AF566939

Evolutionary Biology of Transient Unstable Populations

Edited by Antonio Fontdevila

With 63 Figures

Springer-Verlag Berlin Heidelberg New York
London Paris Tokyo Hong Kong

Professor Dr. ANTONIO FONTDEVILA
Universidad Autónoma de Barcelona
Departamento de Genética y Microbiología
08193 Bellaterra (Barcelona)
Spain

ISBN 3-540-50837-6 Springer-Verlag Berlin Heidelberg New York
ISBN 0-387-50837-6 Springer-Verlag New York Berlin Heidelberg

Library of Congress Cataloging-in-Publication Data. Evolutionary biology of transient unstable populations/edited by Antonio Fontdevila. p. cm. ISBN 0-387-50837-6 (U.S.) 1. Evolution. 2. Population biology. I. Fontdevila, Antonio, 1941– . QH371.E9274 1989 574.5'248 – dc20 89-19689

Printing: Druckhaus Beltz, Hemsbach/Bergstraße
Binding: J. Schäffer GmbH + Co. KG., Grünstadt
2131/3145-543210 – Printed on acid-free paper

Preface

An overview of speciation theory reveals an increasingly held view that many events leading to the origin of new species occur in transient, unstable populations. A transient, unstable population should be understood as a fast episodic phase in a population subjected to genetic and environmental factors that tend to disrupt its cohesive, balanced genome architecure, thus enhancing its probability to produce a new species. Striking the core of Darwinian thought, some authors claim that these processes may be non-adaptive. Among the environmental factors one may cite biotic (e.g. resource availability) and abiotic (e.g. temperature) stress conditions that break up the population stability producing random, unpredictable changes in population size, population trait distribution, breeding structure, inter- and/or intrapopulational hybridization, etc. Genetic factors consist of those events that induce rapid changes in genetic expression and/or that determine reproductive isolation, such as substitutions, insertions, deletions, duplications, transpositions, gross chromosomal rearrangements, recombination and, in general, any mechanism that changes the regulatory pattern of the organism or the balance of its meiotic system. Both kinds of factors are often intertwined in a complex net and may influence each other.

These reflections induced me to propose that the participants of the Sixth European Conference on Population Biology and Evolution center their presentations on theoretical and experimental papers dealing with transient, unstable populations. This volume is representative of the majority, but not all, of the papers presented at this meeting, held in Banyoles (Spain) in July 1988. The first part (A) of the book deals with founder, colonizing and bottleneck populations and consists of a series of chapters ranging from theoretical models to study cases. The results presented by Prevosti and his co-workers show that conventional, natural selection on chromosomal polymorphism may operate very fast in *Drosophila subobscura* colonizing populations during their expansion, a conclusion equally reached by David and collaborators for some allozyme loci in populations of the cosmopolitan species *D. melanogaster*, where short-range genetic variation is attributed to strong differential selection in patchy and transient environments. This large response to selection during the expansion phase of colonization contrasts to the barely detectable intensity of directional selection during the stabilization phase of some colonizing populations, as shown by the work of Ruiz and Santos. This does not contradict the operation of other non-selective factors in colonization, nor does it demonstrate that selection is the main leading force towards speciation. Many authors find suggestive evidence that resource partitioning and seasonality may be responsible for population patchiness. Loeschke, for example, shows that natural selection is too weak to explain the amount of genetic variation in marginal populations, and that drift and migration are mainly responsible for their genetic structure. In fact, some of the research works establish that founder effects are present in terms of incipient prezygotic reproductive isolation (Galiana et al.), in changes in genetic diversity (Brakefield, Fontdevila), and in the reweighting of antagonistic pleiotropic effects of fitness components (Fontdevila). Yet, it remains to be seen whether or not these cases of founder populations lead to speciation events, as discussed by Fontdevila in his work on *Drosophila buzzatii* colonization.

In this respect, an important point is the probability of the establishment of a founder population and the promotion of speciation. Akçakaya and Ginzburg emphasize that patterns of species coexistence, as shown by the fossil record, are better explained by the evolution of resource utilization than by an interspecific competitive model. This may apply to the asymmetric clades or to the hollow curves discussed by Reig. Apparently, the establishment of a new marginal population into a coevolved species guild is, according to Loeschke's work, more likely if it occurs at a resource marginal position. However, these marginal invaders are very vulnerable to resource fluctuations. A conclusion that agrees well with de Jong's view that founders subjected to optimizing selection in a narrow niche show a phenotypic plasticity that is maladaptive in deviating (marginal) environments.

Perhaps the crucial point is how to explain the loss of genetic diversity in bottlenecks postulated by many founder speciation models, and, more importantly, where the new additive genetic variation comes from in the evolving founder populations. Bryant's report suggests that founder effects may change the covariance relationships among morphometric traits. In traits governed by additive processes this change does not promote phenotypic divergence, but in epistatic traits it may convert non-additive variance into additive variance, providing a source of genetic variability of speciogenic value. Nevertheless, this population approach is not the only one to explain the origin of new variability in bottleneck events. The second part (B) of the book deals with an alternative approach. During the last years, chromosomal mechanisms of speciation have been critically revisited by several authors working with different organismal models. The works by Vorontzov and Lyapunova and by Reig emphasize the role of chromosome repatterning in explaining cases of non-gradual speciation as evidenced by population sampling and the fossil record, respectively. Significantly enough, new molecular mechanisms are being incorporated recently to explain the structure and the dynamics of the eukaryotic genome (Hancock and Dover). These mechanisms may not only have evolutionary implications, but they may be operating and/or eliciting genome instabilities and impinging on chromosomal changes. This seems to be the case of mobile genetic elements whose transposition has been related to unusual episodes of enhanced mutation rates promoting new chromosomal rearrangements and/or changes in expression of quantitative characters, as Ratner and Vasilyeva demonstrate in their contribution. Moreover, some evolutionists have linked ecological stress (see McDonald's report) to this genome instability. Interestingly, transposition-mediated mutations could occur under population marginal conditions reminiscent of those that induce speciation in the founder theory. Since any genetic change must pass the population test of natural selection to be incorporated into the gene pool, there must be a way to bridge the gap between these new molecular processes and some classic speciation mechanisms using the unstable, transient populations as the evolutionary stage. This endeavour may be quixotic, but it has an appealing charm at the moment.

I wish to extend my warmest thanks to my colleagues Dr. Mauro Santos and Dr. Alfredo Ruiz who helped me with the organization of this conference and to Mr. Antonio Barbadilla who, using his mastering of computer processing, patiently and cheerfully produced the camera-ready version of many of the book chapters. Special gratitude is dedicated to my secretary Ms. Julia Provecho who assisted me with patience and efficiency in both, the organization of the meeting and the typing of many manuscripts. I also wish to thank the following Spanish funding agencies that made this conference economically possible: Dirreción General de Investigación Científica y Técnica (Ministerio de Educación y Ciencia, España), Comissió Interdepartamental de Recerca i Innovació Tecnològica (Generalitat de Catalunya), Diputació de Girona, and several offices at the Universitat Autònoma de Barcelona (Vicerrectorat d'Investigació, Vicerrectorat de Relaciones Exteriors i Campus, and Institut de Ciències de l'Educació) whose encouragement is also appreciated. The meeting was held at the Casa d'Espiritualitat and the Centre Excursionista of Banyoles, and I wish to thank their staff for their assistance during the meeting.

Bellaterra, July 1989 Antonio Fontdevila

Contents

Part A Founder, Colonizing and Bottleneck Populations

A.1. Theoretical Framework

A.2. Experimental

Part B Evolutionary Mechanisms

B.1. Molecular

B.2. Chromosomal

List of Contributors

You will find the addresses at the beginning of the respective contribution.

Part A Founder, Colonizing and Bottleneck Populations

Phenotypically Plastic Characters in Isolated Populations

G. de Jong

Department of Population Biology and Evolution, University of Utrecht,
Padualaan 8, 3584 CH Utrecht, The Netherlands

Phenotypic plasticity is often held to be an adaptive strategy (Sultan 1987); and while it might be doubtful whether it is warranted to include adaptivity in its evolutionary sense in the definition of phenotypic plasticity, at least many good examples exist that plastic responses are adaptive (Schlichting 1986). The environment seems to cause the physiology of plant or animal to come up with a phenotype that seems not only appropriate to the circumstances, but is actually that phenotype that leads to highest fitness. Often though, the only clear phenomenon is the physiological reaction to the environment; whether this physiological reaction leads to the optimum phenotype, that with the highest fitness, might actually often have been assumed.

If phenotypic plasticity is an adaptive strategy to cope with a spatially varying environment, the implication seems to be that in a constant environment, or an environment that shows little variation, phenotypic plasticity would not be needed, and would not be selected for. This would lead to the assumption that populations that are very local and experience little spatial environmental variation would not show any phenotypic plasticity. The observation is however that such populations, if brought into the laboratory, are often capable of much phenotypic plasticity. Examples can be found in frogs (Berven, Gill and Smith-Gill 1979, Berven and Gill 1983); different frog populations differed in the plasticity shown. This difference in phenotypic plasticity was found too by Dingle *et al* (1982) in the milkweed bug *Oncopeltus fasciatus*: Iowa and Puerto Rico populations differed in sensitivity of size to temperature. Concomitantly, Iowa and Puerto Rico populations of *Oncopeltus fasciatus* showed different genetic covariance patterns between life history characters. Dingle, Evans and Palmer (1988) interpreted the different covariance patterns for life history characters between the island and continental populations in terms of a difference between a non-migratory and a migratory life history strategy. This interpretation, while cogently argued, did not take any influence of phenotypic plasticity on the genetic covariances into account. A background zero hypothesis about the relation between phenotypic plasticity and genetic variances and covariances was not stated.

Such a background is provided here. Given phenotypic plasticity of genotypes that is identical between populations, what will the genetic variances and covariances be like in each environment? A life history strategy interpretation put upon the sign of the genetic covariance between life history characters will only make sense if this covariance has the same sign in all environments. But if a sign change in the genetic covariance is a possible or likely consequence of different environments, given phenotypic plasticity, one should be careful with interpreting the sign of the genetic covariance within one environment in terms of an adaptive life history strategy for that environment.

Populations that experience a narrow environmental range might yet exhibit phenotypic plasticity; possibly with unexpected consequences when one compares populations. But populations might live in a spatially varying environment, and plasticity might seem adaptive to local circumstances. Again, some basic background on how selection effects phenotypic plasticity seems lacking.

The model to be described deals with a continuous environment; the complement of a continuous environment is the genotype as a continuous function. For any given genotype, a systematic change in the environmental variable leads to a systematic change of the phenotype of individuals growing up under that environmental variable. The genotype acts as a mapping function from environment to phenotype (Scharloo 1987); the most common name for this type of mapping function is "reaction norm" (Woltereck 1909).

In nature, reaction norms are widespread, and easily visible from *Daphnia* (from which the idea derives) to clonal plants to size in humans under different nutrition. The secular increase in length in Western Europe, certainly in the Netherlands, is in fact due to an improved early diet; what means that length in humans is a reaction norm, or phenotypically plastic character. Length in humans too serves to remind us that genetic variation, both within and between populations, and phenotypic plasticity do go together.

Reaction norms can have many shapes in nature, from linear functions of the environment to threshold phenomena. A linear reaction norm was for instance found for wing length in *Drosophila melanogaster* (Coyne and Beecham 1987). The model here will be primarily concerned with linear reaction norms: the phenotype is related to the environment by way of a linear function representing the genotype.

These then are the ingredients of the model: an environment with a continuous environmental variable, like temperature, that influences the eventual phenotype of the individual growing up in a given environment; and different genotypes, all giving the phenotype as a (linear) function of the environment.

What to do with it? Two things. First, genetic variance and genetic covariance as a function of the environment: looking for patterns in genetic variance and genetic covariance over populations that only differ in their environment. Second, selection. Selection can happen in populations that have a wide distribution over the environment, or a narrow distribution. Selection can be directional or optimizing. In all these cases, does selection really lead to phenotypic plasticity that is an adaptation? Or becomes phenotypic plasticity a constraint, and perhaps a handicap for the population rather than an asset?

SEPARATE POPULATIONS OVER A RANGE OF ENVIRONMENTS.

Model of genetically variable phenotypic plasticity

We'll only take a simple case here, and leave more formal treatment elsewhere. The model is about one or two quantitative characters, that are phenotypically plastic. To start out, we will not deal with one population in which the individuals experience different environments, but compare populations that each experience a very narrow range of environmental values. Each population is in a different environment. The genotypes in the populations are the same, and present in the same frequencies. Apart from the environment, the populations are identical. We will look into the effect of different environments on the phenotypic variance and covariance of phenotypically plastic characters.

As usual in quantitative genetics, the phenotypic value will be formally represented by the sum of the genotypic value and the environmental value: $P=G+E$, but here we will suppose that there is no environmental noise, and therefore that the genotypic value fully represents the phenotypic value. Phenotypic plasticity means that "the genotype is the reaction norm", and this will mean here that a genotype is represented by a linear function of the environment. The genotypic and phenotypic value in a certain environment x become the function value: $P = g(x) = a+cx$ (Fig. 1).

Genetic variation in reaction norms can be modelled by supposing each genotype to be represented by a separate straight line (Fig. 2). For two alleles at one locus, the three genotypes can be represented by

$$g_{11}(x) = a_{11} + c_{11}\, x$$
$$g_{12}(x) = a_{12} + c_{12}\, x$$
$$g_{22}(x) = a_{22} + c_{22}\, x$$

and if the heterozygote is intermediate between the two homozygotes for all values x of the environment, the genotypic values can be thought to arise

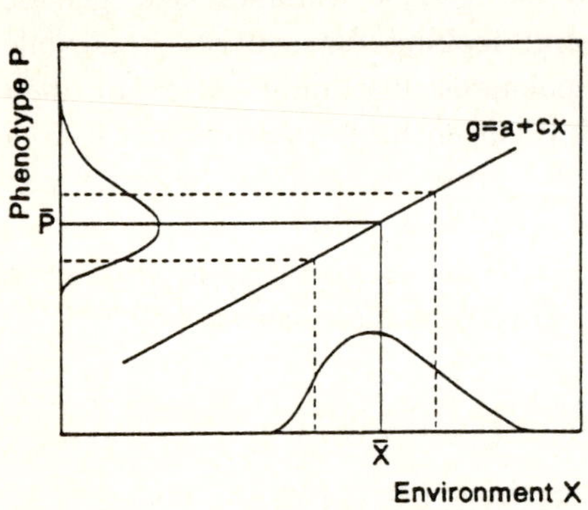

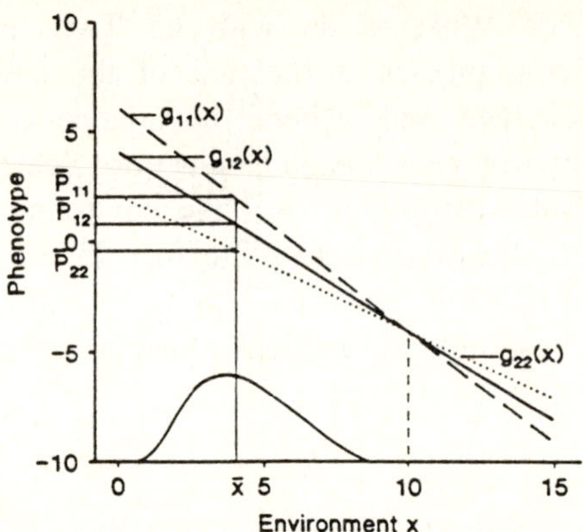

Figure 1. A reaction norm g=a+cx maps the distribution of the environment as the individuals perceive it into a distribution of phenotypic values.

Figure 2. Genetic variation in reaction norms: here $g_{11}(x)=6-x$; $g_{12}(x)=4-.8x$; $g_{22}(x)=2-.6x$. At the mean environmental value of x=4, the mean genotypic and phenotypic values become $P_{11}=2$; $P_{12}=.8$; $P_{22}=-.4$.

additively from linear allelic effects $g_1(x)$ and $g_2(x)$:

$$g_1(x) = a_1 + c_1 x$$
$$g_2(x) = a_2 + c_2 x$$

giving

$$g_{ij}(x) = g_i(x) + g_j(x) = (a_i + a_j) + (c_i + c_j) x \qquad (i=1,2;\ j=1,2)$$

Additivity of allelic effects within and between loci is the only case we'll be concerned with.

The other quantities usual in quantitative genetics, the average effect of a gene substitution and the additive genetic variance can now be defined. The average effect of a gene substitution becomes:

$$\Gamma(x) = g_1(x) - g_2(x) = (a_1 - a_2) + (c_1 - c_2) x$$

This shows two things. 1. The average effect of a gene substitution is a linear function of the environment if the allelic effects are linear and if the slopes c_1 and c_2 of the allelic effects, and therefore the slopes c_{11}, c_{12} and c_{22} of the reaction norms, are unequal. This implicates that the average effect of a gene substitution is a constant ($a_1 - a_2$), if the reaction norms are equidistant. In that case, the mean phenotype changes with the environment, but not the genetic variance. The case we are interested in is however that neither the allelic effects nor the reaction norms are equidistant, and that the lines representing the allelic effects cross at some environmental value x'. The lines representing the reaction norms for the three additive genotypes cross at this environmental value x' too, - and the average effect of a gene substitution is zero at x' (Fig. 3).

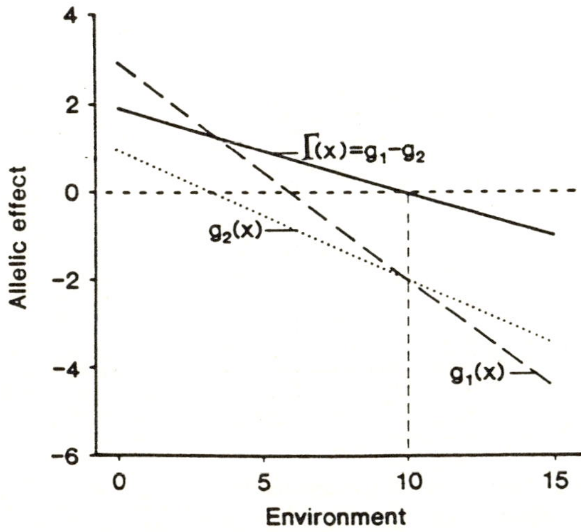

Figure 3. Allelic effects g_1, g_2 and average effect Γ of a gene substitution belonging to the additive reaction norms in figure 2. $g_1(x)=3-.5x$; $g_2(x)=1-.3x$; $\Gamma(x)=g_1(x)-g_2(x)=2-.2x$; $x'=10$.

Crossing of the reaction norms and of the allelic effects means a sign change for the average effect of a gene substitution. These sign changes, and at which environmental value they occur, are of primary importance in the total pattern of genetic variances and covariances over the environment.

2. The second point is that the average effect of a gene substitution $\Gamma(x)$ can be thought of as composed of an average effect of a gene substitution for the intercept, $\alpha = a_1 - a_2$, and an average effect of a gene substitution for the slope, $\gamma = c_1 - c_2$, and the environmental value x: $\Gamma(x) = \alpha + \gamma x$.

The two average effects α and γ are both constants, for any given trait and locus. The environmental value x' where the reaction norms cross is therefore equal to $x' = -\alpha/\gamma$. At the crossing point $x'= -\alpha/\gamma$, $\Gamma(x)=0$, and the contribution of this locus to the additive genetic variance of this trait and to the additive genetic covariance of this trait with any other trait is zero. This is the same as to say that for this one locus, V_A has its minimum, of zero, at x', while COV_A has a zero point and sign change.

This means that some sort of basic pattern exists in comparing the additive genetic variances of two traits and their additive genetic covariance, between populations that only differ in the environmental value x. A graph of the additive genetic variances and covariance over environments, between populations, is given in Fig. 4. This shows the basic pattern in a two locus example. The roots of the additive genetic covariance are found at the minima of the additive genetic variances. It implies that the sign of the additive genetic covariance is not a fixed feature if we deal with phenotypically plastic traits that show genotype-environment interaction. The additive genetic covariance might change sign with change of environment without any further change in

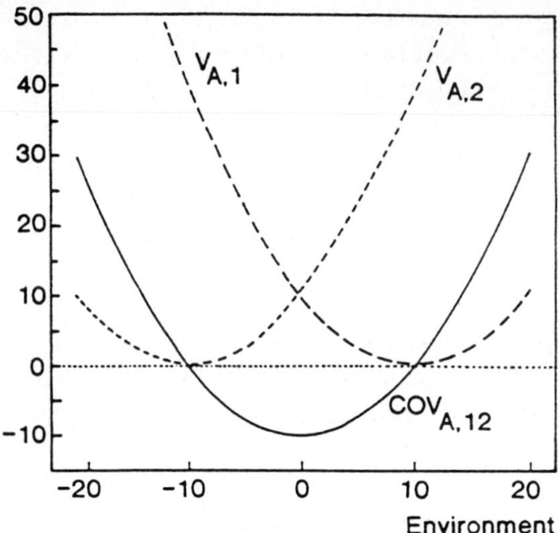

Figure 4. Basic pattern in additive genetic variances for two traits and their additive genetic covariance as determined by <u>two loci</u>. Linear allelic effects and linear reaction norms resulting in linear average effect of a gene substitution lead to quadratic functions for genetic variances and covariances. Here the allelic effects are: locus A, trait 1: $a_1(x)= 6 - .6x$; $g_2(x)= 2 - .2x$; locus B, trait 1: $b_1(x)= 3 - .5x$; $b_2(x)= 1 - .3x$; locus A, trait 2: $a_1(x)= -6 - .6x$; $a_2(x)= -2 - .2x$; locus B, trait 2: $b_1(x)= -3 - .5x$; $b_2(x)= -1 - .3x$; For both loci, $x'=10$ for trait 1 and $x'= -10$ for trait 2, giving the positions of the minima of the variances and the roots of the covariances.

the characters within the population or the gene frequencies within the population. This means that the additive genetic covariance is not a quantity that should be interpreted lightly, just on the basis of its value in one environment.

Polygenes

For each locus, there is a pattern between the additive genetic variance of a trait and its additive genetic covariance with any other trait, if we compare populations with identical reaction norms and identical genotype frequencies over a range of environments. This pattern can be distorted when more loci are present or allelic effects for two traits are functionally related. Yet, even with polygenes, the pattern persists. In general, there is a sign change for the additive genetic covariance, in the region between the minima of the additive genetic variances of the two traits.

The positions of the minima of the additive genetic variances and of the roots of the additive genetic covariances can actually be found for polygenic inheritance of two phenotypically plastic traits. Let the average effects of a gene substitution for intercepts themselves be distributed with a mean $\bar{\alpha}\{1\}$ and a variance $\sigma^2\{\alpha,1\}$ for trait 1. Let the average effects of a gene substitution for slopes for trait 1 themselves be distributed with mean $\bar{\gamma}\{1\}$ and variance

$\sigma^2\{\gamma,1\}$. Let all covariances and cross-covariances between intercepts and slopes be zero: $cov\{\alpha,1,\gamma,1\}= cov\{\alpha,1,\gamma,2\}= cov\{\alpha,2,\gamma,1\}= cov\{\alpha,2,\gamma,2\}= 0$. And, most importantly, let the covariance between intercepts $cov\{\alpha,1,\alpha,2\}$ and the covariance between slopes $cov\{\gamma,1,\gamma,2\}$ be zero too.

It should be clear what these conditions mean. Allelic effects on intercepts are independent of allelic effects on slopes, both within and between traits. But too, allelic effects on intercepts for one trait are independent of allelic effects on intercepts for the other trait, over all loci and genotypes; and allelic effects on slopes for one trait are independent of allelic effects on slopes for the other trait, for all loci and genotypes. These independences presumably mean that the traits show no developmental or functional relationship. Pleiotropy is supposed to be present, all loci influencing both traits. But the pleiotropic effects are independent, unstructured, and genetic covariation for the traits need not betray a functional relationship. A zero covariance between the average effects of a gene substitution for the two traits means that both traits are the result of many converging developmental and physiological pathways. Some loci will have alleles boosting both traits and alleles lowering both traits, along part of one pathway. Some loci will have alleles that boost one trait to the detriment of others, somewhere else in the development. The loci will have to be scattered over the developmental pathways leading to the two traits for the average effects of a gene substitution to result in overall independence. Yet, there might be an evolved functional relationship indicated by the means of the average effects of a gene substitution in slopes and intercepts. The covariances for the average effects are after all attributable to the deviations of the average effects from their means.

The information about the distribution of average effects makes it possible to take the expectation of the additive genetic variance and of the additive genetic covariance. Using this expectation as the additive genetic variance and additive genetic covariance in a large population in linkage equilibrium gives

$$V_{A,1 \text{ at } x} = [\ \Sigma_{loci}\ 2p_iq_i\ ]\ .\ [\ (\ \bar{\alpha}\{1\}+\bar{\gamma}\{1\}x\ )^2 + \sigma^2\{\alpha,1\} + x^2\sigma^2\{\gamma,1\}\ ]$$

$$COV_{A,1,2 \text{ at } x} = [\ \Sigma_{loci}\ 2p_iq_i\ ]\ .\ [\ (\ \bar{\alpha}\{1\}+\bar{\gamma}\{1\}x\ )\ \ (\ \bar{\alpha}\{2\}+\bar{\gamma}\{2\}x\ )\ ]$$

The minimum of the additive genetic variance is found at

$$x'_1 \quad = \quad -\ [\ \bar{\alpha}\{1\}\ \bar{\gamma}\{1\}\ ]\ /\ [\ \bar{\gamma}^2\{1\} + \sigma^2\{\gamma,1\}\ ]$$

In the polygenic case, the gene frequencies have no influence on the expectation of the position of the minimum of the additive genetic variance. Similarly, the gene frequencies have no influence on the roots and the position of the extremum of the additive genetic covariance.

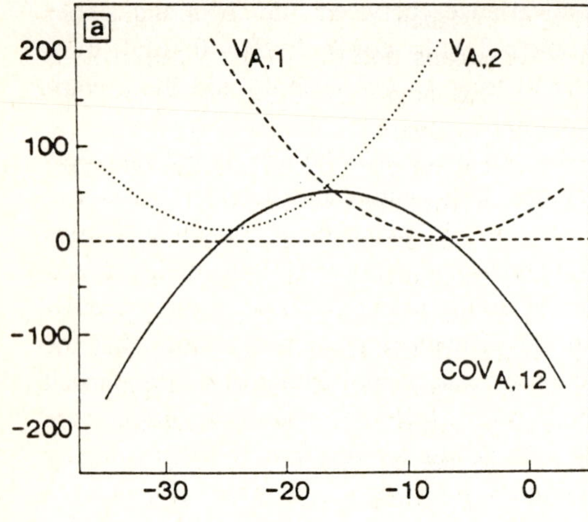

Figure 5. Basic pattern in additive genetic variances for two traits and their additive genetic covariance as determined by polygenes. Linear average effects of a gene substitution for each locus and both traits. All covariances between intercepts α and slopes γ of the average effects are zero within and between traits. Independence between intercepts and slopes always leads to a sign change in the additive genetic covariance between traits, taken over a range of environments. The roots of the additive genetic covariance $-\bar{\alpha}\{1\}/\bar{\gamma}\{1\}$ and $-\bar{\alpha}\{2\}/\bar{\gamma}\{2\}$ are "near" the positions of the minima of the additive genetic variances $-[\bar{\alpha}\{1\}\ \bar{\gamma}\{1\}]/[\bar{\gamma}^2\{1\} + \sigma^2\{\gamma,1\}]$ and $-[\bar{\alpha}\{2\}\ \bar{\gamma}\{2\}]/[\bar{\gamma}^2\{2\} + \sigma^2\{\gamma,2\}]$. Here $\alpha\{1\}=5.0$; $\alpha\{2\}=-20.0$; $\gamma\{1\}=.75$; $\gamma\{2\}=-.80$; $\sigma^2\{\alpha,1\}=2.0$; $\sigma^2\{\gamma,1\}=.0064$; $\sigma^2\{\alpha,2\}=9.0$; $\sigma^2\{\gamma,2\}=.0036$.

In this case, of independence of underlying average effects, the additive genetic covariance has always two roots; the discriminant is a perfect square. This means that the additive genetic covariance will always change sign (Fig. 5). The roots of the additive genetic covariance are found at $x'_1= -\bar{\alpha}_1/\bar{\gamma}_1$ and $x'_2= -\bar{\alpha}_2/\bar{\gamma}_2$. The roots of the covariance are therefore "near" the positions of the minima of the additive genetic variances, "near" depending on the magnitude of the variances $\sigma^2\{\gamma,1\}$.

Independence in pleiotropic effects means a sign change in the additive genetic covariance over the full range of algebraically possible values for the environmental variable. Of course, it is not necessary that the sign change occurs in the range of environmental values that is physiologically possible or is ecologically present. Sign changes in the additive genetic covariance have been found in studies of environmental effects on plastic traits (Schlichting 1986, Giesel, Murphy and Manlove 1982, Gebhardt and Stearns 1988). What arises is the problem of interpretation.

An empirical sign change in the additive genetic covariance over a range of environments can occur if the average effects of a gene substitution for intercepts and for slopes are uncorrelated or very slightly correlated between traits. The best way to interpret this might be to abstain from ascribing biological importance to any deviation from the mean average effect of a gene substitution over loci. The deviations from the mean average effect of a gene substitution would seem to be just mutational noise. The mean average effects of a gene substitution might however be due to selection, and could be interpreted. The coefficient of x^2 in the additive genetic variance is $\bar{\gamma}\{1\}\bar{\gamma}\{2\}$; a mountain parabola for the additive genetic covariance means that $\bar{\gamma}\{1\}\bar{\gamma}\{2\}$ is negative, a valley parabola that $\bar{\gamma}\{1\}\bar{\gamma}\{2\}$ is positive. But negative $\bar{\gamma}\{1\}\bar{\gamma}\{2\}$ means that $\bar{\gamma}\{1\}$ and $\bar{\gamma}\{2\}$ differ in sign, and that there is overall a trade-off in effect of alleles on slopes for the two traits: allelic effect for steep slope for the one trait going on average together with allelic effect for shallow slope for the other trait. Positive $\bar{\gamma}\{1\}\bar{\gamma}\{2\}$ means that $\bar{\gamma}\{1\}$ and $\bar{\gamma}\{2\}$ have the the same sign, that overall allelic effects for slopes go together for the two traits: allelic effect for steep slope for the one trait going on average with allelic effect for steep slope for the other trait. It might be called a trade-off for slope means for negative $\bar{\gamma}\{1\}\bar{\gamma}\{2\}$ and a synergistic effect for slope means for positive $\bar{\gamma}\{1\}\bar{\gamma}\{2]$. For negative $\bar{\gamma}\{1\}\bar{\gamma}\{2\}$, the interpretation is not in terms of a trade-off within each individual (such a trade-off would emerge in the covariances $cov\{\alpha,1,\alpha,2\}$ or $cov\{\gamma,1,\gamma,2\}$) nor in a trade-off between character values, but rather in a genetic trade-off between steep and shallow slopes, between high plasticity and low plasticity. Similarly, a valley parabola would indicate that high plasticity for one trait goes together with high plasticity for the other trait. Interpretation of the sign of the additive genetic covariance separately for each environment, and interpretation of the sign of the overall pattern in the additive genetic covariance point to different conclusions.

A POPULATION WITH A DISTRIBUTION OVER THE ENVIRONMENT

Total, genetic and plasticity variance

Up to now, we have compared populations that each lived in one environment x. A population might however contain individuals that live in different environments, so that a distribution of individuals over the environmental values is the result. The distribution of the environmental values as the population experiences is converted by a reaction norm in a distribution of phenotypic values (Fig. 1). If the mean environmental value is $\bar{x}$, and the variance of the environmental values is $\sigma^2\{x\}$, the mean phenotypic value for a given reaction norm is P=a+cx, and the phenotypic variance is $c^2\sigma^2\{x\}$.

For any number of genotypes within the population, the total variance - excluding noise - becomes

$$V_{total} = V_{G\ at\ \bar{x}} + \sigma^2\{x\}\ [\ V_{G,c} + \bar{c}^2\ ]$$

where $V_{G\ at\ \bar{x}}$ is the genetic variance as found in the mean environment $\bar{x}$

$V_{G,c}$ is the genetic variance in slopes

$\bar{c}$ is the mean slope for the character. That is, the average value of the slopes weighted by the frequency of the genotypes exhibiting them. For one locus with two alleles:

$$\bar{c} = p^2c_{11} + 2pqc_{12} + q^2c_{22}$$

With equidistant reaction norms, there is no genetic variance in the slopes. Falconer (1981) includes the systematic environmental deviation due to phenotypic plasticity in the general environmental deviation. Without genetic variance in slopes,

$$V_{total} = V_{G\ at\ x} + \sigma^2\{x\}\ c^2$$

corresponding to a decompostion of the phenotypic variation in a genetic variation at the mean environment and an environmental component depending upon the variance of the environment. If the reaction norms are non-equidistant, and the genetic variance in slopes exists, the additional variance component $\sigma^2\{x\}\ V_{G,c}$ corresponds to the genotype-environment interaction variance. In that case, both environmental variance and genotype-environment interaction variance contain genetic variation. Scheiner and Goodnight (1984) propose to call $\sigma^2\{x\}\ [\ V_{G,c} + \bar{c}^2\ ]$ the plasticity variance. Their proposal is meant to divide the total variance into a component that can be ascribed to plasticity and a component that can be ascribed to genetic variation; but actually, the split is not clean, as the genetic variance depends upon the value of the mean environment, and the plasticity variance contains genetic variation.

Directional selection

Falconer (1981) does not give any special attention to selection on phenotypically plastic characters. The implication seems to be that the environmental variance due to phenotypic plasticity does not play any special role in selection, anyway as long as the environmental values have the same distribution in parents and offspring. The absence of a genotype-environment correlation does not necessarily remove all correlation between parental and offspring environment, however, and phenotypic plasticity might come to play a role in selection, and be selected on.

Let the environment be stable over generations, and the mean and variance of the environmental distribution as the population perceives it be the same between generations. Yet different life-history patterns are possible. Consider two extremes. The first situation is rather animal-like: males and females grow up in their individual environment, reach a phenotypic value as determined by their genotype for that environment, and then form one mating pool over the whole population. Mating is at random for every zygote. The zygote offspring of every female is distributed according to the environmental distribution over the whole environmental range. Another situation is rather like plants. Females remain in the environment they grew up in, and male gametes are distributed at random over the whole population. All offspring of a female grow up in the same environment as the mother. The mean phenotypic value of the offspring of a female is the phenotypic value at the maternal environment averaged over the genotype frequencies in the offspring. The mean phenotypic value of the offspring of a male is again related to the mean environment. In both the first and second situation the overall distribution of genotypes remains independent of the environment. This permits us to compute the parent-offspring regressions. These differ with the life history, and in the second life history the parent- offsprint regressions depend upon the level of subdivision of the population.

In the first case, and the male version of the second case, there is no correlation between the parental and offspring environment, and the parent - offspring covariance becomes

$$COV(\mathrm{O,P}) = 1/2\ [\ V_{\mathrm{A,a}} + 2\bar{\mathrm{x}}\ COV_{\mathrm{A,a,c}} + \bar{\mathrm{x}}^2\ V_{\mathrm{A,c}}\ ] = 1/2\ V_{\mathrm{A,a+c\bar{x}}}$$

That is, the parent - offspring covariance as if both parents and offspring in fact lived all in the same environment, the mean environment of the population.
When all offspring remain in the maternal environment, the mother - offspring covariance becomes

$$COV(\mathrm{O,M}) = 1/2\ V_{\mathrm{A,a+c\bar{x}}} + 1/2\ V_{\mathrm{A,c}}\ \sigma^2\{\mathrm{x}\} + \bar{\mathrm{c}}^2\ \sigma^2\{\mathrm{x}\}$$

In fact, the two situations are extremes. When the covariance between the environment of the mother, x_0, and the mean environment of her offspring, x_1, is $\mathrm{cov}\{x_0,x_1\}$, the mother - offspring covariance becomes

$$COV(\mathrm{O,M}) = 1/2\ V_{\mathrm{A,a+c\bar{x}}} + 1/2\ V_{\mathrm{A,c}}\ \mathrm{cov}\{x_0,x_1\} + \bar{\mathrm{c}}^2\ \mathrm{cov}\{x_0,x_1\}$$

The heritability as estimated by the mother-offspring covariance becomes

$$h^2 = [V_{\mathrm{A,a+c\bar{x}}} + (V_{\mathrm{A,c}}+2\bar{\mathrm{c}}^2)\ \mathrm{cov}\{x_0,x_1\}]\ /\ [V_{\mathrm{G,a+c\bar{x}}} + (V_{\mathrm{G,c}}+\bar{\mathrm{c}}^2)\ \sigma^2\{\mathrm{x}\}]$$

Directional selection over the total population might have different result depending upon the life history. The simplest case is $\mathrm{cov}\{x_0,x_1\} = 0$. In the numerator of the heritability, all indications of phenotypic plasticity are lost. Phenotypic plasticity only surfaces as part of the environmental variation, in the denominator. As a consequence, the phenotypic plasticity is not at all "seen" by selection; in $R=h^2S$, it is only the additive genetic variance at the mean environment that is of importance. Selection therefore acts as if only the genotypic values at the mean environment exist. The reaction norm that has highest value at x is fixed by selection for high values, independent of the slope. The slope of the reaction norm is not selected on. Slopes of reaction norms originating by directional selection are therefore not adaptive. Rather, they bring a physiological constraint.

In the general case, there is different selection in males and females. The general quantitative genetics formulation for different selection in males and females is

$$R_d = b_{md} S_m + b_{fd} S_f$$
$$R_s = b_{ms} S_m + b_{fs} S_f$$

where d=daughter, s=son, m=mother, f=father, and b is the regression coefficient between a specified parent and a specified offspring type. Using this, while supposing the selection differential S the same in males and females, and all difference between the b's only to derive from the covariance between maternal and offspring environment would give for selection response both in sons and in daughters $R = (b_{m.} + b_{f.})\, S$, being

$$R = (1/\overline{w})\ [V_{A,a+c\bar{x}} + (\ V_{A,c} + 2\bar{c}^2)\ \mathrm{cov}\{x_0,x_1\}]\ .\ (\partial\overline{w}/\partial\overline{P})$$

Supposing selection for higher phenotypic values this indicates that a tight correlation between maternal and offspring environment would lead to a higher selection response. Given positive covariance between maternal and offspring environment, a high genetic variance in slope leads to increased speed of selection; and a high speed of selection is promoted by high plasticity too.

But as most important point, the slopes are now *visible* to selection. Net selection depends both on the genotypic value at the mean environment and on the slope. What weighs most depends upon the actual values, but the tendency seems to be to select for high slope at positive $\mathrm{cov}\{x_0,x_1\}$. Strong plasticity could then be a consequence of directional selection for higher phenotypic values. As with the first case, there is no way in which the resulting plasticity can be called adaptive, however.

Optimizing selection

Selection might not be for just higher (or lower) phenotypic value, but for a fixed phenotypic value per environment. An optimality analysis might point to an optimum phenotypic value, depending upon the environmental value; an example of such an optimality analysis is given by Stearns and Koella (1986). Within each environment selection would be optimizing, and if environments are separate one would expect a version of the model of Via and Lande (1985) to apply. But now consider an optimum phenotypic mean that is a function of the environment, while the population is continuously distributed over the environment. A fitness function describing optimizing selection is

$$w(P)= 1 - a' (P_{opt} - P)^2$$

fitness depending upon the phenotype independent of the genotype.

Let $P_{opt}= b' + b''x$, a linear function of the environment, just as the reaction norms give the genotypic values as linear functions of the environment. Genotypic fitness becomes in a one locus model

$$w_{ij} = 1 - a' [(b' + b'' \bar{x})-(a_{ij} + c_{ij} \bar{x})]^2 - a' \sigma^2\{x\} [b''-c_{ij}]^2$$

Phenotypic quadratic selection leads to a genotypic fitness that includes two components. One component, $[(b'+b''x)-(a_{ij} + c_{ij} x)]^2$, depends upon the difference between the genotypic value and the value of the phenotypic optimum at the environmental mean. The other component, $\sigma^2\{x\} [b''-c_{ij}]^2$, depends upon the difference in slope between the reaction norm for a genotype and the optimum phenotype; the difference in slope being weighted for importance by the variance of the environmental values. This generates separate selection components on average genotypic value and on slope. The relative importance of the two is related to niche width, by way of the variance of the environmental values as the population perceives it, $\sigma^2\{x\}$.

Given the intercepts, what slope would give highest fitness? Differentiating genotypic fitness w_{ij} towards slope c_{ij} gives that the highest fitness is for a slope of

$$\hat{c}_{ij} = [b'- a_{ij}] \bar{x} / (\sigma^2\{x\} + \bar{x}^2) + b''$$

Only if x=0 or if a_{ij} =b' is $\hat{c}_{ij}$ =b"! That is, only if no selection component at the mean environment exists, or if the intercept of a reaction norm happens to

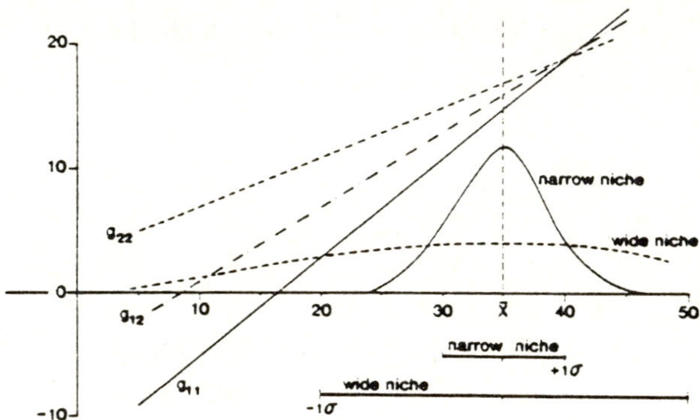

Figure 6. Optimizing selection; the optimum phenotype P_{opt}=0 in all environments x. At x=35 g_{11} is selected for in a population with a narrow niche, g_{22} is selected for in a population with a wide niche. $g_{11}(x)$=-13+.8x; $g_{12}(x)$=-5+.6x; $g_{22}(x)$=3+.4x.

Table 1

Equilibrium gene frequency in the case of optimizing selection illustrated in figure 6.

$\sigma^2\{x\}$	40	60	80	100	120	140	160	180	200
x									
32	1	1	1	1	1	1	.89	.61	.37
33	1	1	1	1	1	1	.68	.40	.17
34	1	1	1	1	1	.73	.41	.14	0
35	1	1	1	1	.78	.38	.07	0	0
36	1	1	1	.78	.29	0	0	0	0
37	1	1	.70	.11	0	0	0	0	0
38	1	.47	0	0	0	0	0	0	0

be the intercept of the optimum phenotype, is the slope of the optimum phenotype the slope giving highest fitness to a reaction norm. Otherwise, the "best" slope of a given set of reaction norms depends upon the variance of the environment.

An example is given in Fig. 6 and table I. The optimum phenotype $P_{opt}=0$ for all environments. A wide environmental range for the population, that is large $\sigma^2\{x\}$, can select for the genotype with the shallow reaction norm even if this is the genotype that at the average environment has a genotypic value further from the optimum phenotype. Variance prevails. A narrow environmental range for the population makes the selection component in the mean environment prevail. The genotype with the steep reaction norm is here selected for. Selection needs *not* lead to the optimal slope - as local values at the mean environment interfere.

From the expression for c_{ij} follows that at $\sigma^2\{x\}=0$ the slope selected for is going to depend upon the weighted difference in intercepts between reaction norm and optimum phenotype. A high variance leads to a "best" slope that approximates the slope of the optimum phenotype; the result of selection will not necessarily be the optimum phenotype itself, but any reaction norm parallel to it.

Optimizing selection per environment for a wide ranging population would lead to phenotypic plasticity that more or less follows the optimum phenotype, and must be called adaptive. A narrow range of environments for the population leads to little selection on the slope itself, and the plasticity actually found might be a consequence of the mean genotypic value it is associated with. That is, it would have to be called a constraint, rather than an adaptation. Only research into the actual fitness profiles over environments can decide whether plasticity would be an adaptation or a constraint.

CONCLUSION

Phenotypic plasticity is not delimited to wide ranging populations. Populations showing a narrow niche in nature might show a large capacity for phenotypic plasticity under experimental circumstances. In case genotype-environment interaction exists, this generates patterns of genetic variances and covariances when the same population is raised in different environments. Caution is needed then in the interpretation of the sign of the additive genetic covariance, as the additive genetic covariance as a function of the environment in general changes sign. It is the pattern of covariances over the whole range of the environment rather than the values of the covariances in a specific environment that will need interpretation. This applies too to populations that have a narrow environmental range under natural circumstances.

Optimizing selection in a population occupying a wide ranging environment would lead to phenotypic plasticity that approaches the optimum slope for the character over the range of environments the population experiences. But selection on a phenotypically plastic character in a population occupying a narrow niche might as well lead to phenotypic plasticity that leads to severe maladaptation in environments deviating from the environmental mean. It is at least possible that a generalist species, experiencing a wide range of environments, closely follows the optimum phenotype and easily extends its range, while it is at least possible too that a specialist, experiencing a narrow range of environments, so closely is adapted to the environmental mean under selective disregard for phenotypic plasticity, that it experiences a strong maladaptive change in phenotype with any change in environment, and cannot easily adapt to a change in environment.

REFERENCES

Berven KA (1987) The heritable basis of variation in larval developmental patterns within populations of wood frogs (*Rana sylvatica*). Evolution 41:1088-1097

Berven KA, Gill DE (1983) Interpreting geographic variation in life history traits. Amer. Zool. 23:85-97

Berven KA, Gill DE and Smith-Gill SJ (1979) Countergradient selection in the green frog, *Rana clamitans* Evolution 33:609-623

Coyne JA, Beecham E (1987) Heritability of two morphological characters within and between natural populations of *Drosophila melanogaster* Genetics 117:727-737

Dingle H, Blau WS, Brown CK, and Hegmann JP (1982) Population crosses and the genetic structure of milkweed bug life histories. Pp 209-229 in: H.Dingle and J.P.Hegmann (eds), Evolution and genetics of life histories. Springer Verlag Heidelberg

Dingle H, Evans KE, and Palmer JO (1988) Responses to selection amoung life history traits in a non-migratory population of milkwood bugs (*Oncopeltus fasciatus*). Evolution 42:79-92

Falconer DS (1981) Introduction to quantitative genetics Longman

Gebhardt MD, Stearns SC (1988) Reaction norms for developmental time and weight at eclosion in *Drosophila mercatorum* J Evol Biol 1:335-354

Giesel JT, Murphy PA and Manlove MN (1982) The influence of temperature on genetic interrelationships of life history traits in a population of *Drosophila melanogaster*: what tangled datasets we weave. Amer Natur 119:464-479

Scharloo W (1987) Constraints in selection response in: V.Loeschcke (ed) Genetic constraints on adaptive evolution Springer Heidelberg

Scheiner SM, Goodnight CJ (1984) The comparison of phenotypic plasticity and genetic variantion in populations of the grass *Danthonia spicata* Evolution 38:845-855

Schlichting CD (1986) The evolution of phenotypic plasticity in plants Annu Rev Ecol Syst 17:667-693

Stearns SC, Koella JC (1986) The evolution of phenotypic plasticity in life history traits: predictions of reaction norms for age and size at maturity Evolution 40:893-913

Sultan SE (1987) Evolutionary implications of phenotypic plasticity in plants Evolutionary Biology 21:127-178

Via S, Lande R (1985) Genotype-environment interaction and the evolution of phenotypic plasticity Evolution 39:505-522

Woltereck R (1909) Weitere experimentelle Untersuchungen über Artveränderung, speziell über das Wesen quantitativer Artunterschiede bei Daphniden. Verhandlungen der Deutschen Zoologische Gesellschaft 19:110-172

Multivariate Morphometrics of Bottlenecked Populations

E.H. Bryant

Department of Biology, University of Houston, Houston, TX 77004, USA

Introduction

In one way or another population bottlenecks and/or founder events have played critical roles in theories of speciation, including those of Mayr (1954, 1970, 1982), Carson (1968, 1975, 1982) and Templeton (1980a,b). In Mayr's view (1954) the internal genetic change fostered by these founder events is "the most drastic change (except for polyploidy and hybridization) which may occur in a natural population, since it may affect all loci at once." Basically, these speciation models envisage founder events as disruptive to established genetic relationships among polygenic traits, thereby creating opportunity for establishing new polygenic balances. In Carson's founder flush theory, for example, he suggests that the genetic architecture can be viewed as containing "open" and "closed" portions. The "open" portion would largely have an additive genetic basis, be responsive to selection, and account for variation within species, while the "closed" portion would largely have a nonadditive genetic basis, be unresponsive to selection, and account for variation between species. Bottlenecks would then "unlock" the closed genetic system and allow for divergence of polygeneic traits in ways that would not normally occur within species.

These theories have not been well tested, in part because the concepts of "genetic balance" and "open" and "closed" genetic system are difficult to operationalize. One possible way to interpret altered genetic environment is in terms of the genetic covariance relationships among polygenic traits. These genetic relationships (i.e., the additive genetic covariance structure) govern the directions and magnitudes of multivariate phenotypic evolution (Lande 1979; Lande and Arnold 1983), so whether or not bottlenecks create opportunity for novel differentiation would depend on whether they alter the genetic covariance structure among traits. The putative action of bottlenecks can then be evaluated directly by comparing genetic relationships among polygenic traits in bottleneck lines with an ancestral (pre-bottleneck) population. We have addressed this problem during the past several years using experimental bottleneck populations of the housefly, *Musca domestica* L., that originated from a single outbred line. I synthesize a number of our findings here that bear upon altered multivariate relationships among morphometric traits within and among our bottleneck lines. Specifically, I address two related questions:

1. Do bottlenecks alter the additive genetic covariance relationships among traits?

2. Do bottlenecks promote significant phenotypic divergence from an ancestral population and if so, in what multivariate directions does this occur?

In brief, our experimental protocol was as follows (details can be found in Bryant et al. 1986 or Bryant and Meffert 1988a). Individual virgin pairs of flies were sampled from an outbred laboratory line of the housefly recently established from a natural population, to create four replicated bottleneck (founder) lines, each with either one, four, or 16 pairs of flies (total of 12 bottleneck lines plus an unbottlenecked control line). Additive genetic variances and covariances for eight morphometric traits were estimated on all lines and the control (ancestral) population using parent/offspring covariances (Falconer 1981). The morphometric traits were (for a complete description see Bryant 1977): 1) wing length 2) wing width 3) inner-eye separation 4) head width 5) scutellum length 6) scutellum width 7) metafemur length and 8) length of thoracic suture. All traits were measured on parents and offspring using an ocular micrometer and converted natural logarithms. Significance levels for all measures were determined by bootstrap simulations (Efron 1982; Efron and Gong 1983) using the additive genetic covariance matrix for the control population as the sampling universe. Details of these procedures can be found in Bryant and Meffert (1988a,b).

Changes in Covariance Structure

The primary concern here is whether or not there are changes in the multivariate genetic relationships among traits as a result of the experimental bottleneck, and if so, do these occur in similar ways for the different bottleneck sizes. These concerns can be subdivided into two questions with respect to changes in the multivariate genetic ellipsoid compared with the control:

1) Are there changes in the overall level of eccentricity of the multivariate ellipsoids for bottleneck lines?
2) Are there changes in the orientation of these ellipsoids?

Measures of genetic integration among the traits for the additive genetic correlation and covariance matrices are summarized in Figure 1. Two measures of integration are given: 1) the average correlation among traits (panel a), and 2) the coefficient of integration devised by Cheverud et al. (1983) as

$$C.I. = [\ \Sigma\ (\lambda_i - \lambda)/(n^2 - n)]^{1/2},$$

where λ_i is the i th eigenvalue of the additive genetic correlation (or covariance) matrix, λ is the average of the eigenvalues, and n is the number of traits (panel b). A higher C.I. indicates greater integration among traits. Both measures were significantly greater in the four-pair and 16-pair lines than in either the control or single-pair lines, indicating that the general level of genetic integration of the genotype was increased for the intermediate bottleneck size lines in relation to the control but not for the single-pair lines.

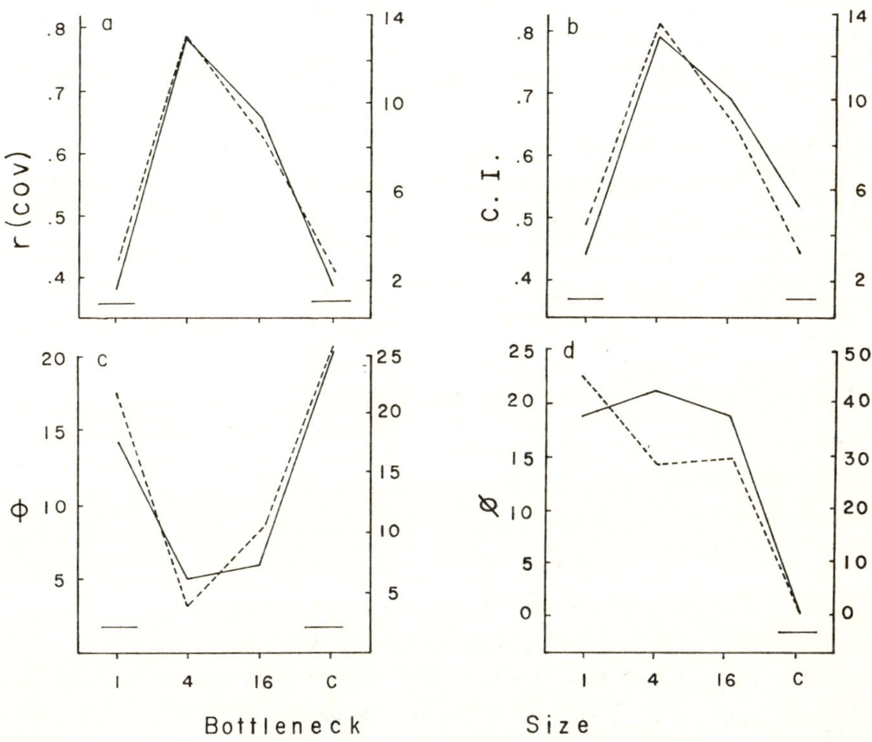

Figure 1. Measures of morphological integration for the single-pair, four-pair, and 16-pair bottleneck lines, based upon additive correlation (solid lines and left hand scale) or additive covariance (dashed lines and right hand scale) matrices. a) levels of genetic correlation and genetic covariance among the eight traits, averaged over all n(n-1)/2 combinations of traits. b) coefficients of integration. c) angles in degrees of the principal axes for bottleneck lines to the isometry vector. d) angles in degrees of the first principal axes for the bottleneck lines to that for the control.

The second question can be addressed by looking at the orientation of the major or principal axes of the ellipsoids for the bottleneck lines. The orientation of these major axes for the bottleneck lines can be viewed in two ways: 1) What is the orientation of the major axes for the bottleneck lines in relation to that for the control? and 2) What is the orientation of these axes in relation to a vector of general size (i.e., an isometry vector)?

These results are summarized in panels c and d of Figure 1, where again the replicate lines have been pooled to obtain an average correlation structure for a bottleneck size. The deviation of major axes for the bottleneck lines from that of the control was nearly the same for all covariance matrices, respectively (Fig. 1d). But in spite of this similarity among angles in relation to the control, the major axes for the bottleneck lines were not the same, since the major axes of the four- and 16-pair lines were nearly collinear with the isometry vector (approximately 5° deviation from isometry for these axes), while those for either the single-pair or control lines

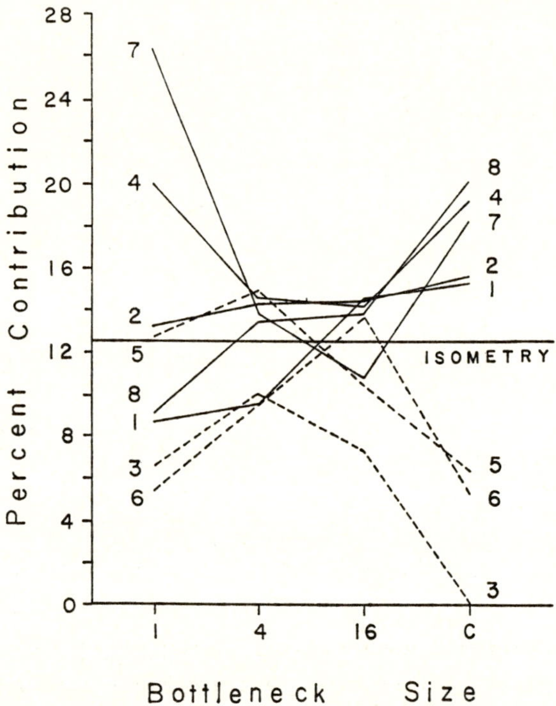

Figure 2. Percentage contribution of individual traits to the first principal axes of the additive genetic correlation matrices for the single-pair, four-pair, and 16-pair bottleneck lines. Equal contributions of traits are represented by an isometry vector, so deviations from isometry indicate variations in the contributions of traits to the respective principal axes. Numbers identifying traits are given in the text.

were 15° and 25° deviant from isometry vector for the correlation and covariance matrices, respectively (Fig.1c). That the major axes for the bottleneck lines were different from that of the control is clearly evident when we look at the percentage contribution of individual traits to the total variance associated with this axis (i.e., its eigenvalue; note that a perfect isometry vector would have equal percent contributions from all traits). This is illustrated in Fig. 2 for the first principal axis of the additive correlation matrices. The two major suites of traits (group factors) evident in the control population (designated with solid vs. dashed lines in Figure 2) are not evident in any of the bottleneck lines: the major axes for the four- and 16-pair lines were very close to an isometry vector and that for the single-pair lines represents a remixing of the trait relationships so that the original two suites of traits are no longer evident.

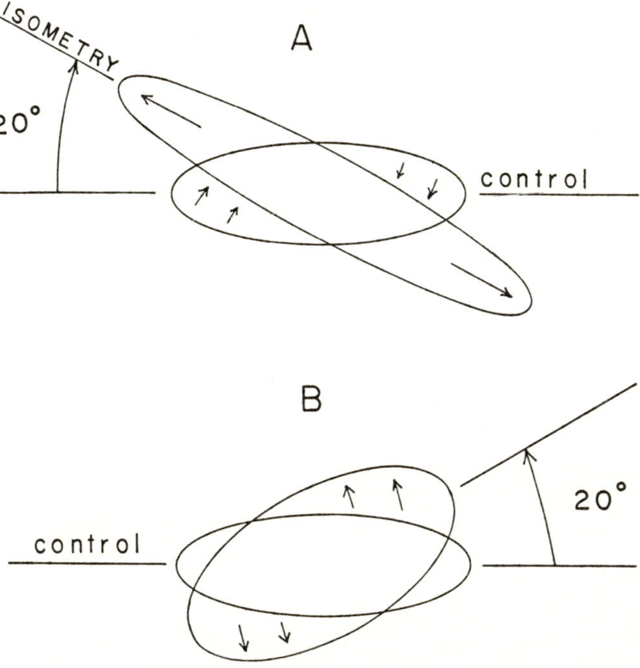

Figure 3. Summary representation of the changes in the covariance ellipsoids for the bottleneck lines compared to the control. a) the covariance ellipsoids for the four-pair and 16-pair lines showing greater elongation of the ellipsoids in the direction of the isometry vector. b) the covariance ellipsoids for the single-pair lines indicating lessening of their eccentricities and reorientation from the control in a direction away from the isometry vector.

The major change in eccentricity and orientation of these ellipsoids is summarized in Fig. 3 (modified from Bryant and Meffert 1988a). The additive genetic ellipsoids for the four- and 16-pair lines were more eccentric than that of the control and nearly collinear with an isometry vector. On the other hand, that for the single-pair lines was slightly less eccentric than that of the control and deviated from the control major axis in a direction away from isometry. Analyses of major axes for individual single-pair lines showed considerable variation in their multivariate orientations (average angle separating major axes of any two replicate single-pair lines was 53°, for example, for the additive correlation ellipsoids). Hence, some of the decrease in eccentricity (integration) of the average ellipsoid for the single-pair lines was due to mixing of disparate ellipsoids.

Divergence of Bottleneck Lines

Turning to phenotypic divergence of bottleneck lines from the control we ask two questions:

1) Is there significant divergence of bottleneck lines from the control?
2) If so, what are the directions of the divergence and do these directions differ among bottleneck sizes?

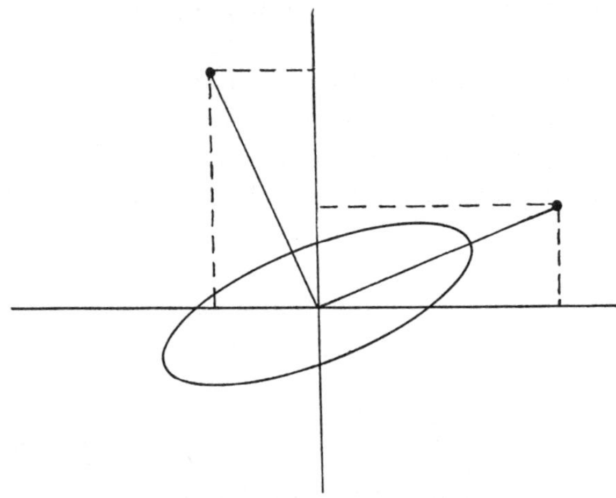

Figure 4. Phenotypic divergence of bottleneck lines from the control in relation to the control covariance ellipsoid. The two vectors are of equal Euclidean distance but differ in their generalized genetic distance due to the relative additive genetic variance available in the vector direction.

To answer either of these questions we need a suitable metric which should take into account the genetic relationships among traits in the base population. Because a given Euclidean distance would involve different numbers of genetic standard deviation units depending on its direction in relation to the covariance ellipsoid (Fig. 4), a suitable metric should accommodate the additive genetic variance along the vector of observed phenotypic change. One such metric is the generalized genetic distance or Mahalanobis' distance measured in units of genetic covariation of the ancestral population (Lande 1979; Bryant and Meffert 1988b). In addition this generalized genetic distance can be easily referenced to the successive major axes of the ancestral covariance ellipsoid, so that all changes can be compared with a single ancestral covariance structure.

To convert a mean Euclidean shift (Δz) to generalized genetic distances along the respective principal axes of the control covariance ellipsoid (Δp), we compute $\Delta p = \Delta z^t A \Lambda^{-1/2}$, where A is a matrix of eigenvectors and $\Lambda^{-1/2}$ is a diagonal matrix of reciprocals of square roots of eigenvalues of the control additive genetic covariance matrix. The squares of the elements of the vector Δp give the contribution of each principal axis to total generalized genetic distance, which can be obtained as the sum of all squared elements.

The contributions of the first four principal axes of the control covariance matrix (accounting for 96% of the total additive genetic variance) to total generalized genetic distance between a bottleneck line and the control line are given in Table 1. Based upon bootstrap simulations all bottleneck lines diverged significantly from the control (shown by an underscore); the greatest divergence form the control occurred for the 16-pair lines and the least in the four-pair lines. But these generalized distances were partitioned differently among the principal axes for the

Table 1. Phenotypic divergence in generalized genetic distance units (D^2) of bottleneck lines from the control along the first four principal axes of the control additive genetic covariance matrix. Independent contributions of the individual axes sum to the total squared distance of lines from the control. An underscore indicates a significant divergence of a bottleneck line from the control at $P < .05$.

	Principal Axis D^2				
Bottleneck Size	I	II	III	IV	Total D^2
1	<u>2.61</u>	<u>1.61</u>	<u>1.95</u>	<u>3.56</u>	<u>9.73</u>
4	<u>1.94</u>	0.31	0.51	<u>2.00</u>	<u>4.76</u>
16	<u>6.95</u>	0.47	0.97	<u>4.31</u>	<u>12.71</u>

three bottleneck sizes. The four- and 16-pair lines diverged predominantly along axes I and IV of control, whereas the single-pair lines diverged more nearly equally along all four axes of the control. The likely explanation for the differences in divergence pattern among the bottleneck sizes rests in differential shifts in size and shape among them, since principal axes of covariance matrices usually represent different levels of size and shape variation. Size and shape can be partitioned as in Fig. 5: size is taken to be a phenotypic shift in the direction of the control vector with no change in orientation, whereas shape is taken to be a shift in direction without a change in length of the mean phenotypic vector. Using this definition the contributions of individual principal axes to size, shape, or total divergence are represented by the squared coordinates of the respective vectors onto the principal axes as in Figure 5. Therefore, these contributions can be greater than the total divergence along a particular axis, but these contributions for size and shape summed over all axes must be less than or equal to the contributions for total divergence summed over all axes.

The partitions of generalized distances into size and shape components are shown in Table 2, separated further into contributions of individual axes to these components. The divergence

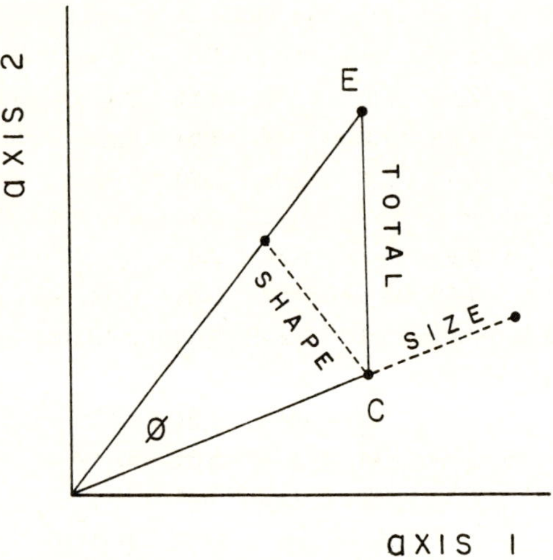

Figure 5. Partitioning of total phenotypic divergence between a bottleneck line (E) and the control (C) into components of size and shape. Size is the difference in lengths of the vectors in the direction of the control vector, whereas shape is the distance between vectors of equal length and depends upon the angle separating the vectors. The squares of the coordinates of the total, size, and shape vectors yield the contributions of the azes to these vector lengths, as given in Table 2.

along axes I and IV of the control for the 16-pair lines clearly occurred because of a shift in size in the direction of the mean phenotypic vector for the control. On the other hand, divergence for the single-pair lines occurred principally in shape along principal axes II, III, and IV of the control additive genetic covariance matrix. The four-pair lines were intermediate in the contributions of size and shape to total divergence.

A final question about phenotypic divergence of these lines from the control concerns the variation in directions of phenotypic divergence among the replicate lines within a bottleneck size: Do all replicate lines tend to diverge in similar directions? In particular, is there a concerted shape divergence among the single-pair lines? Table 3 displays the variances (in Δp) among the four replicated lines computed first within each principal axes and then summed over axes to yield a total variance. The greatest variance occurred among the single-pair lines (more than

Table 2. Contributions of size and shape to the squared phenotypic divergence along the first four principal axes of the control additive genetic covariance matrix. Independent contributions of axes to size and shape divergence sum to the total size and shape divergence but components for individual axes are not additive. An underscore indicates a significant divergence of a particular D^2 from the control at $P < .05$.

Bottleneck Size		Principal Axis I	II	III	IV	Total D^2
	D^2	<u>2.61</u>	<u>1.61</u>	<u>1.95</u>	<u>3.56</u>	<u>9.73</u>
1	D^2:Size	<u>1.33</u>	0.07	0.04	0.64	<u>2.08</u>
	D^2:Shape	0.97	<u>1.85</u>	<u>2.33</u>	<u>2.46</u>	<u>7.62</u>
	D^2	<u>1.94</u>	0.31	0.51	<u>2.00</u>	<u>4.76</u>
4	D^2:Size	<u>1.37</u>	0.08	0.04	0.66	<u>2.14</u>
	D^2:Shape	0.66	0.27	0.61	<u>1.08</u>	<u>2.62</u>
	D^2	<u>6.95</u>	0.47	<u>0.97</u>	<u>4.31</u>	<u>12.71</u>
16	D^2:Size	<u>7.39</u>	0.41	0.21	<u>3.56</u>	<u>11.57</u>
	D^2:Shape	0.16	0.22	0.33	0.41	<u>1.12</u>

twice the variance among the lines of the other bottleneck sizes). Hence there was considerable variation in the directions of shape divergence among these single-pair lines, so that the divergence in shape for the lines occurred along a number of directions all of which represented phenotypic changes that were different from the major axis of genetic variation observed for the ancestral population.

Table 3. Variance in phenotypic divergence among replicate lines within a bottleneck size category, separated into contributions of size and shape and then summed over the four principal axes. Variances are for ΔP, the signed generalized genetic distance.

	Bottleneck Size 1	4	16
Total Variance	6.76	3.14	2.47
Size Variance	1.22	1.88	1.78
Shape Variance	5.51	1.24	0.68

Discussion and Conclusions

I believe our results confirm a potential role for bottlenecks to alter covariance structure and thereby open new avenues of evolutionary change unavailable or at least less available to ancestral populations. In our experiments this was true for both size as well as shape variation. In the control line the major axis of genetic variation did not represent isomorphism, so that a change in pure size could not easily occur without invoking heterochrony. But in the altered genetic environment within these lines a pure size change could easily occur since the major axes of the four- and 16-pair lines were nearly collinear with an isomorphism axis, and indeed a large size component of divergence was observed for these lines. In contrast, the single-pair lines exhibit high levels of divergence in shape, and within these lines we also observed reorientation of the major axes of variation towards greater shape (i.e., away from an isomorphic axis). In this regard our single-pair lines exhibit a breakdown of the original trait relationships to favor novel changes in phenotype, both by a lessening of the eccentricity of the additive genetic ellipsoid and by a change in its orientation in relation to the control.

How bottlenecks might serve to promote such novel differentiation, particularly in shape from the control can be visualized in the following paradigm. We can interpret the relative "openness" of a particular multivariate vector in relation to the level of additive genetic variance along it:

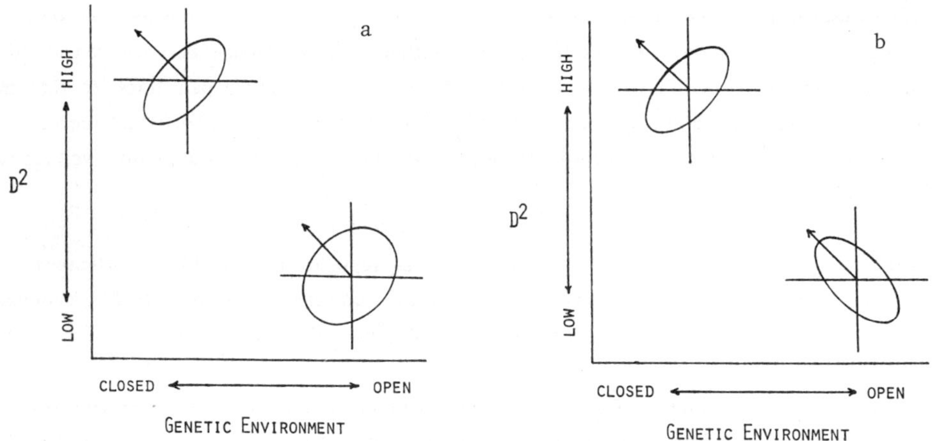

Figure 6. Changes in the genetic covariance ellipsoid rendering a phenotypic change more or less open to evolution. A vector with less additive genetic variance (a high D^2) would be less open to evolutionary change than one with more additive genetic variance available (a low D^2). a) additive genetic variance along the vector altered by the changes in the eccentricity of the genetic covariance ellipsoid. b) additive genetic variance along the vector altered by a reorientation of the genetic covariance ellipsoid.

the more additive genetic variance available along a vector, the faster will be the evolutionary response and hence, the more "open" the population would be to potential evolutionary change along the vector. Conversely, the smaller the generalized genetic distance (D^2), the greater resistance there would be to evolutionary change in along the vector.

The potential change along a particular Euclidean vector could then be altered (i.e., rendered more open or closed) by two processes of realignment in the covariance structure. First, as illustrated in Figure 6a, there could be a lessening of the genetic integration among the traits in question, to cause a broadening of the additive genetic ellipsoid in the direction of the particular vector, and consequently yield a lower D^2 distance. Secondly, as in Figure 6b, a lower evolutionary distance (D^2) could be achieved by a reorientation of the additive genetic ellipsoid. Both types of changes were apparent in our experiment, and were most evident in the single-pair lines where considerable realignment of shape axes occurred.

If bottlenecks can cause such shifts in the eccentricities or in the directions of the additive genetic ellipsoids, what genetic processes could be involved? Such shifts are not expected to occur when genetic processes underlying the traits are purely additive (e.g., Lande 1979), because

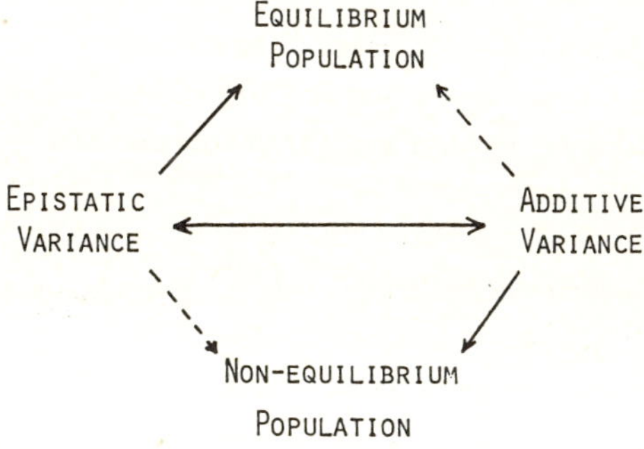

Figure 7. The ebb and flow of additive genetic variance in equilibrium and non-equilibrium populations. Much of the genetic variance in equilibrium populations may be locked up in epistatic interactions, some of which may be transformed to additive components of variance available for immediate evolutionary change as a result of a population bottleneck or founder event.

this yields a linear relationship of genotype to phenotype. On the other hand, when traits are affected by nonadditive genetic processes, such as epistasis, there is unlikely to be a simple linear mapping of genotype to phenotype. One reason is that additive genetic variation for a particular trait (or multivariate vector) in a population stems from additive as well as nonadditive genetic processes. We have previously shown that sampling of a small number of individuals from a population with epistatic effects (specifically multiplicative dominance) will tend to increase additive genetic variance in lines derived from these samples (Bryant et al. 1986). More importantly, Goodnight (1987, 1988) has recently developed models showing explicitly how epistatic variance in an ancestral population can be transformed into additive genetic variance in a derived bottleneck population. As he demonstrated, the level of additive genetic variation in a bottleneck population can be several times higher than that in the original population when there is epistasis.

The ebb and flow of variance between additive and nonadditive (epistatic) components in equilibrium and non-equilibrium populations is shown in Fig. 7. The epistatic component of variation, not being responsive to selection, can serve as a reservoir of additive genetic variation. A bottleneck can then transform some of this variation to the additive component responsive to selection. If different levels of epistasis affect the different axes of the covariance ellipsoid, a bottleneck would likely influence additive genetic variances along them unequally. In this way, a bottleneck could serve to alter the orientation as well as the shape of the additive genetic covariance ellipsoid and thereby open new avenues to selection unavailable in the source population. This paradigm suggests a mechanism by which bottlenecks can promote evolutionary change in the way envisaged by speciation models.

Acknowledgments. This work was supported by grants from the National Science Foundation (BSR-8198128) and the University of Houston Coastal Center. I thank Dr. Trevor Price for valuable discussions during the formulation of many of these ideas.

References

Bryant EH (1977). Morphometric adaptation of the housefly, Musca domestica L, in the United States. Evolution 31:580-596.

Bryant EH, McCommas SM, Combs LM (1986). The effect of an experimental bottleneck upon quantitative genetic variation in the housefly. Genetics 114:1191-1211.

Bryant EH, Meffert LM (1988a). Effect of an experimental bottleneck on morphological integration in the housefly. Evolution 42:698-707.

Bryant EH, Meffert LM (1988b). Multivariate phenotypic differentiation among bottleneck lines of the housefly. Evolution, in press.

Carson HL (1968). The population flush and its consequences. In: Lewontin RC (ed) Population Biology and Evolution. Syracuse Univ. Press, Syracuse, N.Y., pp 123-137.

Carson HL (1975). The genetics of speciation at the diploid level. Amer. natur. 109:73-92.

Carson HL (1982). Speciation as a major reorganization of polygenic balances. In:Barigozzi C (ed) Mechanisms of Speciation. Alan R Liss, N.Y., pp 411-433.

Cheverud JM, Rutledge JJ, Atchley WR (1983). Quantitative genetics of development: Genetic correlations among age-specific trait values and the evolution of ontogeny. Evolution 37:895-905.

Efron B (1982). The jackknife, the bootstrap and other resampling plans. Soc. Indust. Appl. Math., Philadelphia, PA.

Efron B, Gong G (1983). A leisurely look at the bootstrap, the jackknife and cross-validation. Amer. Stat. 37:36-48.

Falconer DS (1981). Introduction to Quantitative Genetics. Longman, N.Y.

Goodnight CJ (1987). On the effect of founder events on epistatic genetic variance. Evolution 41:80-91.

Goodnight CJ (1988). Epistasis and the effect of founder events on the additive genetic variance. Evolution 42:441-454.

Lande R (1979). Quantitative genetic analysis of multivariate evolution, applied to brain:body size allometry. Evolution 33:402-416.

Lande R, Arnold SJ (1983). The measurement of selection on correlated character. Evolution 37:1210-1226.

Mayr E (1954). Changes of genetic environment and evolution. In: Huxley J (ed) Evolution as a Process. Allen and Unwin, London, U.K., pp 156-180.

Mayr E (1970). Population, Species and Evolution. Belknap, Cambridge, MA.

Mayr E (1982). Processes of speciation in animals. In: Barigossi C (ed) Mechanisms of Speciation. Liss, N.Y., pp 1-19.

Templeton AR (1980a). The theory of speciation via the founder principle. Genetics 94:1011-1038.

Templeton AR (1980b). Modes of speciation and inferences based upon genetic distances. Evolution 41:719-729.

Templeton AR (1982). Mechanisms of speciation - a population genetic approach. Ann. Rev. Ecol. Syst. 12:23-48.

Niche Overlaps and the Evolution of Competitive Interactions

H. REŞIT AKÇAKAYA and L.R. GINZBURG

Department of Ecology and Evolution, State University of New York, Stony Brook, NY 11794, USA

Introduction

Several theoretical studies have suggested that the nature of interactions between species effect the structure of communities. Most of these studies were based on the analysis of an interaction matrix A, and its stability properties. The elements of this matrix describe the interactions between species: diagonal elements a_{ii} describe the effect of the population of species i on itself (i.e., self-limitation as a result of density-dependent factors), and off-diagonal elements a_{ij} describe the effect of species j on the growth rate of species i. The signs of the two corresponding elements of the matrix, a_{ij} and a_{ji} together define the type of interaction between species i and species j: (0,0) for no interaction, (–,–) for competition, (–,+) for predation/parasitism, (0,–) for amensalism, (0,+) for commensalism, and (+,+) for mutualism. The number of species (i.e., the dimension of the matrix), m, and the proportion of non-zero elements of the matrix (connectance), C, were used as measures of the complexity of the system (May, 1972).

In this section, we briefly review studies on the complexity and stability of ecological systems related to the evolution of community structure. We then present a model based on a new approach and summarize the results of this model as presented in detail in Ginzburg, Akçakaya & Kim (1988). The third section combines this approach with the niche concept and considers the relation between resource utilization and competition.

Gardner & Ashby (1970) analyzed the stability of randomly constructed matrices with off-diagonal elements −1, 0, or +1, and diagonal elements uniformly distributed between −1.0 and −0.1. They found that for this kind of matrix, stability decreased as either m or C increased. In a similar study, May (1972) used random matrices with coefficients $a_{ij}=0$ (with probability 1–C) and $a_{ij}=R$ (with probability C), where R is a random number with mean of 0 and variance of s^2, s being a measure of the strength of non-zero interactions. He found a sharp transition from stability to instability as the quantity s increased over 1.0. These and similar studies (Roberts, 1974; Gilpin & Case, 1976) showed that complexity (measured as an increase in m, C or s) usually results in instability, which seemed to contradict the observed complexity of natural systems.

Parallel ideas were developed in population genetics, where the question of the maintenance of genetic variation was instigated by the observation that polymorphisms were more common than they had been thought. Theoretical models of selection showed that polymorphisms with more than 5 alleles were extremely difficult to maintain by heterosis (Gillespie, 1977; Lewontin, Ginzburg & Tuljapurkar, 1978). These studies analyzed the stability of randomly constructed fitness matrices, similar to the interaction matrices described above. Such an analysis is more readily applicable in genetics because the matrix is symmetric (making the analysis simpler) and because a fitness matrix is a more realistic representation of a panmictic diploid population than an interaction matrix is of an ecosystem. Fitness values (the elements of the matrix) are meaningfully restricted to the interval (0,1), whereas using the same method in ecology means making assumptions about the population growth rates, r_i's (they are assumed to be equal in some studies, given arbitrary values in others), interaction functions (all are linear, Lotka-Volterra type), and interaction strengths (taken from arbitrarily assigned distributions).

Given the difficulty in constructing stable, complex systems with randomly chosen matrix elements, several authors looked for mathematical conditions that will give stability (May, 1973; Siljak, 1975; McMurtie, 1975; Tregonning & Roberts, 1978), or searched for biological mechanisms that may restrict the parameter values to subsets containing more stable combinations (Gillespie, 1977; DeAngelis, 1975; DeAngelis *et al.*, 1975; Deakin, 1975; see also Karlin, 1981; Karlin & Feldman, 1981; Clark & Feldman, 1986 for similar attempts in population genetics). One of the common results of these studies is that self-limitation is a stabilizing factor, whereas strong competition and mutualism are destabilizing. There is also some empirical evidence that species interact strongly with only a few others, or weakly with many (Margalef, 1968).

Another way of constructing complex and stable systems was explored by Tregonning & Roberts (1979, see also Roberts & Tregonning, 1980; and Taylor, 1985). These authors used a generalized Lotka-Volterra model:

$$\frac{dN_i}{dt} = N_i\left(r_i + \sum_{j=1}^{m} a_{ij}N_j\right) \tag{1}$$

where N_i is the size and r_i is the intrinsic growth rate of population *i*; and a_{ij} are the interaction coefficients as described above. They constructed random systems with dimension $m=50$, and eliminated the species with the most negative "equilibrium" population size, N_i^*, at each step until the matrix became stable. In this way, they were able to construct stable systems with a mean number of species m of the order of 25. In other words, although it is difficult to obtain a random 25×25 matrix supporting a stable and feasible equilibrium, such a matrix could be sampled with a relatively high probability as a subset of a larger unstable matrix.

Other authors used a more "evolutionary" process, gradually increasing the complexity (i.e., dimension) of the system by adding species to a system of coexisting species (Robinson & Valentine, 1979; Post & Pimm, 1983; Taylor, 1985; see also Aoki, 1980; Spencer & Marks, 1988; Marks & Spencer, 1989, for similar population genetics models using fitness matrices as described earlier). In these studies, it was possible to construct stable systems with a large number of variables (species/alleles). The number of species increased rapidly with initial additions (sometimes "collapsing" to a lower dimension as the result of an addition), and then stabilized at a level depending on the pool of parameters from which new species (alleles) were sampled. Colwell & Winkler (1984) introduced another "evolutionary" process, based on mutating a set of characters, rather than interaction coefficients. Their species did not compete until they were sampled and assembled in island biotas, after which a constant proportion of species on each island was eliminated by competition according to the similarities of their characters to those of the other members of the same biota.

In summary, methods to find the small stability regions in the parameter spaces of complex systems can be devised. Whether analogous mechanisms which result in the stable structuring of ecological systems exist in nature is, however, an open question.

One problem with the above models is that specific values for interaction parameters were always selected from an arbitrarily chosen distribution. The distribution may have a discrete character, with a probability of zero and non-zero coefficients, and a continuous character for non-zero elements (usually uniform distribution). Most importantly, new species have always been sampled completely independently of the properties of the species existing in the community at the time of introduction. This would be a realistic method if we did not have any knowledge of the characteristics of the invading species, such as in the case of species introduced from other continents. However, in the course of evolution of communities, species naturally arise from already existing ones (or they may arise artificially in the case of genetically engineered species). In both cases, the characteristics of the new species are similar to those of its progenitor. This process —which may effect the structure of the resulting community, as well as the conditions for the invading species to be successful—can be simulated by generating the parameters of the new species as deviations from parameters of one of the existing species, instead of sampling them independently from a separate preassigned ensemble. Building a complex, stable community by adding new species with this process is a more realistic approach if we are studying the natural evolution of a closed community, excluding the invasion of unrelated species.

Model 1: Evolution of competition matrix

Our approach concerns the effect of repeated invasions on the community structure. The basic question is how the interactions between the species in a community are shaped as a result of species turnover caused by the addition and extinction of species that evolved "continuously" out of the existing ones.

The invasion-extinction process we used in our simulations differed from the models used by the other authors in one important respect: instead of adding a species characterized by random coefficients taken out of some arbitrary distribution assigned independently of the existing community, we made the assumption that the new (mutant) species will be ecologically similar to its progenitor. The added row and column in the interaction matrix are picked to be close to the row and column that correspond to the progenitor species. This is a reasonable assumption in mimicking natural evolution. It is also reasonable for newly created recombinant organism of known origin.

All simulations started with a single species ($r_1=1.0$, $a_{11}=-0.01$). We assigned the coefficients of the new species as independent random normal deviates (with given standard deviations) from those of an existing species. Since the scope of this study was limited to competition communities, all values of a_{ij} were restricted to be non-positive. The progenitor species was selected with equal probability from the existing ones. We introduced the new species to the community and simulated the growth of all populations with the population dynamic model described in eq. 1 until a new stable equilibrium was reached. These "invasion attempts" were repeated 1000 times for each simulation, and 5 simulations were run at each of the two levels of standard deviation which characterize the closeness of a mutant species to its progenitor. These levels were .0001 and .0002 for a_{ij} and .01 and .02 for r_i, corresponding to 1% and 2% of the initial values of a_{11} and r_1, respectively. The value of standard deviation did not have an important effect on the results.

We assumed that the evolutionary time scale in which speciation events occur is much slower than the ecological time scale of population dynamics. Consequently, we measured the time in terms of speciations and let the community reaçh an equilibrium before a new speciation occurred. This makes our results interpretable as a description of the direction of evolution. If speciation rates vary widely (e.g., if larger number of species results in higher rate of speciation per unit of actual time), our "time" axis cannot be translated linearly into the actual time scale.

The simulation at each step resulted in either the extinction of the invading species (repulsion), or in its invading the community with or without driving one or more of the existing species to extinction. The proportion of these outcomes, the observed frequency of transitions from *m* to *n* species (i.e., number of species before and after the invasion), as well as changes in the number of species and strength of interactions through time are given in Ginzburg *et al.* (1988).

There was a gradual increase in the number of species, to about 6 species after 1000 speciation events. The largest community observed during the simulation had 9 species. An interesting result of the simulation was the change in the strength of the interactions through time. In the final communities, the average value of the diagonal elements of the interaction matrix (intraspecific competition) has decreased to about half of its initial value, whereas the average value of the off-diagonal elements (interspecific competition) has decreased to about one-tenth. Thus the ratio of the average strength of interspecific competition to the average strength of intraspecific competition decreased from 1.0 to 0.2.

Most theoretical studies of competition assume symmetric interactions, $a_{ij}=a_{ji}$ (e.g., Roughgarden, 1979). Since we did not impose such a restriction on the competition coefficients, it is interesting to observe the change in the degree of asymmetry of interactions. Our results have shown that for this model, the interactions become more and more asymmetric as new species entered the community.

The results of this study can be summarized in four main points.

(1) a process simulating the speciation events as ecologically continuous mutations (deviations) in the strength of competitive interactions results in stable communities with many coexisting species;

(2) with this process, weak competitive interactions between species evolve. As a result, the community contains a few (usually two) strongly interacting and several weakly interacting species. Strong interactions are usually a result of recent invasions resulting in coexistence of the new species with its progenitor;
(3) during the process, the interactions become increasingly asymmetric. Interactions in which one of the species has a stronger effect on the other appear more and more often as new species enter the community;
(4) the probability that a new species will invade the community is approximately independent of the number of species in the community at the time of speciation.

Model 2: Niche structure and competition

Changes in the ecological niche of the progenitor species is a generally accepted mechanism underlying changes in the interaction coefficients. Although these coefficients are defined as the effect of population sizes on growth rates, designing experiments to measure changes in growth rates is very difficult. Hence several authors have suggested formulations of interaction coefficients as functions of resource utilizations (niches) of the species in the community. Most of these formulations utilize the quantity p_{ia}, the relative utilization of the ath resource category by the ith species. In the case of two resources, the quantity p_{iab} refers to the relative utilization of the ath category of resource 1 and the bth category of resource 2 by species i. For one-dimensional niches, Levins (1968) defined the competition coefficients as

$$a_{ij} = \frac{\sum_a p_{ia} p_{ja}}{\sum_a p_{ia}^2} \tag{2}$$

Pianka (1973) proposed a modification which gives symmetric matrices; Colwell & Futuyma (1971), Schoener (1974) and Abrams (1980) discussed problems associated with measuring a_{ij} in the field with this formulation; and several studies have attempted to estimate niche overlap in one or few resource dimensions as a measure of the strength of competition (e.g., Schoener, 1968; Culver, 1970; Brown & Lieberman, 1973; Cody, 1974).

We can restate the result of this study in terms of niches becoming increasingly non-overlapping except for an occasional recent speciation which results in the coexistence of the new species with its progenitor. In order to incorporate niche structures into our model, we simulated speciations as changes in the resource utilization of the progenitor species and calculated the interaction coefficients as a function of the breadth and overlap of resource utilization. For this, we used a formulation similar to the one discussed above in its general behavior, but much simpler than it, in order to make thousands of a_{ij} calculations based on multidimensional niches possible. We defined

$$a_{ij} = \frac{o_{ij}}{b_i} \tag{3}$$

where o_{ij} is the amount of overlap of niches of species i and j, and b_i is the niche breadth of species i.

As pointed out by May (1975), an interaction matrix defined by Levins' formulation (2) (or by the simpler formulation that we used), always gives a stable system, no matter what the number of species or the strength of interactions are. This means that every new species may coexist with its progenitor, no matter how small the difference between their niches. In reality however, such two populations with minimal differences in their niches will not be classified as separate species, either because they have not developed mechanisms for genetic isolation, or because they do not have enough morphological differences to be recognized as different species. For this reason, we defined a "threshold niche difference"; and did not accept deviations smaller than this threshold as new species. Two other assumptions we made are uniformity of resources and independence of resource dimensions to permit rectangular niches, which makes calculations much faster.

In this second model, we simulated speciation as deviations in the resource utilization of one of the existed species which was selected as the progenitor species. Then we calculated the interaction matrix as described above and projected the new equilibrium community as in the first model. We made simulations with 3 different resource (niche) spaces, with 1, 2 and 3 dimensions. The results of these simulations are given in Table 1 and Figures 1, 2 and 3. Figure 1 shows the number of species after each invasion attempt, for simulations with 1, 2 and 3 resource dimensions. These figures are averaged over periods of 10 invasion attempts.

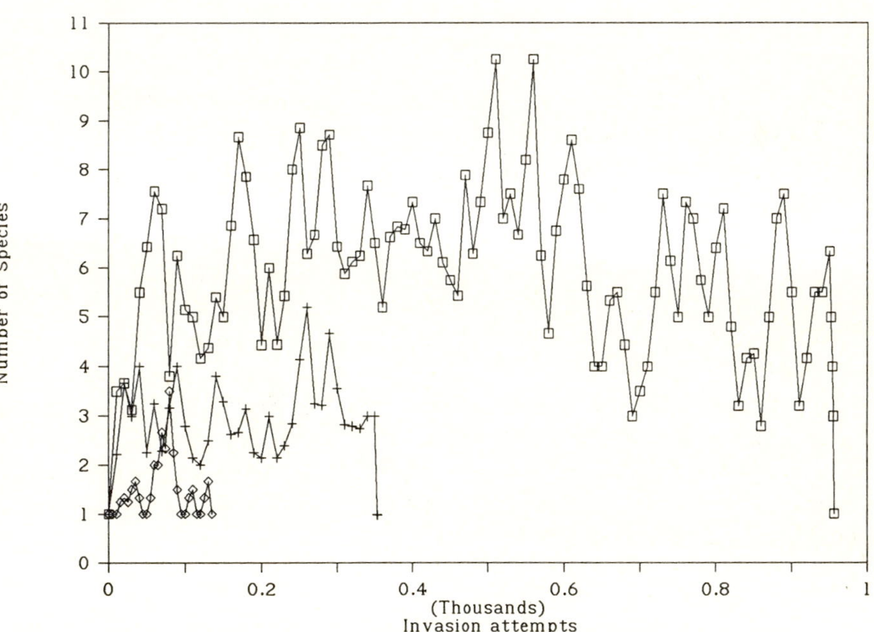

Figure 1. Number of species after each invasion attempt
Square: 3 dimensions; Plus: 2 dimensions; Diamond: 1 dimension

The maximum number of species in the simulations were 4 species for 1 dimension, 6 species for 2 dimensions, and 12 species for 3 dimensions of resources. We terminated the simulations when the number of species at the end of the simulation decreased to 1 species which occupied more than 90% of the total niche space. This "super-species" occurred after 131 invasion attempts in 1 dimensional niche space; after 356 attempts in 2 dimensional, and 956 attempts in 3 dimensional niche spaces.

TABLE 1: Probabilities of various outcomes of an invasion attempt in simulations with 1, 2 and 3 resource dimensions

Outcome of Invasion Attempt	**1**	**2**	**3**
Repulsion	0.435	0.371	0.406
Invasion and no extinctions	0.153	0.213	0.229
Invasion and 1 extinction (Replacement)	0.267	0.253	0.196
Invasion and 2 extinctions	0.137	0.126	0.126
Invasion and 3 extinctions	0.008	0.028	0.036
Invasion and 4 extinctions	0.0	0.006	0.004
Invasion and 5 extinctions	0.0	0.003	0.003

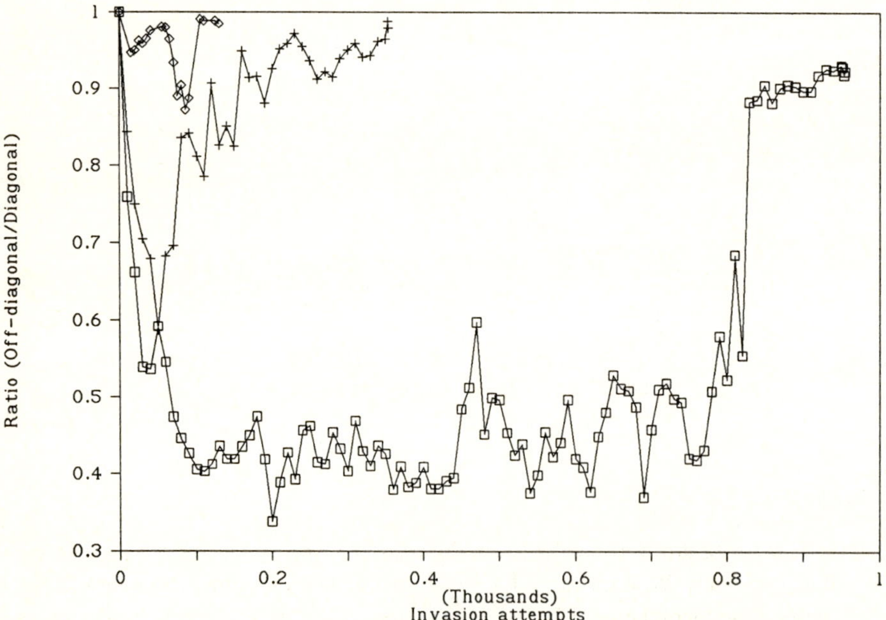

Figure 2. Ratio of interspecific to intraspecific competition coefficients
Square: 3 dimensions; Plus: 2 dimensions; Diamond: 1 dimension

Although the number of species is very different in simulations with different number of resource dimensions, the probabilities of various outcomes of an invasion attempt is not very different. These probabilities are given in Table 1. Repulsion is the failure of a new species to invade the community. Invasion may result in the extinction of the progenitor species (replacement), or the extinction of more than one species (one is almost always the progenitor).

Another important difference among simulations with different number of niche dimensions relates to the ratio of off-diagonal elements to diagonal elements (strength of interspecific competition relative to the strength of intraspecific competition). This ratio reaches much lower values in 3 dimensional niche spaces (Figure 2).

Figure 3 shows the longevity distribution of species in the 2 simulations with 2 and 3 niche dimensions. The simulation with 1 dimension did not have sufficient number of species to construct a distribution. Longevities were measured as the number of invasion attempts during which the species remained in the community. The mean longevities were 2.6, 4.8 and 10.2 invasion attempts (with standard deviations of 2.0, 3.8 and 10.2) in simulations with 1, 2 and 3 niche dimensions, respectively. Mean longevity thus seems to increase rapidly with the dimension of the resource space.

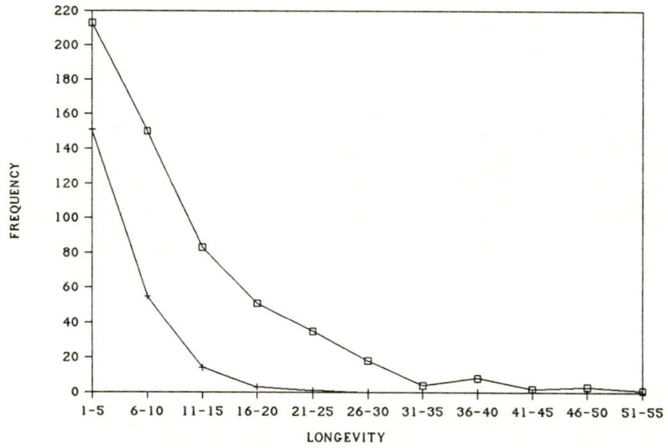

Fig 3: Longevity Distributions
Square: 3 dimensions; Plus: 2 dimensions

Discussion

The most important difference between the results of the two models is the change in the number of species. Whereas the first model results in an indefinite increase in the number of species, the second model, based on resource utilization of species, gives a more realistic picture: the number of species reaches a maximum, or a plateau, and then declines.

This is exactly what is observed in the fossil record. The evolution of Ordovician invertebrates is an example of this pattern (Rosenzweig & Taylor, 1980). Symmetry of the speciation pattern is another characteristic of clades in the fossil record. Gould *et al.* (1977) claimed that early speciations result in asymmetric clades with a rapid increase in the number of species in the beginning, and a slow decrease at the end. Our simulation with 3 dimensional resource space seems to follow this pattern.

Another pattern that emerges from our simulations and is similar to the pattern observed in the fossil record is the distribution of longevities. Although we measure time as the number of invasion attempts *per species*, which cannot be transformed linearly to real time units, our simulations nevertheless result in an approximately exponential distribution of species longevities as is observed in the fossil record.

These two patterns correspond to observations in nature, demonstrating that our approach is a realistic model for the evolution of competitive interactions between species. In this model, we are not concerned with the specific mechanism of speciation. Ecological continuity in our simulations imply that origination of new species results from speciation rather than immigration from other areas. Our model therefore simulates speciation in a large area that has recently become hospitable for colonization and was colonized by a single species which in turn originated other species. This scenario, together with characterizing species by their respective niches, results in a process similar to the mechanism of "competitive speciation" proposed by Endler (1977), Tauber & Tauber (1977), Rosenzweig (1978) and Gibbons (1979).

Both models result in a decrease in the ratio of interspecific to intraspecific competition coefficients. In other words, niche overlaps decrease and interspecific competition becomes weaker a result of the species turnover caused by repeated invasions and extinctions. This trend is reversed at the end of the simulation in the case of our second, niche-based, model. When there are very few species left in the community at the end of the simulation, they interact strongly, and soon there is only one species left. In the fossil record, however, usually the whole clade goes extinct. This does not, in our opinion, invalidate our model. Besides the difficulties in detecting a single species, and the possibility that it may go extinct because of environmental factors, the last species to evolve may be so different from the rest of the species in the clade that, it can be classified in a different family. Disappearance of some groups has been attributed to such taxonomic artefacts and are thought to be pseudoextinctions (Smith, 1988).

If our methodology reasonably represents natural dynamics, this approach may explain the differences observed between fossil records of different taxonomic groups based on the dimensionality of their niche space. As our figures demonstrate, different number of resources (niches with different number of dimensions) impose distinctive speciation dynamics (speciation rates, mean and maximum longevities, etc.) that can be recognized in the fossil record.

Acknowledgments. We would like to thank Jeffrey Levinton for helpful discussions and Geoff Jacquez for useful comments on the manuscript. This is contribution number 699 in Ecology and Evolution, State University of New York at Stony Brook. The simulations were made using the Cornell National Supercomputer Facility, a resource of the Center for Theory and Simulation in Science and Engineering at Cornell University, which is funded in part by the National Science Foundation, New York State, and the IBM Corporation. Support to both authors was provided by grants from the U.S. Environmental Protection Agency (Gulf Breeze Laboratory, Florida), and the Electric Power Research Institute (Palo Alto, California).

References

Abrams P (1980) Comments on measuring niche overlap. Ecology 61:44-49

Aoki K (1980) A criterion for the establishment of a stable polymorphism of a higher order with an application to the evolution of polymorphism. J Math Biol 9:133-146

Brown JH, Lieberman GA (1973) Resource utilization and coexistence of seed-eating desert rodents in sand dune habitats. Ecology 54:788-797

Clark AG, Feldman MV (1986) A numerical simulation of the one-locus multiple-allele fertility model. Genetics 113:161-176

Cody ML (1974) Competition and community structure. Princeton University Press New Jersey

Colwell RK, Futuyma DJ (1971) On the measurement of niche breadth and overlap. Ecology 52:567-576

Colwell RK, Winkler DW (1984) A null model for null models in biogeography. Pp. 344-359 in: DR Strong, D Simberloff, LB Abele, AB Thistle (eds) Ecological communities: conceptual issues and the evidence. Princeton University Press New Jersey

Connor EF, Simberloff D (1979) The assembly of species communities: chance or competition? Ecology 60:1132-1140

Culver DC (1970) Analysis of simple cave communities: niche separation and species packing. Ecology 51:949-958

Deakin MAB (1975) The steady state of ecosystems. Math BioSci 24:319-331

DeAngelis DL (1975) Stability and connectance in food web models. Ecology 56:238-243

DeAngelis DL, Goldstein RA, O'Neill RV (1975) A model for trophic interactions. Ecology 56:881-892

Diamond JM, Gilpin ME (1982) Examination of the "null" model of Connor and Simberloff for species co-occurances on islands. Oecologia 52:64-74

Endler JA (1977) Geographic variation, speciation and clines. Princeton University Press New Jersey

Gardner MR, Ashby WR (1970) Connectance of large dynamic (cybernetic) systems: critical values for stability. Nature 228:784

Gibbons JRH (1979) A model for sympatric speciation in *Megarhyssa*. (Hymenoptera: Ichneumonidae): competitive speciation. Am Natur 114:719-741

Gillespie JH (1977) A general model for enzyme variation in natural populations. III. Multiple alleles. Evolution 31:85-90

Gilpin ME, Case TJ (1976) Multiple domains of attraction in competition communities Nature 61:40-42

Ginzburg LR, Akçakaya HR, Kim J (1988) Evolution of community structure: Competition. J theor Biol 133:513-523

Gould SJ, Raup DM, Sepkoski JJ, Schopf TJM, Simberloff DS (1977) The shape of evolution: a comparison of real and random clades. Paleobiology 3:23-40

Karlin S (1981) Some natural viability systems for a multiallelic locus: a theoretical study. Genetics 97:457-473

Karlin S, Feldman MW (1981) A theoretical and numerical assessment of genetic variability. Genetics 97:475-493

Levins R (1968) Evolution in changing environments. Princeton University Press New Jersey

Lewontin RC, Ginzburg LR, Tuljapurkar SD (1978) Heterosis as an explanation for large amounts of polymorphism. Genetics 88:149-170

Margalef R (1968) Perspectives in ecological theory. University of Chicago Press Chicago

Marks RW, Spencer HG (1989) The maintenance of single-locus polymorphism. II. The evolution of fitnesses and allele frequencies Amer Natur (in press)

May RM (1972) Will a large complex system be stable? Nature 238:413-414

May RM (1973) Stability and complexity in model ecosystems. Princeton University Press New Jersey

May RM (1975) Some notes on estimating the competition matrix, *a*. Ecology 56:737-741

McMurtie RE (1975) Determinants of stability of large randomly connected systems. J theor Biol 50:1-11

Pianka ER (1973) The structure of lizard communities. Ann Rev Ecol Syst 4:53-74

Post WM, Pimm SL (1983) Community assembly and food web stability. Math BioSci 64:169-192

Roberts A (1974) The stability of a feasible random ecosystem. Nature 251:607-608

Roberts A, Tregonning K (1980) The robustness of natural systems. Nature 288:265-266

Robinson JV, Valentine WD (1979) The concepts of elasticity, invulnerability and invadability. J theor Biol 81:91-104

Roughgarden J (1979) Theory of population genetics and evolutionary ecology: an introduction. MacMillan New York

Rosenzweig ML (1978) Competitive speciation. Biol J Linn Soc 10:275-289

Rosenzweig ML, Taylor JA (1980) Speciation and diversity in Ordovician invertebrates: filling niches quickly and carefully. Oikos 35:236-243

Schoener TW (1968) The Anolis lizards of Bimini: resource partitioning in a complex fauna. Ecology 49:704-726

Schoener TW (1974) Some methods for calculating competition coefficients from resource-utilization spectra. Amer Natur 104:73-83

Siljak DD (1975) When is a complex ecosystem stable? Math BioSci 25:25-50

Smith AB (1988) Patterns of diversification and extinction in early palaeozoic echinoderms. Palaeontology 31:799-828

Spencer HG, Marks RW (1988) The maintenance of single-locus polymorphism. I. Numerical Studies of a viability selection model. Genetics 120:605-613

Strong DR, Simberloff D, Abele LB, Thistle AB (eds) (1984) Ecological communities: conceptual issues and the evidence. Princeton University Press New Jersey

Tauber CA, Tauber MJ (1977) Sympatric speciation based on allelic changes at three loci: evidence from natural populations in two habitats. Science 197:1298-1299

Taylor PJ (1985) Construction and turnover of multispecies communities: a critique of approaches to ecological complexity Ph.D. Thesis. Harvard University Cambridge MA

Tregonning K, Roberts A (1978) Ecosystem-like behavior of a random interaction model I. Bull Math Biol 40:513-524

Tregonning K, Roberts A (1979) Complex systems which evolve towards homeostasis. Nature 281:563-564

A.1. Theoretical Framework

Marginal Populations in Competitive Guilds

V. Loeschke

Department of Ecology and Genetics, University of Aarhus, Ny Munkegade, 8000 Aarhus C, Denmark

Introduction: A scenario for marginal populations

The term marginal populations has been used in a geographical sense as well as in describing a population adapted to a harsh type of environment or an ill-adapted population which persistence is crucially dependent on the inflow of migrant individuals. At least the latter concept implies that marginal populations share properties with transient or unstable populations.

In the following a somewhat different characterization of being marginal will be used, though it contains properties of the geographical as well as the adaptive concept of marginality. Consider a guild of resource limited populations that utilize resource spectra which consist of items of different size, but similar type (seeds of different plants, insects of different size etc.). Let us assume that a gradient exists from North to South in the mean of the available resources (for instance, further south seeds are on average larger). Further assume that there is a temporary change in the resource distribution at each location along the North South gradient (short-term shifts in the mean and/or in the variance). Our species of concern finds its northern populations located in a position rather central concerning its utilization of resource items. Further south it is becoming increasingly marginal in the utilization of available resources, even though it might have adapted to a certain degree to

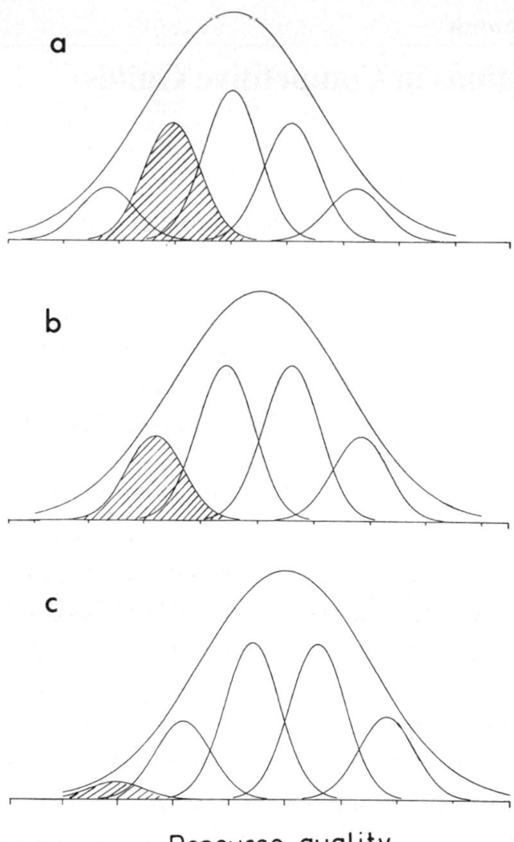

Figure 1: A scenario for a marginal population. (a) Our species has its resource utilization (the dashed area) at a rather central position in the northern guild of competitors which is at coevolutionary equilibrium for five resident species with a relative resource width $\sigma/w = 3$. The height of the utilization functions correspond to the relative densities of the competing populations. (b) Further south the resource spectrum has shifted its mean consisting of on average larger resource items. The left marginal competitor of the northern guild has gone extinct. The guild consists now of four competitors at coevolutionary equilibrium. Our species has a slightly different resource utilization function when compared to the northern population. (c) Still further south the resource spectrum has further shifted its mean towards still larger items. Our species has now a rather marginal location of its utilization function. Here the equilibrium location and density of its utilization function (height) are given assuming that the resident species respond to the marginal invader in their population sizes but not by a change in the utilization means.

a different resource abundance distribution (Fig. 1) through selection favoring traits that get a larger share of available resources. An intuitive conceptual basis of this scenario presupposes that a character like body length, beak depth or head size is directly related to the way resources are exploited.

The Model

The description of intra- and interspecific competitive interactions within this scenario uses a model based on the niche concept of MacArthur & Levins (1967; Roughgarden 1976; Christiansen and Fenchel 1977; Christiansen and Loeschcke 1980; Matessi and Jayakar 1981; Case 1982; Loeschcke 1984, 1985).

Let us consider a guild of u species competing for renewable resources and assume a genetically controlled variable character in these species to be directly related to the exploitation of resources. For simplicity assume a one locus m allele autosomal inheritance of the character in a diploid, random mating population with nonoverlapping generations. The resource utilization by genotype mn in species i (i = 1, 2, .., u) is specified by Gaussian utilization functions U_{imn} with mean D_{imn} and variance W^2. The variance W^2 is assumed independent of the mean and equal among species. The utilization mean is given as $D_{imn} = D_i + d_{im} + d_{in}$, where D_i is a scaling constant and d_{im} and d_{in} are the contributions of alleles m and n in species i, i.e, we assume additivity in allele contributions.

Competition coefficients $\alpha_{imn,jkl}$ between genotypes mn in species i and genotype kl in species j are defined through the overlap of the utilization functions

$$\alpha_{imn,jkl} = \exp\left[-(D_{imn} - D_{jkl})^2 / 4W^2\right].$$

We further assume a continuously renewed biotic resource with a short generation time compared to that of the competitors (predators) that share the resources. In the absence of the competitors and on a logarithmic scale the resource spectrum $S(\tau)$ is assumed to be proportional to a Gaussian distribution with mean M and variance σ^2. Then we get the carrying capacity of genotype mn in species i as

$$K_{imn} = \exp\left[-(M-D_{imn})^2 / (2\sigma^2 + 2W^2)\right].$$

The recursions determining population dynamics and evolution in the system are for the population sizes

$$x'_i = x_i \left[1 + V\left(K_i - \sum_{j=1}^{u} \alpha_{i,j} x_j\right)\right], \quad i = 1, 2, \quad , u,$$

and for the allele frequency of allele m in species i

$$p'_{im} = p_{im} \left[1+V\left(K_{im} - \sum_{j=1}^{u} \alpha_{im,j} x_j\right)\right] / \left[1+V\left(K_i - \sum_{j=1}^{u} \alpha_{i,j} x_j\right)\right],$$

where V is a proportionally constant related to the efficiency of the conversion of resources into individuals and K_i, K_{im}, $\alpha_{i,j}$ and $\alpha_{im,j}$ are averages (for details of parameter definition see Loeschcke 1984).

Results

First, I will assume a constant resource spectrum and give the condition for invasion of marginal populations into coevolved competitive guilds, characterize equilibrium sizes and the initial rate of increase of colonizing marginal populations (Loeschcke 1985). Then I will characterize genetic variation in marginal populations, conditions for polymorphism and whether the polymorphism is protected, and the time to reach genetic equilibrium. Thereafter, I will look at the implications a temporary changing resource spectrum can have on marginal populations. Finally, I will discuss the expected amount of genetic variation in marginal populations, the likelihood of chance extinction, and the role of historical aspects within the given scenario for marginal populations.

A constant resource spectrum

Assume a guild of competitors being at coevolutionary equilibrium (Matessi and Jayakar 1981; Loeschcke 1985) and consider species i to invade the u-species community. This is possible if $K_i > \sum_{j=1}^{u} \alpha_{ij} \hat{x}_j$ for $i = u+1$ (Loeschcke 1985). If invasion is successful species i will eventually reach its equilibrium size, x_i, in a guild of now u+1 species when $K_i = \sum_{j=1}^{u+1} \alpha_{ij} \hat{x}_j$ and $\hat{x}_j \geq 0$. The rate at which the population size of the invader will initially increase is $\triangle_i = (x_i'-x_i)/x_i$ V disregarding intraspecific competition (Loeschcke 1985).

Figure 2 shows for coevolved guilds of three and four monomorphic resident species that equilibrium size and, especially, the rate of increase is considerably higher for marginal invaders than for populations invading within the range of resident species. Hereby, it is assumed that the resident species respond to the new exploiter by a change in population sizes (ecological time scale), but not in their mode of resource utilization (evolutionary time scale). The result holds independent of the number of resident species in the coevolved guild, and the probability of an invasion to be initially successful therefore is generally higher for species invading at the margin of the resource spectrum than between residents. This difference gets even more pronounced when the resident species are genetically variable.

Let us now consider the pattern of genetic variation in a marginal population. From an equilibrium oriented approach the

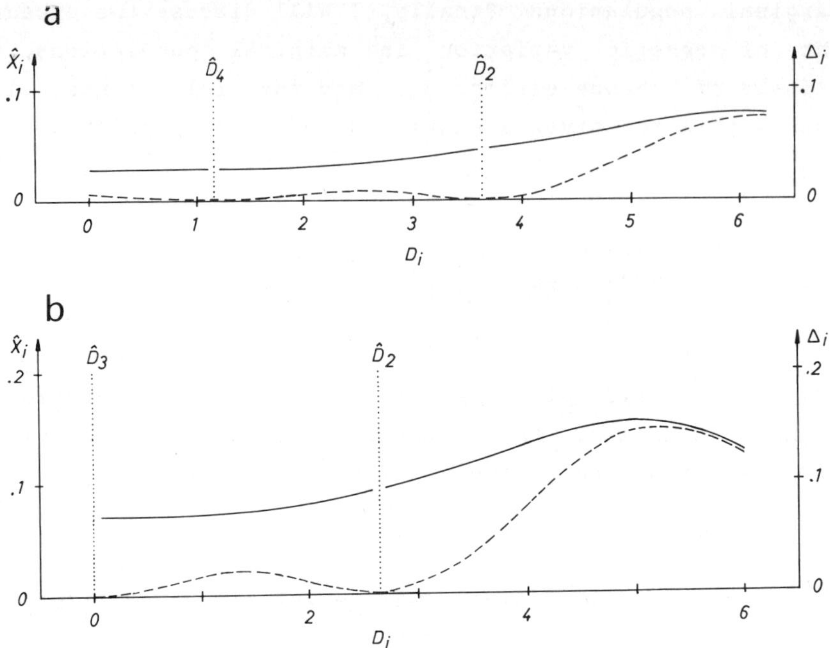

Figure 2: Equilibrium population size $\hat{x}_i$ (solid lines) and initial rate of increase Δ_i (dashed lines) as a function of the utilization mean D_i of an immigrant population invading communities of (a) three and (b) four resident species. The relative resource width, σ/W, is set at 3. Dotted lines indicate the equilibrium utilization means of the resident species. Because curves are symmetrical, left margins are omitted. (After Loeschcke 1985).

structure of genetic variation in marginal populations can be characterized like with pure intraspecific competition as described by Christiansen and Loeschcke 1980 (see also Loeschcke 1983; Christiansen and Loeschcke 1989). If resource utilization takes place away from the evolutionary equilibrium location of the invading population, directional selection will favor genotypes that utilize the resource closest to the equilibrium location and the population will eventually become monomorphic for the allele m with the smallest contribution in

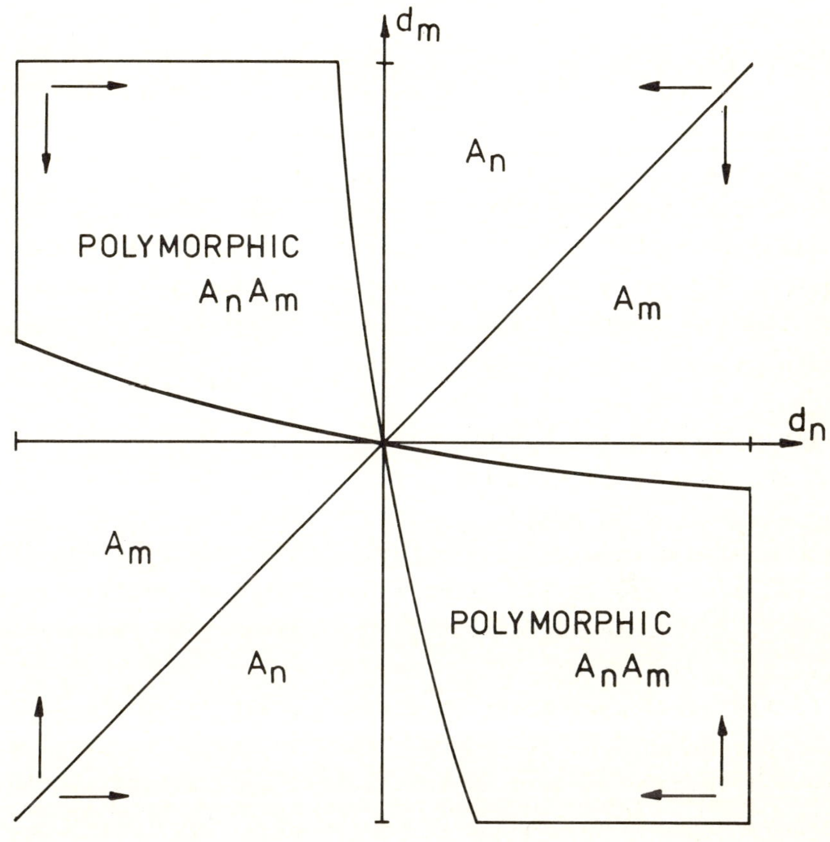

Figure 3: The qualitative pattern of genetic variation at equilibrium in a marginal population as a function of the contributions of alleles A_n and A_m, d_n and d_m ($d_k = 0$ at the evolutionary equilibrium location). If the contributions are of the same sign, the allele with the smaller absolute contribution will become fixed at equilibrium. It will become substituted, however, by any new mutant with a smaller contribution of the same sign as indicated by the arrows. If alleles exist with contributions of different sign and not too different magnitudes, equilibria will be polymorphic. Again, every new mutant with a smaller contribution will substitute the one of the same sign. Thus polymorphism is transient and evolution will eventually lead to monomorphism.

absolute value, $|d_m|$, if the scaling constant is equal with the evolutionary equilibrium. With the resource utilization close to the equilibrium location, two-allele polymorphism is possible, if alleles with contributions of different sign exist, i.e if $d_m < 0 < d_n$, and if the contributions are not too different in magnitude (Figure 3). In contrast to the pure intraspecific model of Christiansen and Loeschcke (1980), the mode of selection in the marginal population close to the equilibrium location is not a simple function of the relative resource width, $\delta = \sigma/w$. Even for a rather large relative resource width as $\delta = 3$, the mode of selection is always stabilizing selection in the example with four resident species and one marginal population, i.e. polymorphism is transient and alleles with smaller contributions will replace the ones with the same sign until eventually a monomorphic equilibrium is reached (Figure 3). From an equilibrium oriented approach we will therefore not expect polymorphism to be maintained in marginal populations.

Let us now consider the time scale at which equilibria are approached, and let us do this by using a three allele example which is depicted in figure 4 by a DeFinetti diagram. We have set σ/w at 3 and assumed the guild to consist of four coevolved resident species and a marginal invader (see Figure 1c and 2b). We start the recursions for the gene frequencies in the marginal population close to each of the three vertices representing monomorphic populations with utilization means set to $D_{i11} = 5.9$ ($d_1 = -0.14$), $D_{i22} = 6.14$ ($d_2 = -0.02$), and $D_{i33} = 6.3$ ($d_3 = +0.06$) respectively, assuming the two rare alleles to have initial frequencies $p_k = p_l = 0.01$ (generation 0). Starting from the two lower vertices at the base of the triangle allele frequencies do not change much between about generation 10000 and generation 30000 and they are still far from equilibrium after more than 100000 generations. At equilibrium $\hat{p}_1 = 0$, $\hat{p}_2 = 0.78$ and $\hat{p}_3 = 0.22$. Still, when starting monomorphic the recursions close to the vertix characterized by $p_3 = 1$, p_1 will increase for about 10000 generations before it slowly starts to decrease until it gets close to zero after about 200000 generations. The time scale

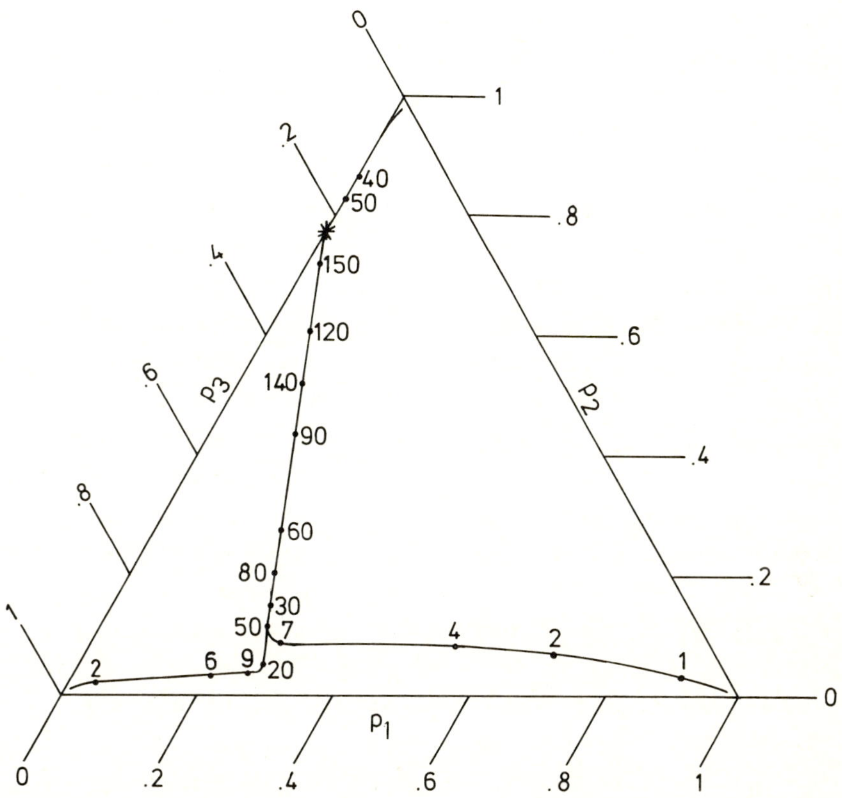

Figure 4: Schematic representation of allele frequency changes in an equilateral diagram (DeFinetti diagram) as a function of the number of generations. The initial frequencies of the three alleles are close to the vertices of the triangle representing monomorphic equilibria. The allele frequency changes over generations are depicted by curves within the triangle with the number of generations (in thousands) given besides the curves. The two curves starting close to the lower vertices fuse after some tens of thousands generations. After fusion the generation number for the curve starting at the right lower vertices is given to the right side of the curve and that for the other curve at the left side. The star on the left side represents the genetic equilibrium. For further explanation see text.

at which equilibria are approached is not so much dependent on the relative resource width, $\delta = \sigma/w$ (on average slightly slower for $\delta = 2$), but more on the size of the allele contributions and the mode of selection. The smaller the contributions the weaker is selection and the longer it will take to approach the equilibrium. Only in case of directional selection, i.e. with $d_m < 0$ or $d_m > 0$ for all m ($m = 1$, 2 and 3), the approach to genetic equilibrium is considerably faster than with stabilizing selection, but it is still in the order of magnitude of thousands of generations. Natural selection therefore does not play any role for the amount of genetic variation that prevails in any marginal population in time spans of relevance for such populations, as selection generally is weak and equilibria are unlikely to be attained within such time spans.

A temporary changing resource spectrum

Temporary or short-term changes in the availability are likely to occur in a biotic resource. Such changes may result in an increase or decrease in the resource width, σ/W, in a shift in the resource mean M, or in a change in the absolute amount of available resources (super- or subabundance). As the population dynamics of the resources in the model is assumed to occur on a much faster time scale than that of the guild of competitors, changes in the resource distribution appear as sudden on the consumer level.

Temporary changes in the resource distribution can have drastical effects on marginal populations. A temporary reduction in the variance of the resource abundance, a shift of the resource mean away from the utilization mean of the marginal population, as well as a general decline or shortage in resource availability cause a decrease in the initial rate of increase, Δ_i, and the equilibrium size, $\hat{x}_i$. This may endanger the persistence of the marginal population that already is relatively small in size and thereby vulnerable to all kinds of fluctuations. For example a reduction of the relative resource width, σ/W, from 3 to 2, leads to negative rates of initial increase of the marginal invader in a coevolved guild of four residents (Loeschcke 1985).

An increase in the variance of the resource, a shift of the resource mean closer to the utilization mean of the marginal population, or a temporary superabundance may help to get a marginal invader established within the guild, as it positively affects the initial rate of increase, but as long as the change in resource availability is temporary, it will not have persisting effects on the composition of the guild or on the genetic structure of the marginal population.

Discussion: Genetic variation, chance extinction and historical events in marginal populations

On basis of a deterministic model we have argued that natural selection is generally too weak to play any decisive role for the amount of genetic variation that prevails in marginal populations, as genetic equilibria will only be attained at an evolutionary time scale. As marginal populations are by our definition finite populations, their genetic variability is dependent on the effective population size (Wright 1969; Kimura 1983; Lande and Barrowclough 1987) and on the genotypes that enter the population by migration.

In populations of small effective size genetic drift becomes of major importance for the genetic structure of a population. An isolated marginal population of finite size is expected to be less polymorphic and to consist of fewer heterozygotes than a large population as the loss of genetic variation by drift is not balanced by mutation. Even rather large isolated populations need thousands of generations to recover from a temporary bottleneck if single locus variation is to be regained (Lande and Barrowclough 1987). Marginal populations, however, will, according to the definition, often experience immigration by individuals from other populations. In contrast to genetic drift, migration will increase the allelic variation in the marginal population. A quantitative assessment of the relative role of the two opposing forces drift and migration and their implications on the amount of genetic variation is not given here, but a few migrants per generation that participate in reproduction will generally be enough to equalize the genetic structure of the marginal population and the source population. One difference that may remain if the marginal population is considerably smaller than the source population, is that the marginal population due to drift will have fewer of the rare alleles than the larger source population.

Marginal populations as defined here are vulnerable to extinction by chance because their population sizes on average are low. Random variation in the growth rate and in its variance as a result of variation in resource abundance and density independent death rates (white noise or environmental stochasticity), in birth- and death rates (demographic stochasticity), and in the time periods of the occurrence of catastrophes affects population dynamics in marginal populations and thus influences the likelihood of extinction (Keiding 1975; Shaffer 1981, 1987; Goodman 1987; Belovsky 1987; Ewens et al. 1987). The combination of random events and other factors, intrinsic to the biology of a population as e.g. social interactions, may also press the population under a minimum size necessary for survival (Jacobs 1984).

Historical events can have a large impact on the species composition of competitive guilds and on the pattern of genetic variation in marginal populations within the guild. Coevolved guilds are difficult to invade at a central position of the resource spectrum (Loeschcke 1985). The likelihood of invasion to be successful is much higher at the margin of the resource spectrum, as is the likelihood of extinction due to chance or fluctuations in the resources. Within the physiological constraints of a species often more than one of the modes of resource utilization is possible that are realized by the different species of a given guild. Then the actual mode of resource utilization of a population can depend on the history of the build-up of the guild (Loeschcke 1987). Historical events play as much a role for the genetic composition of a marginal population if it is rather isolated from other populations of the same species. The genotypes that founded the population and the chance dominated fate of the alleles constituting these genotypes in the initial phase of increase in population size will for a large number of generations determine the genetic composition of the population.

Summary

A scenario for marginal populations is suggested that emphasizes the marginality in resource exploitation within a competitive guild. Using a MacArthur and Levins type of competition model, the condition for invasion of a marginal population into a coevolved guild, its initial rate of increase and its equilibrium size are given. It is shown that invasion is more likely to be successful if it occurs at a marginal position of the resource spectrum, but that marginal invaders are very vulnerable to temporary fluctuations in the resources.

The pattern of genetic variation in marginal populations and the time to reach genetic equilibrium are characterized for constant and temporary changing resources. It is concluded that natural selection is too weak to play any role for the amount of genetic variation that prevails in marginal populations in time spans of relevance for such populations, and that genetic drift and migration are of major importance for the genetic structure in marginal populations.

Finally, chance extinction and the impact of historical events on the species composition in competitive guilds and on the genetic composition of marginal populations are discussed.

Acknowledgement: I am grateful to Freddy B. Christiansen and Jørn Madsen for critical reading of the manuscript, to Kirsten Petersen for collating computer outputs and typing some versions of the manuscript, to Arno Jensen for drawing the figures, and the Research Foundation of the University of Aarhus (Aarhus Universitets Forskningsfond) for financial support.

References:

Belovsky GE (1987) Extinction models and mammalian persistence. In: Soulé ME (ed) Viable populations for conservation. Cambridge Univ Press, Cambridge, pp 35-38

Case TJ (1982) Coevolution in resource-limited competition communities. Theor Popul Biol 21: 69-91

Christiansen FB, Fenchel TM (1977) Theories of populations in biological communities. Springer Verlag, Berlin Heidelberg New York

Christiansen FB, Loeschcke V (1980) Evolution and intraspecific exploitative competition. I. One-locus theory for small additive effects. Theor Popul Biol 18:297-313

Christiansen FB, Loeschcke V (1989) Evolution and competition. In: Wöhrmann K, Jain SK (eds) Topics in population biology and evolution. Springer Verlag, Berlin Heidelberg New York (in press)

Ewens WJ, Brockwell PJ, Gani JM, Resnick SI (1987) Minimum viable population sizes in the presence of catastrophes. In: Soulé ME (ed) Viable populations for conservation. Cambridge Univ Press, Cambridge, pp 54-68

Goodman D (1987) The demography of chance extinctions. In: Soulé ME (ed.) Viable populations for conservation. Cambridge Univ Press, Cambridge, pp 11-34

Jacobs J (1984) Cooperation, optimal density and low-density thresholds: yet another modification of the logistic model. Oecologia (Berlin) 64: 389-395

Keiding N (1975) Extinction and growth in random environments. Theor Popul Biol 8:49-63

Kimura M (1983) The neutral theory of molecular evolution. Cambridge Univ Press, Cambridge

Lande R, Barrowclough F (1987) Effective population size, genetic variation, and their use in population management, In: Soulé ME (ed) Viable populations for conservation. Cambridge Univ Press, Cambridge, pp 87-124

Loeschcke V (1983) Species diversity and genetic polymorphism: dual aspects of an exploitative competition model. Atti Assoc Genet Ital (suppl.) 29:53-66

Loeschcke V (1984) The interplay between genetic composition, species number and population sizes under exploitative competition. In: Wöhrmann K, Loeschcke V (eds) Population biology and evolution. Springer Verlag, Berlin Heidelberg New York, pp 235-246

Loeschcke V (1985) Coevolution and invasion in competitive guilds. Am Nat 126:505 -520

Loeschcke V (1987) Niche structure and evolution in ecosystems. In: Schulze ED , Zwölfer H (eds) Ecological Studies Vol 61. Springer Verlag, Berlin Heidelberg New York, pp 320-332

MacArthur RH, Levins R (1967) The limiting similarity, convergence, and divergence of competing species. Am Nat 101:377-385

Matessi C, Jayakar SD (1981) Coevolution of species in competition: a theoretical study. Proc Natl Acad Sci USA 78:1081-1084

Roughgarden J (1976) Resource partitioning among competing species - a coevolutionary approach. Theor Popul Biol 9:388-424

Shaffer M (1981) Minimum population sizes for species conservation. Bioscience 31: 131-134

Shaffer M (1987) Minimum viable populations: coping with uncertainty. In: Soulé ME (ed) Viable populations for conservation. Cambridge Univ Press, Cambridge, pp 69-86

Wright S (1969) Evolution and the genetics of populations Vol 2. Univ of Chicago Press, Chicago

Flush-Crash Experiments in *Drosophila*

A. Galiana, F.J. Ayala[1], and A. Moya

Laboratori de Genética, Universitat de València, c/Dr. Moliner, 50,
46100 Burjassot, Spain

1. INTRODUCTION

In 1968, Hampton Carson (1968) proposed the founder-flush-crash model, a new model of speciation, which readily lends itself to experimental testing in the laboratory (Carson 1971).

Carson's model was motivated by the conditions presumed to have facilitated the extensive adaptative radiations experienced by drosophilids in the Hawaiian archipelago (Carson 1975; Carson and Kaneshiro 1976).

Some authors have questioned whether the founder-flush-crash model is relevant to circumstances other than those of the Hawaiian drosophilids or even there (Barton and Charlesworth 1984, Ringo 1977). The conditions under which the model applies are, indeed, quite restrictive (Templeton 1981; Carson and Templeton 1984); but the model has received substantial theoretical elaboration (Carson 1975; Carson and Templeton 1984; Templeton 1980, 1981) and has been explored in a number of laboratory experiments (Bryant et al. 1986; Dodd and Powell 1985; Powell 1978; Ringo 1987; Ringo et al. 1985, 1986). We shall herein review these experiments and report the results of an extensive study conducted in our laboratories over the last five years.

1 Department of Ecology and Evolutionary Biology, University of California, Irvine, 92717 California, U.S.A.

2. THE FOUNDER-FLUSH-CRASH MODEL

Carson's model assumes that there is an ancestral population from which colonizers arrive to new habitats, empty of competitors. In the laboratory, these empty habitats are culture bottles or population cages. Exponential population growth would occur in a new habitat for a number of generations while abundant resources persist: this is the flush phase of the model. We will refer to these "island" populations derived from the colonizers as derived lines. There can be a succession of colonization events: after the derived line fills its habitat, a propagule leaves this habitat and colonizes a new empty island. The exponential growth in a newly colonized habitat eventually yields resource saturation, at which point natural selection due to competition for limited resources will occur. Carson's model assumes that a crash in population numbers is likely to occur, which is referred to in the experiments as the crash phase. Thus, we have a cycle of colonization, flush and crash, or simply the "flush-crash cycle". Because inbreeding effects may arise as a consequence of colonizing events and may be important, lines in the laboratory experiments are usually maintained that go through several successive generations at very low numbers: these are the inbred (derived) lines.

The founder-flush-crash model has implications in three dimensions: demographic, ecological, and genetic.

The experimental protocol seeks to reproduce a distinctive demographic pattern by carrying the population through several bottlenecks, every one followed by a fast expansion in population numbers. Population numbers have, obviously, important genetic consequences (Carson 1968; Templeton 1980).

The ecological scenario of the model is accomplished experimentally by reducing competition for resources as much as possible during the beginning of the flush phase, which is eventually followed by substantial competition that occurs towards the end of the cycle (crash phase). The population crash acts like a sickle in that it should permit the survival of only some genotypes that are succesful competitors. These may be quite different from the ancestral ones, since new genetic arrays may arise during the flush phase owing to mutation and recombination under tolerant conditions with little or no natural selection due to competition for resources.

The genetic requirements of the model are the most difficult to monitor in laboratory experiments. The founders must have genetic traits likely to trigger an episode of fast evolution. The best one can do is to conduct as large a number of experiments as possible, as well as to choose advisely both the species and the ancestral population. Genetic considerations suggest that colonizing species should be avoided, for their genotypes are presumably adapted to resist founder events; they are succesful at creating new populations but not new species. It is usually difficult to say whether or not a given species should be considered to be a colonizing species, because we do not have a quantitative, objective way of telling, for instance, where a species might be placed in the gradient between _r_ and _K_ selection strategies.

Templeton (1980) is the author who has explored the most the genetic dimension of the founder-flush-crash model. The conditions that he has identified as favorable include high chromosome number, long recombinational map, abundant offspring, and panmictic mating. Within the _Drosophila_ genus, Templeton indicates that _D. pseudoobscura_ is a good choice, althogh _D. virilis_ might perhaps be even more promising.

It would seem that Hawaiian species would provide the most promising options, but they remain difficult to culture in the laboratory in spite of recent improvements (Spieth 1980).

3. EXPERIMENTAL DESIGN

Our experimental species is _D. pseudoobscura_, as it was for Powell (Powell 1978; Dodd and Powell 1985). Two separate founder populations are used, derived each from a number of strains, collected respectively at Bryce Canyon National Park (Utah, U.S.A.) and Lago Zirahuén (Michoacán, México), about 2000 km apart from each other. The strains were kindly supplied by Dr. Wyatt Anderson. Additional details about the ancestral populations and how they were established, are given also in Moya et al. (in preparation).

The two ancestral populations will be herein symbolized as BCA and MA, for the U.S. and Mexican population, respectively. Through repeated backcrosses for a number of generations, we introduced the orange (or) eye-colour marker in BCA and sepia (se) in MA. Chromosomally, BCA was completely monomorphic, whereas MA was extensively polymorphic for third chromosome rearrangements.

The derived lines were started in December 1984, each with either 1, 3, 5, 7, or 9 pairs of virgin flies. These derived lines are shown in Table 1.

TABLE 1. The 45 derived lines

n	BCA populations	MA populations
1	BC2, BC4, BC32, BC33	M3, M5, M37
3	BC7, BC9, BC10, BC11, BC12, BC34	M7, M8, M10, M11, M12, M36
5	BC13, BC14, BC15, BC17, BC18	M13, M14, M15
7	BC19, BC20, BC21, BC22, BC23, BC24	M19, M20, M21
9	BC25, BC26, BC27, BC28, BC29, BC30	M25 ,M26, M27

n : number of pairs used as founders in each bottleneck. BCA and MA refer to the localities of origin.

3.1. Protocol

The experimental populations were started in 1985 at the University of California at Davis, where the flies were raised in half-pint bottles at 21°C. In December 1985, the populations were carried to the University of Valencia, Spain, where they were cultured in 150ml bottles, first at 25°C and after several months at 21°C.

Each flush phase is as follows: the n pairs of founders are allowed to lay eggs in one culture bottle for one week, and then transferred to a second bottle for another week, and then discarded. 100 adult flies are randomly chosen among the flies emerging from these two bottles. These flies become the parents of the next generation by distributing them among five cultures, at 20 flies per culture bottle; after one week they are moved to a second set of five bottles, and after another week they are discarded. Thus, in this second generation there are 10 culture bottles per experimental population, a condition that is mantained throughout the flush phase. The opportunity for population expansion, without competition for resources (r-selection) is provided every flush generation by taking one-pair of newly emerged flies from each one of the 10 cultures in order to establish each one of the five cultures of the next generation. The 20 flies in each culture are transferred to a new culture after one week, and discarded after one more week -- so that we have 10 cultures with 20 parents each in each generation througout the flush phase. The selection of each pair of flies from each of the ten cultures of the previous generation insures that gene flow is mantained, while at the same time the conditions for population expansion are mimicked by reducing the number of parents to 20 in each of 10 bottles in every generation. (We estimate that if all flies in each generation would have been allowed to reproduce in similar conditions as those selected as parents, each experimental population would have expanded to more than one million flies by the end of each flush phase.)

In each cycle we allow for 4 to 7 flush generations before we begin the crash phase. After the last flush generation of a cycle, five crash cultures are established by putting together in each new bottle all flies emerged from two cultures of the last flush generation, one from the first set of five flush bottles and the other one from the second set, but one with different parents. About 300 flies are thus allowed to become parents in each of these crash cultures, instead of the 20 used during the flush phase. This crowding leads to competition for food and other resources, particularly among the larvae. After one week, these flies are transferred to a second set of five bottles; and they are finally discarded after one more week.

The crash phase consists of a single generation. A new flush-crash cycle is initiated by random selection of n virgin pairs of flies from any one of the ten crash cultures.

Each flush-crash cycle is completed in about half a year. We still maintain all the derived populations and continue to carry them through flush-crash cycles.

3.2. About inbreeding

We have assessed the population size for each line and generation in order to obtain an estimate of the average effective population size, N_e. The average N_e is calculated using Wright's harmonic-mean method. Population sizes are estimated for the second and the following flush generations, as well as for the crash generation, under the assumption that all discarded flies become parents under those same conditions experienced by the flies that are actually chosen to become parents of the following generation in the cultures. Thus, the first two or three flush generations -- those with the lowest estimated numbers -- contribute the most to the N_e estimate. In Table 2 are shown the values of the fraction of reduced heterozygosity after several flush-crash cycles.

As is apparent in the table, founder events do not necessarily lead to massive genetic fixation, even when the number of founders is small or after repeated founder events. Enzyme polymorphism data associated with natural founder events, support this conclusion (Baker and Moeed 1987; Janson 1987;

TABLE 2. Estimates of effective population size, $\underline{N}_e$, and fraction, $\underline{F}_t$, of reduced heterozygosity relative to the ancestral population. t refers to the considered number of cycles

n	N_e estimate	F_4	F_5
1	9.79	0.63	0.71
3	29.06	0.28	0.34
5	47.45	0.18	0.23
7	66.66	0.13	0.17
9	84.56	0.11	0.13

Sene and Carson 1977). However, some theoretical analyses indicate that little variability is likely to remain after extreme bottlenecks (Maruyama and Fuerst 1985, Motro and Thomson 1982; but see Goodnight 1988 and Templeton 1980).

3.3. Sexual isolation experiments

As it has been the case in former flush-crash experiments (Powell 1978; Ringo et al. 1985), we study mating behavior as the main measure of reproductive isolation. The reason is that premating reproductive barriers are thought proner to evolve under the conditions of the model than other reproductive isolating mechanisms. We obtain estimates of ethological isolation by means of multiple-choice mating tests. Four replicates are carried out for each cross between any two given lines. In each replicate, 12 (15 for all experiments after the 4th cycle) six-days-old virgin flies of each sex-line combination are put together in a mating chamber. Flies matings within the first 45 min (60 min after the 4th cycle) are removed from the mating chamber for identification. The mating tests are conducted during the second half of the morning in any given day, so as to avoid any major circadian distortions. In crosses between two lines derived from the same natural population, and hence with the same eye color, we follow the usual practice of clipping the tip of the wing in one of the two lines. We have not detected any significant bias in mating behavior due to this clipping.

We use five measures for analyzing our data: I (joint isolation index), X (which has the distribution of an independence chi square), V, Q (Gilbert and Starmer 1985) and Y (Ringo 1987). We report herein only the results obtained with the first measure, I, because there is good agreement among the results obtained with all five measures, despite their different formulations and distributions.

We have also made sexual isolation tests between inbred derived lines (data not shown). These lines underwent the usual flush-crash cycles, but with three (instead of one) consecutive generations consisting of one pair of flies each at the beginning of each cycle (see Figure 1). None of the inbred lines yields any significant deviation from random mating. This result coincides with Powell's (1978), and is predictable under the genetic transilience model (Templeton 1981). The harmonic mean (N_e) estimate for the inbred lines is 4.63, and the fixation indices (F_t) are 0.87 and 0.92, for the 4 and 5 cycles, respectively.

4. PREVIOUS EXPERIMENTS

Powell's (1978) experiment is the first published response to Carson's (1971) challenge. He chose as ancestral population a mixture of D. pseudoobscura strains from several localities. He reproduced "islands" in cages and allowed there the expansion of populations derived from one pair of flies. The choice of a "polyhybrid" ancestral population was motivated by the goal of starting with as much genetic variability as possible. Figure 1 diagrams his experimental protocol. The results of this first attempt were partially confirmatory of Carson's model; and the reproductive isolation achieved seemed stable enough (Dodd and Powell 1985). Tests between independently evolved populations (as measured by I, the joint isolation index, see below) were in some cases significantly homogamic; many of these significant tests involved one particular line ("1"). Furthermore, no homogamy was found between inbred derived lines that had not been subject to the flush-crash cycle.

Ringo et al. (1985) chose instead *D. simulans* for their experiments, but they also started with a polyhybrid ancestral population, and used population cages derived from one pair of founders (see Figure 1). Their results were neither so positive nor stable as Powell's. The instances of apparent reproductive isolation between lines could be attributed for the most part to lowered mating propensities induced by the bottlenecks (Ringo et al. 1986).

Table 3 identifies some distinctive characteristics for each of the three flush-crash *Drosophila* experiments reported to date.

TABLE 3. Comparison among flush-crash *Drosophila* experiments.

	Powell	Ringo	Moya
DESIGN:			
Species	*pseudoobscura*	*simulans*	*pseudoobscura*
Cultures	cages	cages	bottles
ANCESTRAL POPULATIONS:			
Number	one	one	two
Origin	polyhybrid	polyhybrid	local
Chromosomes	polymorphic	monomorphic	one polymorphic, one monomorphic
FLUSH-CRASH PROTOCOL:			
Cycle length	3-4 months	6 months	5-7 months
Flush generations	3-5*	10-12*	4-7
Crash	progessive	progressive	one generation
Derived lines	8	8	45
Founding couples	1	1	1,3,5,7, or 9
ISOLATION TESTS:			
After cycles	1 and 4	all 6 ones	4 and 5
Mating choice	multiple	multiple	multiple
Flies/line/sex	12	6	12
Homogamic crosses	15/41(36.5%)	12/216(5.5%)	31/276(11.2%)
Stability	good	not good	good

* overlapping generations

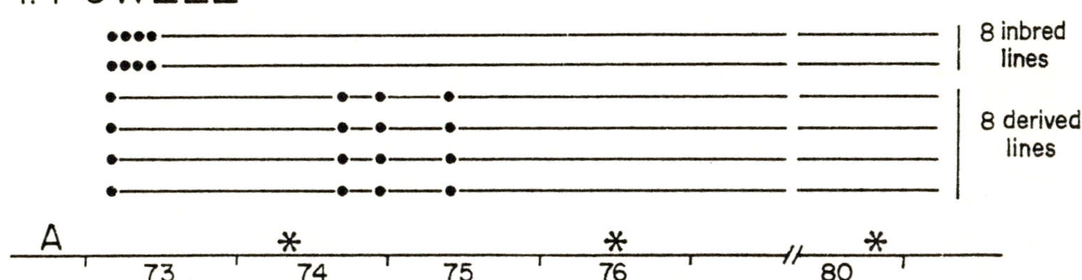

2. RINGO

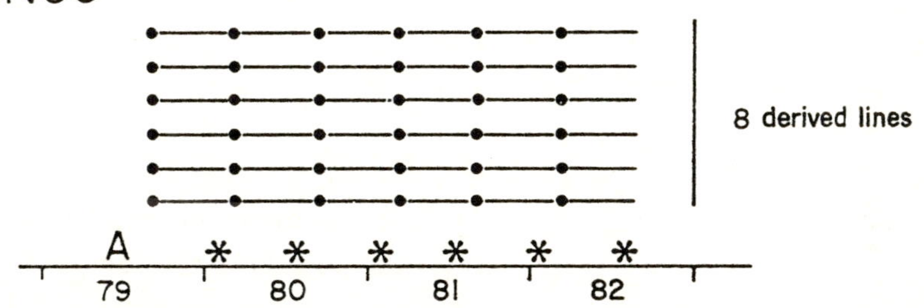

3. MOYA

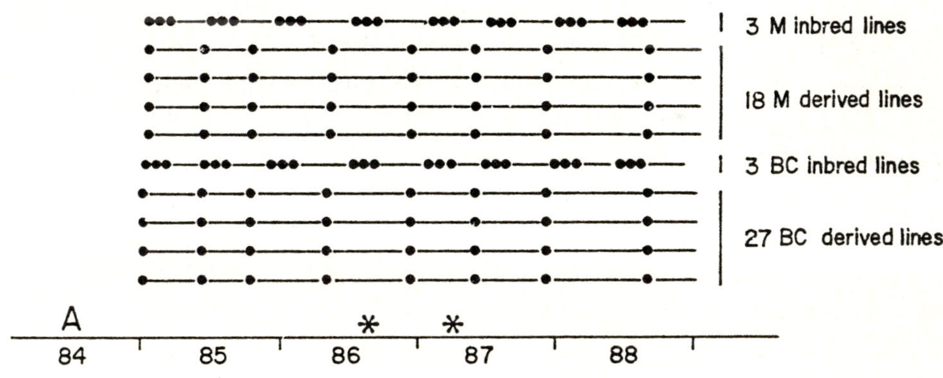

FIGURE 1. Diagram of the flush-crash protocol followed in three Drosophila experiments. The ancestral population is represented by A. Horizontal lines represent derived populations; dots represent bottleneck events. Asteriscs indicate the times at which tests were made for reproductive isolation between populations. M refers to the lines derived from the Mexican locality; BC to those from Bryce Canyon. See text and Table 3 for additional details

Our experiment is distinct in that (1) it uses individuals derived from one or another single local population; (2) it uses bottles for culturing the flies in such a way that the number of parents is known and little or no competition for resources exists during the flush generations; and (3) bottleneck sizes larger than one pair are used as well as those with only two founder flies. These characteristics would seem to fit the demands of the flush-crash model better than the alternatives followed in previous experiments.

5. RESULTS AND DISCUSSION

In a first series of sexual isolation tests carried out in 1986 (Moya et al., in preparation), we chose ten derived lines, one per bottleneck size and origin, together with the two ancestrals populations, and we made all possible pairwise combinations between them (a total of 66 crosses). The most interesting results involved lines with n=1 or 3. Consequently, and given the difficulty of undertaking a complete analysis, our second series of tests, started during the 5th flush-crash cycle in January 1987, and finished in September 1988, has involved all the possible pairwise combinations between 1- and 3-pair lines, as well as between each of them and both ancestral populations (a total of 210 crosses). Deviations from random mating are tested by means of the _I_ index.

In the experiment corresponding to the 4th cycle, 8 out of 66 (12%) of the pairwise combinations tested, showed statistically significant homogamy. In the 5th cycle experiment, 23 out of 210 (11%) combinations are significantly homogamic. The derived line M3 has been the most persistently homogamic, with 4 (out of 9) significantly homogamic crosses in the 4th cycle, and 6 (out of 20) in the 5th cycle. Only two out of the 31 crosses yielding significant excess of homogamic matings involve one of the ancestral populations and a derived line. Powell (1978) and Dodd and Powell (1985) also found that most significantly homogamic crosses occurred between two derived lines rather than between the ancestral population and one derived line. On the other hand, Ringo et al. (1985) observed in Drosophila simulans more instances of significant excess of homogamic matings between their ancestral population and one derived line than between pairs of derived lines.

TABLE 4. Isolation index, I, for crosses that were repeated in both the 4th and the 5th flush-crash cycles.

cross	4th cycle	5th cycle
BCA x MA	0.095	0.068
M3 x BC2	0.148	-0.039
BC7 x BC2	-0.200	0.012
M7 x BC2	0.184	-0.062
BCA x BC2	0.079	-0.157
MA x BC2	-0.280*	0.023
M3 x BC7	0.306**	0.349**
M7 x BC7	0.087	-0.068
BCA x BC7	0.032	-0.238
MA x BC7	-0.024	0.024
M7 x M3	0.269**	0.000
BCA x M3	0.049	0.080
MA x M3	0.068	0.011
BCA x M7	0.139	-0.133
MA x M7	0.000	-0.091

* P<0.05, ** P<0.01

Table 4 gives the joint isolation index, I for those crosses that have been performed after both the 4th and the 5th cycles. They involve all pairwise combinations between six populations: the derived lines BC2, BC7, M3 and M7, and the two ancestral populations BCA and MA. Three crosses corresponding to the 4th cycle give statistically significant deviations from random mating. Two of them (M3xBC7 and M3xM7) show a significant excess of homogamic matings. In the 5th cycle the M3xBC7 cross was still significantly homogamic, which suggests that sexual isolation is in some instances stable through time. Table 5 summarizes the results obtained with each of the lines BC2, M3, BC7 and M7. The data obtained in all sexual isolation tests involving each one of these lines and any other derived lines have been combined, and the I index has been calculated for the accumulated total numbers. In addition, the I values are also calculated using the accumulated data that involve only either the BC or the M populations. The table shows separately the data obtained in each series of experiments.

We have found only one case of significant heterogamy, in the cross MAxBC2 (see Table 4), and only in the first series of experiments.

TABLE 5. Isolation index, I, for each of the four lines tested in both the 4th and 5th flush-crash cycles. The I values are calculated accumulating mating data as explained in the text.

	4th cycle			5th cycle		
line	total	M	BC	total	M	BC
BC2	0.004	-0.025	0.023	0.065*	0.076	0.054
M3	0.162**	0.124*	0.197**	0.125**	0.106**	0.143**
BC7	0.080*	0.121*	0.032	0.110**	0.191**	0.020
M7	0.128**	0.040	0.198**	0.070*	0.086*	0.057

total: all accumulated data
M: accumulated data for M derived lines only.
BC: accumulated data for BC derived lines only.
* P<0.05, ** P<0.01

Flush-crash experiments previously performed by other authors have been criticized on the grounds that the mixture of individuals from different geographic origins used to establish the ancestral population, does not correspond to any likely situation in nature (Barton and Charlesworth 1984), and hence that it does not test Carson's speciation model. Such criticism does not apply to our experiment because our experimental populations derived each from a single ancestral natural population. Bryant et al.'s (1986) experiments with *Musca domestica* also avoided that criticism by using founders collected in one single locality.

It might seem that founder events would be likely to lead to speciation when the number of founders would be intermediate. If the number of founders is too small, there would be too much inbreeding, which leads to homozygosis and multiple deleterious effects. If the number of founders is large, no drift is likely to occur that might lead to the genetic reorganization postulated by the founder-flush-and-crash model. Our results, however, are not consistent with that inference. We have, indeed, obtained a greater incidence of sexual isolation with populations derived from only 1 or 3 pairs of founders than with populations derived from 5, 7, or 9 pairs of founders (data not shown). Bryant et al. (1986) found a larger increase of non-additive variance associated to the most extreme (1-pair) bottlenecks.

We are continuing to expose our populations to flush-crash cycles. It will be critical to ascertain whether the sexual isolation exhibited by some populations relative to others persist in the future. For the present, it seems warranted to conclude that (1) some derived lines show statistically significant homogamy with respect to other lines and that these deviations from random mating are stable (Tables 4 and 5); (2) that M3 is the line most frequently involved in significant homogamic crosses both after the 4th and after the 5th cycle and (3) that the cross that yields the highest sexual isolation after the 5th cycle is M3xBC7 (I=0.349, Table 4), which involves the two lines that also exhibit the highest sexual isolation with resoect to all other lines (see Table 5).

ACKNOWLEDGEMENTS. This work has been supported by grant PB86-0517 from DGICYT (Spain) to AM, a grant from the U.S. Department of Energy to FJA, and a fellowship from the Generalitat Valenciana to AG.

REFERENCES

Baker AJ, Moeed A (1987) Rapid differentiation and founder effect in colonizing populations of common mynas (Acridotheres tristis). Evolution 41:525-538

Barton NH, Charlesworth B (1984) Genetic revolutions, founder effects, and speciation. Annu Rev Ecol Syst 15:133-164

Bryant EH, McCommas SA, Combs LM (1986) The effect of an experimental bottleneck upon quantitative genetic variation on the housefly. Genetics 114:1191-1211

Carson HL (1968) The population flush and its genetic consequences. In: Lewontin RC (ed) Population Biology and Evolution. Syracuse University Press, Syracuse, NY, pp 123-137

Carson HL (1971) Speciation and the founder principle. Stadler Symp 3:51-70

Carson HL (1975) The genetics of speciation at the diploid level. Am Nat 109:73-92

Carson HL, Kaneshiro KY (1976) Drosophila of Hawaii: Carson systematics and ecological genetics. Annu Rev Ecol Syst 7:311-346

Carson HL, Templeton AR (1984) Genetic revolutions in relation to speciation phenomena: the founding of new populations. Annu Rev Ecol Syst 15:97-131

Dodd DMB, Powell JR (1985) Founder-flush speciation: an update of experimental results with Drosophila. Evolution 39:1388-1392

Gilbert DG, Starmer WT (1985) Statistics of sexual selection. Evolution 39:1380-1383

Goodnight JC (1988) Epistasis and the effect of founder events on the additive genetic variance. Evolution 42:441-454

Janson K (1987) Genetic drift in small and recently founded populations of the marine snail Littorina saxatilis. Heredity 58:31-37

Maruyama T, Fuerst PA (1985) Population bottlenecks and Motro nonequilibrium models in population genetics. II. Number of alleles in a small population that was founded by a recent bottleneck. Genetics 111:675-689

Motro U, Thomson G (1982) On heterozygosity and the effective size of population subject to size changes. Evolution 36:1059-1066

Moya A, Galiana A, Castro JA, Ayala FJ (in preparation) Founder-flush speciation in Drosophila pseudoobscura: a new experiment

Powell JR (1978) The founder-flush speciation theory: an experimental approach. Evolution 32:465-474

Ringo JM (1977) Why 300 species of Hawaiian Drosophila? The sexual selection hypothesis. Evolution 31:694-696

Ringo JM (1987) The effect of successive founder events on mating propensity of Drosophila. In Huettel M (ed) Evolutionary genetics of invertebrate behavior. Plenum Press, NY, pp 79-88

Ringo JM, Barton KM, Dowse HB (1986) The effect of genetic drift on mating propensity, courtship behavior, and

postmating fitness in _Drosophila simulans_. Behavior 97:226-233
Ringo JM, Wood D, Rockwell R, Dowse H (1985) An experiment testing two hypotheses of speciation. Am Nat 126:642-661
Sene FM, Carson HL (1977) Genetic variation in Hawaiian _Drosophila_. IV. Close allozymic similarity between _D. silvestris_ and _D. heteroneura_ from the island of Hawaii. Genetics 86:187-198
Spieth HT (1980) Hawaiian Drosophila Project. Proc Hawaiian Entom Soc 23:275-291
Templeton AR (1980) The theory of speciation via the founder principle. Genetics 94:1011-1038
Templeton AR (1981) Mechanisms of speciation - a population genetic approach. Annu Rev Ecol Syst 12:23-48

Founder Effects in Colonizing Populations: The Case of *Drosophila buzzatii*

A. FONTDEVILA

Departamento de Genética y Microbiología, Universidad Autónoma de Barcelona, 08193 Bellaterra (Barcelona), Spain

Colonizing species are of two kinds: those following a defined habitat, usually a human-disturbed environment (weedy species), and those shifting into a new ecological niche, distinct from the ancestral one. It has been argued (Carson, 1965) that the former do not promote speciation, but that the latter do often speciate through founder effects. According to this view, weedy species, the proper colonizing species, have some recognizable properties: a) they are the result of the fixation of genes for a defined niche, e.g. a unique host or a general adaptability; b) their chromosomal polymorphism is rigid; c) their novel genetic constitution is largely fixed in the homozygous condition, and d) they are resistant to changes by the founder effect, throughout complex, balanced systems of heterosis. On the other hand, true founder populations leading to speciation have a high segregation system, able to shift their internal balance (i.e., their coadapted gene complexes) via the founder effect.

The operational definition of a colonizing species is often a tremendous task. Most species originate in a single geographic region where they acquire their basic ecological adaptations. However, with the exception of extreme cases of endemism, the majority of species tend to occupy new adjacent territories in which the ecological conditions differ in some degree. One way by which species try to push their ecological limits further apart is by continuously producing new temporary colonies with high extinction probabilities in the ecological margins . These marginal populations have stirred the interest of evolutionists because in some cases they have been able to report significant genetic changes in them, e.g. the loss of chromosomal polymorphism in marginal *Drosophila* populations (Carson, 1958; Dobzhansky, 1957). Often, these genetic changes have been taken as an indication of a founder effect leading to a speciation event. This linking between a colonizing event and a founder effect has led to much confusion in the speciation theory.

It seems timely to establish a clear-cut distinction between founder effects and colonization, yet this may be a difficult task. Here, I present an introductory study using information produced by my research group with the species *Drosophila buzzatii* during several years. This species, of South American origin, has been qualified as a "weedy species", although a specialist one (Carson, 1965), because it feeds on rots of *Opuntia* plants and follows this plant genus wherever it disseminates. The combined studies of original and colonized populations of *D. buzzatii* have shown some interesting features that allow one to give preliminary answers to relevant questions, such as: a) Are weedy colonizing species resistant to founder effects? b) Is their chromosomal polymorphism rigid? c) How much polymorphism is lost (gained) in colonization? d) Are the antagonistic pleiotropic fitness effects repatterned in colonization? The theory of the founder effect has gone through a series of avatars (compare Carson, 1973 vs Carson, 1982) and reached an ultimate form that embodies some statements that can be tested in well documented cases of colonization. The case of *D. buzzatii* might represent one of these study cases.

THE BIOGEOGRAPHICAL FRAME

The historical discovery of America by Spain in 1492 and its subsequent large scale colonization carried along the artificial destruction of a geographical intercontinental barrier that had lasted since the Cretacious times. The continuous travels by sea allowed many species from one hemisphere to be able to reach the other hemisphere by means of conscious or unconscious transportation. Among these species, the *Cactaceae*, probably of South American origin, have experimented a worldwide spread. In particular, different species of the genus *Opuntia* were transported early to Spain and other Mediterranean countries (Font Quer, 1973), the species *Opuntia ficus-indica* being widely introduced. Other introductions were performed much later, such as to Australia in 1788 (Mann, 1970).

In spite of the early *Opuntia* introduction in the Mediterranean area, we have given reasons for believing that *D. buzzatii* colonized the Old World no earlier than in the eighteenth century (Fontdevila et al., 1981, 1982). This late arrival of *D. buzzatii* might have been crucial to the success in the establishment of new colonies because this species was faced, upon its arrival, with an abundant and open niche to be occupied. Also, it seems an indisputable fact that the many imports of *Opuntia* material, specially to the Canary Islands, for different purposes, ranging from ornamental or foodstuff uses to cochineal explotation, may have enhanced the probability of establishement.

The endemic Argentinian species of *Opuntia* (*O.quimilo*, *O.pampeana*, *O.cordobensis*, *O.aurantiaca*, *O.sulphurea*, etc.) are rarely present in the Old World area of colonization. In this colonized area *O. ficus-indica* is by far the most abundant host species, but other species (*O.maxima*, *O.dillenii*, etc.) can be found sporadically. *D. buzzatii* utilizes most of these *Opuntia* species in Argentina, and this host-plant distribution supports the idea that *D.buzzatii* may have experienced an important host shift during colonization. However, *O.ficus-indica* was also introduced in Argentina by the Spaniards during the colonial times and the *D. buzzatii* shift has also been produced there. The ability of *D. buzzatii* to change host plants has been recently confirmed by the observation that, under natural conditions, this species can also emerge from columnar cacti such as *Cereus validus* and *Trichocereus pasacana*, two abundant species in Argentina.

THE CHROMOSOMAL POLYMORPHISM

A total of 70 localities have been analyzed so far in the Old and the New Worlds (36 and 34 localities, respectively) for their chromosomal polymorphism. Figures 1 and 2 depict a representative sample of the second and fourth chromosomal polymorphisms. In Argentina, the original populations show a high geographical diversity for both polymorphisms. In the second chromosome, frequency of standard (*st*) arrangement ranges from almost nil to 0.5; jz^3 arrangement is also extremely variable (0-0.4) and y^3 is only present in San Luis. Finally, jq^7 arrangement has only been found in very low frequency in Arroyo Escobar. A similar situation applies to the fourth chromosome, where the *s* arrangement is absent in many localities and highly frequent in others. There is no latitudinal or macrogeographical pattern that could explain the actual distribution. As an example, the Quilmes population that could be considered marginal according to its low second chromosomal polymorphism is the most poly-

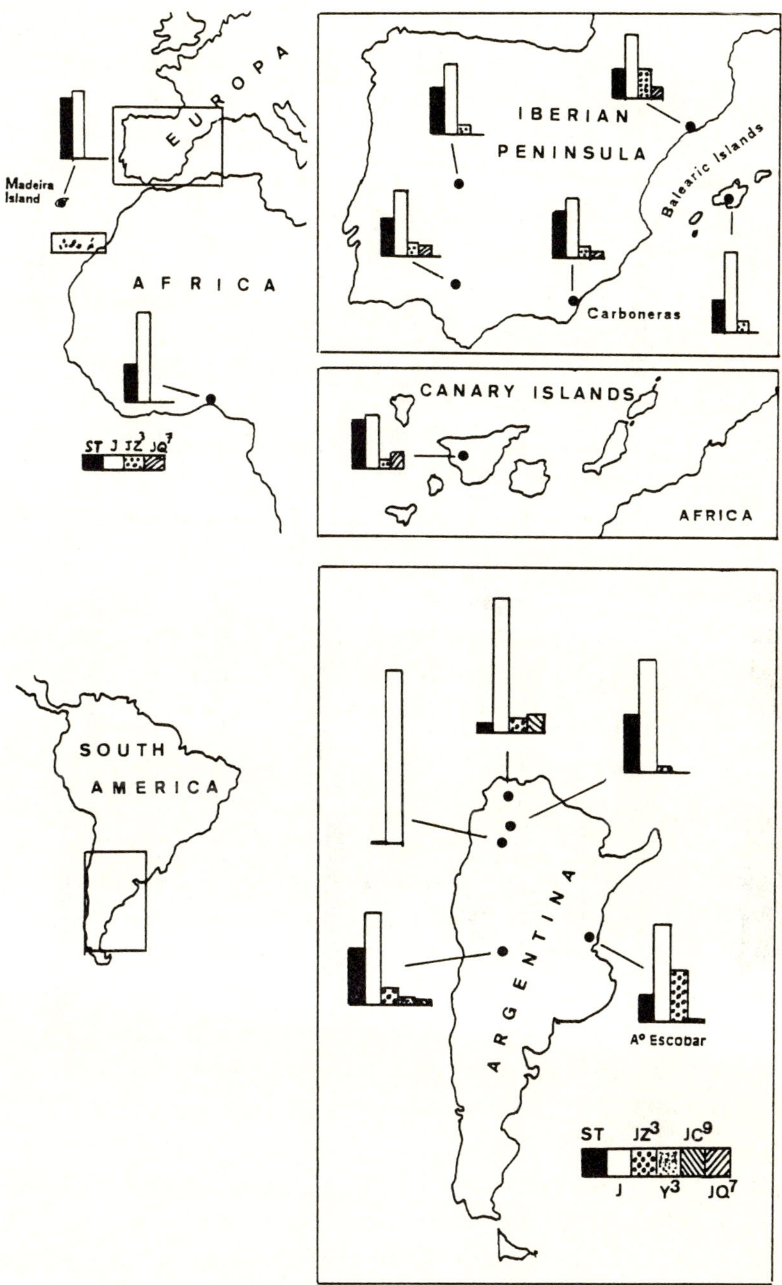

Figure 1: Frequency distributions of second chromosome polymorphism in several original (New World) and colonizing (Old World) populations of *Drosophila buzzatii*. See text for explanation.

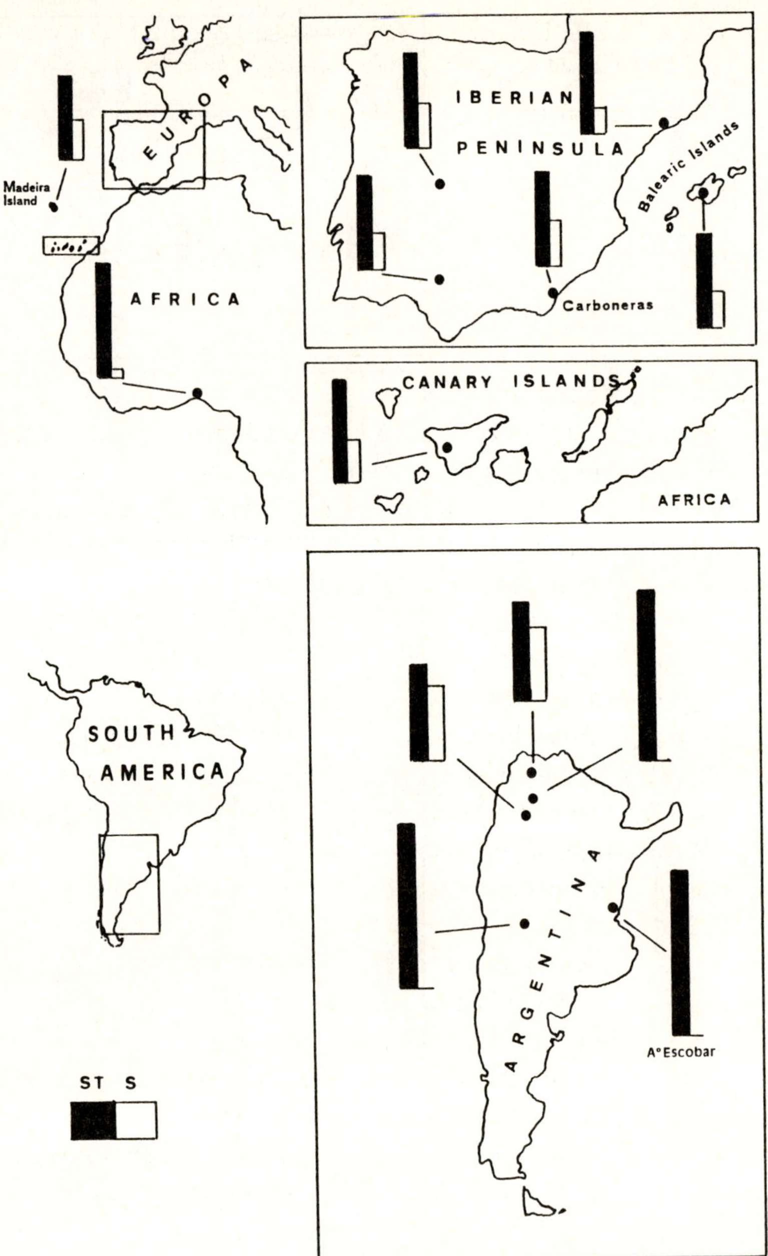

Figure 2: Frequency distributions of fourth chromosome polymorphism in several original (New World) and colonizing (Old World) populations of *Drosophila buzzatii*. See text for explanation.

Table 1: Mean expected heterozygosities (± standard errors) of the chromosomal polymorphism of original and colonized populations of *D. buzzatii*.

	Chromosome		
	Second	Fourth	Total
Original	0.5569 ± 0.0222	0.0079 ± 0.0052	0.2824 ± 0.0115
Colonized	0.5538 ± 0.0239	0.2912 ± 0.0410	0.4225 ± 0.0351

(From Fontdevila et al., 1982)

morphic for the fourth chromosome.

Chromosomal polymorphism in the Old World is much more uniform, yet it shows some geographical variability. Arrangement *j* is also the most frequent, *st* is rather frequent and never absent and jz^3 shows a high variability ranging from nil to 0.25. The presence of jq^7 in many populations at low frequencies is a characteristic of colonized populations, since this arrangement is practically absent in the New World. Chromosome 4 polymorphism shows a moderate to high frequency (up to 0.5) of the *s* arrangement in most of the Old World populations, which contrasts to its absence in most original populations. There is a macrogeographical trend in the colonized populations in that rare inversion frequencies decrease eastwards in the Mediterranean Basin. We have interpreted this trend as a reflection of the losing of rare inversions that accompanies the founder effects in succesive episodes of colonization (Fontdevila et al., 1981, 1982).

Three main conclusions can be inferred from these data. First, original chromosomal polymorphism cannot be considered rigid since it shows a high degree of geographical variability. However, the adaptive factors to which it responds are still unknown, but may be operating differently in each chromosomal polymorphism. Second, there is evidence of founder effects due to the losing of low frequency arrangements in small isolated populations and in populations founded after several episodes of colonization. Third, there is no losing of polymorphism as evidenced by the high, expected heterozygosities in colonized populations (Table 1).

Table 2: Compared allozyme polymorphism between original (A) and colonized (B) populations of several species

Species	Alleles per locus	Polymorphism (%)	Heterozygosity
D.buzzatii [1]			
A: S. America	1.6	32.7	0.098
B· Old World	1.2	17.1	0.068
D.pseudoobscura [2]			
A: N. America	2.0	43.0	0.129
B: Colombia	1.4	25.0	0.051
Passer montanus [3]			
A: Germany	1.5	35.9	0.098
B: N. America	1.3	25.6	0.078
Musca autumnalis [4]			
A: Europe	1.6	36.0	0.053
B: N. America	1.3	36.0	0.030
Acridotheres tristis [5]			
A: India	1.4	17.9	0.060
B: Africa and Oceania	1.2	14.3	0.046

(1) Sanchez, 1986. (2) Lewontin, 1974. (3) St.Louis and Barlow, 1988.
(4) Bryant et al., 1981. (5) Baker and Moeed, 1987.

THE ALLOZYME POLYMORPHISM

A total of 18 populations (13 from the Old World and 5 from the New World) have been analyzed for 22 putative allozyme loci (Sánchez 1986). There are striking differences in allozyme polymorphism between original and colonized populations. The proportion of polymorphic loci (P) is much lower in colonizing populations (0.17) than in original populations (0.33). The proportion of heterozygous loci (H : heterozygosity) as a measure of gene diversity has also diminished significantly in colonizing populations versus original populations (0.068 vs 0.098; $F=17.965$, $P<0.01$), and so it has the mean number of alleles per locus (1.2 vs 1.6). In Table 2, I have summarized these parameters for several historically documented cases of colonization. They show a pattern very similar to that found in *D. buzzatii.*

Table 3: Mean expected (P_e) and observed (P_o) percentage of polymorphic loci under the infinite-alleles model. H stands for mean gene diversity (or expected heterozygosity) and q the criterion of polymorphism. Standard deviation in parenthesis.

Origin of populations	q	H	P_e	P_o
Original	0.05	0.098	27.86 (3.80)	32.72 (4.97)
	0.01	0.098	39.41 (4.89)	42.74 (10.92)
	0.003	0.098	46.80 (5.39)	46.38 (15.86)
Colonizing	0.05	0.068	19.60 (3.64)	17.12 (3.79)
	0.01	0.068	28.44 (4.90)	18.87 (4.49)
	0.003	0.068	34.39 (5.60)	19.92 (5.43)

Changes in polymorphism during colonization have preserved most of the original variability and have mostly eliminated rare alleles. This can be seen by using the infinite-allele model (Kimura, 1971). In a population in mutation-drift equilibrium, expected P values can be calculated from H for each criterion of polymorphism (q), such that $P = 1-q^{H(1-H)}$. Table 3 reports expected and observed P values for criteria of 0.05, 0.01 and 0.003 (including any polymorphism). Differences between P values in original populations are statistically non significant at each criterion. On the other hand, colonizing populations show a good agreement between both P values only at low restrictive criterion (q = 0.05), but at moderate and high restrictive criteria (q = 0.01 and 0.003) the expected polymorphism is much higher than that observed. This can be accounted for by the losing of less frequent alleles in colonization. Obviously, there has not been enough time to replace the lost variability by mutation to lost alleles.

We have also investigated the genetic differentiation among populations using the Wright Fst statistics (Sánchez, 1986). Most of the population differentiation can be accounted for by differentiation within colonizing (Fst = 0.129) or within original (Fst = 0.128) populations, although a slightly higher level of differentiation (Fst = 0.154) is obtained when considering the total ensemble of populations together. A hierarchical analysis was performed in which three levels have been considered: subdivision (Old-New World), region (primary and secondary colonization, Argentina-Bolivia) and deme. The highest differentiation is attained among demes when considering the total number of populations

(Fst = 0.140) or inside subdivisions (Fst = 0.139); however, differentiation among demes within regions is lower (Fst = 0.114). This indicates a certain effect of the region in the population differentiation.

In summary, allozyme variation gives proof of changes in genetic polymorphism during the colonization process. Main changes are of two kinds: 1) Levels of polymorphism and genetic diversity are lower in colonizing populations due mainly to the elimination of rare alleles in the founder event, although some allozyme loci (e.g. Est-1 and Adh-1) have changed their polymorphism in other ways. 2) Colonizing and original populations show similar levels of demic differentiation, but some regions can be distinguished inside each kind of subdivision. These regions are biogeographical areas in the original populations and represent colonizing episodes in the colonized populations.

MEASURING FITNESS COMPONENTS

Changes in chromosomal polymorphism during colonization may be either the outcome of an ecological shift or the result of a founder effect or both. This is a crucial issue for understanding the relationship between colonizing and founder events. Total fitness is the resultant of complex interactions between several components that can be summarized in two: early and late components. Early components are operating in early stages of the life cycle, such as zygotic or gametic viability. Late components operate in the mature adult throughout sexual selection or fecundity. Due to antagonistic pleiotropic effects in the genetic systems, some phenotypes may be diferentially favored through stages in relation to other phenotypes, leading to opposed fitness coefficients in different life stages. This has been called endocyclic selection. Repatterning of the genetic architecture during founder events may lead to rapid changes in this antagonistic pleiotropy.

One way to measure fitness components is to compare trait frequency distributions among different life stages of age classes. The hypothesis says that significant differences among stages is an indication of selection. In order to avoid the effects of random processes in frequency changes, samples of life stages must be taken at the same locality and time, and it is advisable to repeat sampling from several localities and different times, when possible. This method can be used to measure selection in natural populations, provided that enough of the species population ecology is known. Traditionally it has been employed to detect natural selection associated with allozyme markers (electromorphs) in the deer mouse *Peromyscus maniculatus* (Nadeau and Baccus, 1981, 1983), in the eelpout *Zoarces viviparus* (Christiansen, 1977), and in plants such as *Hordeum vulgare* (Clegg et al., 1978a, b) and the oat *Avena barbata* (Allard et al., 1977; Clegg et al., 1978b).

Inferences about the fitness components of the chromosomal polymorphism in *Drosophila* natural populations using the method of life stage comparisons were made in *D. pseudoobscura* for viability (Dobzhansky and Levene, 1948) and for male mating success (Anderson et al., 1979). However, the measurement of natural selection through the whole life cycle has met with tremendous difficulties due to the ecological ignorance surrounding most of the *Drosophila* species amenable to genetic studies. Crumpacker et al., (1977) made the only serious attempt in this respect with *D. pseudoobscura.* Later on, the work in my laboratory with *D. buzzatii* has demonstrated the potentiality of this species to measure fitness components in natural populations (Ruiz et al., 1986).

The cactophilic niche of *D. buzzatii* has been studied by several investigators (Vacek, 1982; Fontdevila, 1982; Barker et al., 1984; Ruiz, Naveira and Fontdevila, 1984; Peris, 1989). With a reasonable effort, field work with this species provides the opportunity to sample individuals at most, if not all, life-history stages in natural populations. We have devised an experimental scheme of sampling and of chromosomal analyses through the entire life cycle (Figure 3). To sum up, three samples are taken directly from the field: a) bait collected adults; b) larvae from *Opuntia* rots, and c) adults emerged from *Opuntia* rots. Collected adult females were analyzed for chromosomal polymorphism in two different ways: 1) Individual females were allowed to lay eggs and one larva from each individual progeny (eggs-1) was analyzed (individual analysis); 2) A group of females were allowed to lay eggs and a sample of this progeny (eggs-2) was analyzed (mass analysis). Paired comparisons between chromosomal inversion distributions of two consecutive life stages allow one to detect fitness components (for a detailed account of the theoretical framework see Ruiz et al., 1986).

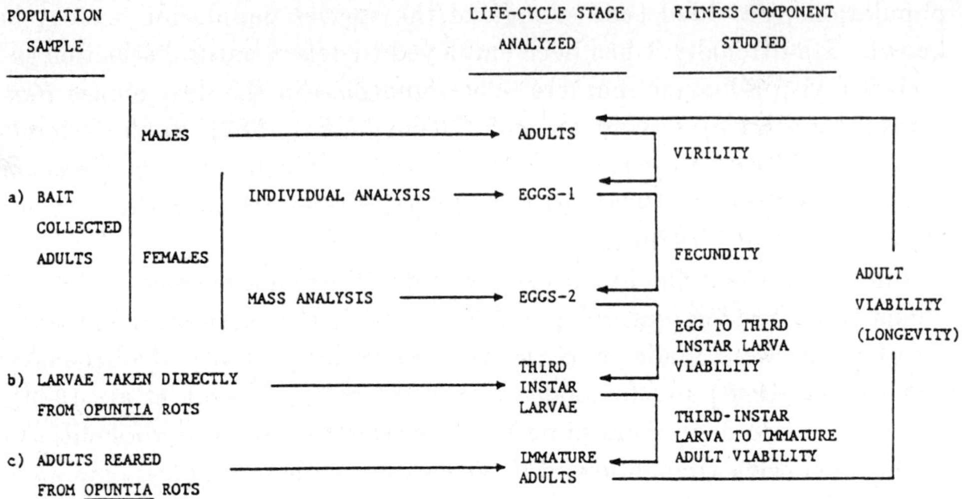

Figure 3: Experimental scheme for the analysis of fitness components carried out in a natural population of *Drosophila buzzatii*. Selection is detected through the comparison of gene and genotypic frequencies between consecutive life-cycle stages (connected by arrows). See text for more explanation (From Ruiz et al., 1986).

CHANGES IN FITNESS COMPONENTS THROUGH COLONIZATION

Several criteria were used to choose the colonizing and original populations to compare. First, presence and absence of rare electromorphs and of chromosomal rearrangements were considered. Arroyo Escobar and Carboneras are probably the best candidates for an original and a derived populations, respectively, since they share the majority of genetic markers. In particular, $2jq^7$ rearrangement is present only in Arroyo Escobar among the original, studied populations. Under the asumption that inversions are unique events in the species history, Arroyo Escobar is the sole possible founder population. The possibility of several founder events cannot be ruled out. As an example, the presence of Pgm^{103} and Em^{98} electromorphs in colonizing populations can only be accounted for by two founder events from Los Negros and Quilmes populations, respectively. However, electromorph back-mutation is a common event that could explain this apparent ambiguity.

Second, Sánchez (1986) has performed principal component analyses with both the allozyme and the chromosomal polymorphisms, showing

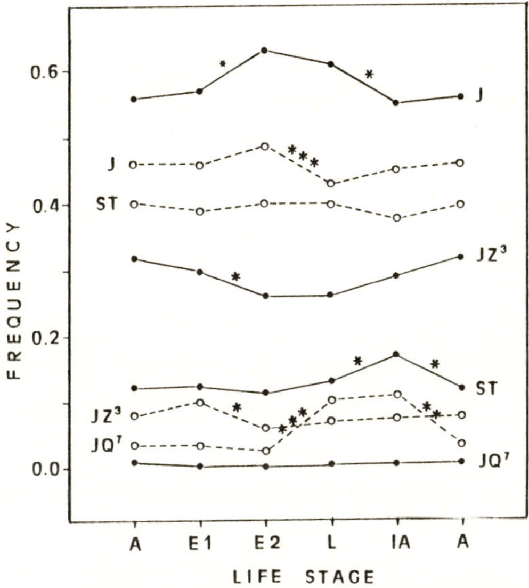

Figure 4: Cyclic frequency changes of second chromosome rearrangements in five consecutive life stages of one original (Arroyo Escobar, Argentina; continuous line) and one colonizing (Carboneras, Spain: dotted line) population. A: collected adults; E1: Egg-1 sample; E-2: Egg-2 sample; L: larvae from rots; IA: immature emerged adults. Statistical significance between stage frequencies: *: $p<0.05$; **: $p<0.01$; ***: $p<0.001$. (Data from Ruiz et al., 1986 and Hasson, 1988).

that Quilmes and Arroyo Escobar are the two original populations that ordinate closest to the colonizing populations. Since Quilmes does not have $2jq^7$ rearrangement, this again leaves Arroyo Escobar as the best original candidate. Third, preliminary work in my laboratory with the *D. melanogaster* copia element has shown that a few repetitive patterns emerge when colonizing and original populations are surveyed in Southern blots. One of them is shared by Arroyo Escobar and Carboneras, giving support to their common heritage.

Figs. 4 and 5 show the relative frequencies of second and fourth chromosome arrangements for each life stage. Two samplings in Arroyo Escobar were performed in two consecutive years at the same season for some life stages and showed very good repeatibility between both samples (Hasson, 1988; Hasson et al.; submitted). These data confirm some of the characteristics of this polymorphism cited above, namely, that arrangement frequencies are quite different when both populations are compared. The colonizing population of Carboneras has significantly increased $2st$,

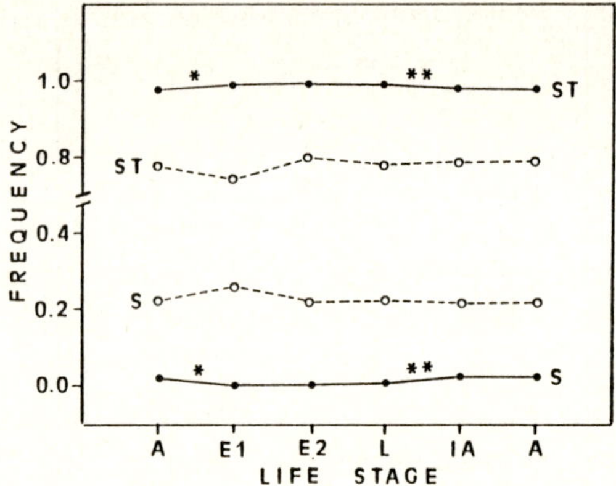

Figure 5: Cyclic frequency changes of fourth chromosome rearrangements in five consecutive life stages of one original (Arroyo Escobar, Argentina; continuous line) and one colonizing (Carboneras, Spain; dotted line) population. Symbols as in Fig. 4. (Data from Ruiz et al.; 1986 and Hasson, 1988).

$2jq^7$ and $4s$ frequencies in detriment to the remaining arrangements. The two most dramatic changes are the decrease of $2jz^3$ and the increase of $4s$.

Comparisons between stage frequencies also give significant differences between original and colonizing populations. Most significant is the absence in Arroyo Escobar of the strong larval viability effect for the second chromosome detected in Carboneras, which was attributed most probably to heterozygote superiority. However, fitness values do not satisfy the conditions for stable equilibrium of a multiple polymorphism and some kind of endocyclic selection must be involved (Ruiz et al., 1986). Not less important are the highly significant effects detected for virility and pupal viability in the fourth chromosome polymorphism of the Arroyo Escobar population (Hasson, 1988). These effects were not found in the colonizing population of Carboneras. A more detailed study of frequency changes can be performed with the second chromosome by using the expression $\Delta p/(Var\ p)^{1/2}$, where $Var\ p = p(1-p)(1/n_1 + 1/n_2)$ (Anderson et al., 1979). This statistical test allows one to detect significant changes in particular arrangement frequencies that might be underestimated by the χ^2 test. Figure 4 shows, in fact, that some significant effects are unveiled for fecundity and longevity in the second chromosome polymorphism. Namely, $2j$ is favored in detriment to $2jz^3$ in both populations.

FOUNDER EFFECTS IN COLONIZATION

In the introductory paragraph I stressed the importance of distinguishing founder from colonizing events. Taking for granted that *D. buzzatii* is a colonist, it is important to analyze in detail each founder effect detected in the colonizing history of this species. This is a unique opportunity, since colonists are good material for the study of the founder effect, but detailed comparative studies are very rare and incomplete. This scarcity of factual data has often justified one to turn to endemic non-colonizing species to the study of founder effects.

Speciation theories of the founder effect have evolved through a series of steps towards a more general theory of disorganization-reorganization in the genome (Carson, 1982, 1985). The details of this neo-founder effect theory are vague, but their ensemble is of interest. Here, I present an incomplete and preliminary experimental work that may clarify some of the past and present premises of the founder theory.

The Fate of Genetic Variability in Founder Events. The latest versions of the founder theory (Carson and Templeton, 1984) agree in that ancestral (original) populations must be outcrossed and polymorphic. This discards true weedy colonizing species as candidates to founder events. *D. buzzatii* cannot be qualified as a weedy species according to its genetic structure. Its original populations show a high degree of allozyme genetic variability, comparable to other species in the *repleta* group and not lower than that of other endemic sibling species, such as *D. koepferae* (Sánchez, 1986). *D. buzzatii* chromosomal polymorphism is rather high and shows geographic variability in the endemic areas, contradicting the idea of chromosomal rigidity in a colonizing species. We do not know the ecophysiological traits associated with the adaptive value of this polymorphism, but differences in resource utilization among geographical localities and differential frequency changes of second chromosomal inversions in experimental populations with different natural substrates (Ruiz, Naveira and Fontdevila, 1985) suggest an association with natural resources.

Genetic changes in colonizing populations are most informative. Allozyme changes (Table 2) show a general decrease in polymorphism that can be easily accounted for by a stochastic founder event. The low levels of polymorphism found according to what one expects from the observed gene diversity in colonizing populations, give support to the recent occurrence of a population foundation in which most rare electromorphs have been lost (Table 3). On the other hand, changes in chromosomal polymorphism are of a different kind. There is no reduction in overall inversion heterozygosity for the second chromosome, but there is an increase of heterozygosity for the fourth chromosome. However, founder effects are present either in some small isolated and island populations or in populations originated in secondary waves of colonization (Fontdevila et al., 1981, 1982). The maintenance of ancestral chromosomal variation can be explained by rapid population expansion (flush) after foundation and the frequency changes by a disruption of coadapted complexes through drift accompanied or followed by selection. These events have been suggested as important in the founder models of speciation (Carson and Templeton, 1984).

The Fitness Reorganization after Founder Events. It is not easy to summarize the reorganization steps that have been proposed to occur after founder events. In summary, chance alterations of genetic arrays during founder events are subjected to natural selection leading to a fitness reweighting of the pleiotropic antagonistic effects. Carson (1986) has lately emphasized the role of sexual selection, based mainly in speciation studies with the Hawaiian picture-wing drosophilids. In *D. buzzatii* changes in fitness values and in their pleiotropic effects are observed for its chromosomal polymorphism. Three kinds of changes can be distinguished:

1. Changes in some fitness components. Among them the larval viability component is most illustrative. The original population does not show any significant egg–to-third-instar larval viability effect, but the colonizing population shows a highly significant directional effect. Interestingly, pupal viability shows directional selection in original but not in colonized populations. This may be explained by the action of parasites present in the endemic ar-

eas and absent in the colonized ones. Finally, longevity selection seems to be operating in both populations, yet with different intensity and direction.

2. Fitness components with no changes, but showing significant directional effects. Fecundity is one of these components and demonstrates that it is not affected by the reorganization phase. The fecundity effects on changes in inversion frequencies are barely significant in our study, but this may be due to the fact that our statistic only estimates half of the frequency change (Ruiz et al., 1986). It may be true that during the population flush, selection would relax only for some components such as viability, but not for fecundity. In this case the original fecundity pattern would be retained because of its optimal performance in fitness.

3. Fitness components showing no significant effects. This is true for virility in the second chromosome, but not in the fourth chromosome polymorphism. However, the method of estimating these virility effects detects only half of them, as in the fecundity component. Some recent sexual selection studies using a more direct method of detecting male mating propensity have shown that in the colonizing population of Carboneras there is a significant virility effect associated both with the second chromosome polymorphism and with the fly body size (see Ruiz and Santos, this volume). The virility component seems to be very important in the genome reorganization of the Hawaiian founder effect, but its role in *D. buzzatii* founder populations remains to be seen.

CONCLUDING REMARKS

The above results suggest that true colonists such as *D. buzzatii*, show the signs of founder effects in terms of their genetic polymorphisms and their fitness components associated to chromosomal polymorphism. Unfortunately, very few detailed studies have been performed in natural colonized populations to reveal changes in morphometric characters. However, in some experimental studies with colonists such as the house-fly, *Musca domestica* (Bryant and Meffert, 1988; see also Bryant, this volume) significant changes in the additive genetic covariance structure for polygenic traits have been detected in bottleneck (founder) lines. Research work in my laboratory (see Ruiz and Santos, this volume) establishes a significant relationship between chromosomal polymorphism and body size in the Carboneras population. Thus, it is reasonable to think that *D. buzzatii* will also show the founder effects in morphometric traits.

At the beginning of this work I asked several questions about colonists. Now I am able to give some answers. Many colonists, if not all, are not resistant to the founder effect, as defined by changes in genetic polymorphism and in pleiotropic fitness effects. They are also most probably, susceptible to changes in genetic covariance relationships among morphometric traits. These results, if true, make it inappropriate to discard colonists as good material for speciation studies. Nonetheless, taking this for granted, this finding does not explain why many colonist species do not speciate.

Two aspects of the theory of speciation by founder effect seem to be crucial, namely, rate of speciation and available genetic variability. In early versions of the theory, it was proposed that "species differences may arise abruptly, frequently as the result of a population flush followed by establishment of a new population by a single founder individual" (Carson, 1973, p. 279). This has been challenged by some population geneticists by stating that not one but a series of episodes of genetic drift are necessary for achieving a given degree of isolation (Barton and Charlesworth, 1984). Some authors (Lewontin, 1965) have emphasized that a single bottleneck may not greatly diminish the genetic variance available for selection and a continuous, small, efective population size has to be postulated in order to explain those cases, as reported in Table 2, in which colonizing populations show a lower degree of heterozygosity and polymorphism. However, bottlenecks produce gene frequency shifts that change genetic polymorphisms qualitatively and then selection may drive populations to new adaptive peaks (Wright, 1978).

Often it is very difficult to separate the bottleneck effect from environmental differences, geographical isolation or, even, continuous genetic drift effects. Nonetheless, bottleneck effects have been demonstrated experimentally (Bryant, this volume; Galiana et al., this volume) and continuous drift effects in chromosomal polymorphism have been documented in secondary colonizations of *D. buzzatii* (Fontdevila et al., 1981). My view is rather different from both population geneticists and founder evolutionists. I envisage the loss of genetic variability not as the result of drift effects, but as the intensive selection after a colonizing event. It is common knowledge that abrupt (accelerated) changes in selection experiments exist and are often interspersed with long periods of stasis (see for example MacBean, McKenzie and Parsons, 1971 or Thoday, Gibson and Spickett, 1964). So it is no wonder that colonists may face new, stressful selective forces that produce accelerated responses. However, each one of these responses is limited by the rapid exhausting of available (additive) variability.

There is no satisfactory explanation for resuming the selective response after a plateau. Recombination among polygenic systems has been advocated. Most sensible is the argument that random genetic changes (e.g. drift) may change the gametic disequilibria and generate new additive genetic variability from former epistatic variance (Bryant et al., 1986; Goodnight, 1987, 1988). The action of genetic drift is unpredictable in time and this generates the time sojourns in the observed plateaus. Mutation may restore the new variability, provided that rates and time are appropriate, yet, mutation rates are considered too low by population geneticists to be taken into account to restore depleted additive variability in a reasonable time. Nevertheless, Barton and Charlesworth (1984) discuss the dilemma of genetic fixation in bottlenecks, a moment in which genetic variability is needed to produce a peak shift. They discard mutation as an agent to restore variability and postulate that selection must be operating in founder events. But, in my view, it is this intensive selection that is the main responsible cause of exhausting variability.

The data presented here strongly suggest that colonizers experiment the founder effects. Since colonizing events do not lead to speciation, other conditions must be invoked to explain speciation. I am proposing here that founder effects do not only induce changes in selective pressures and genetic organizations, but also changes in kind and rate of new mutations due to the stressful conditions of foundation. An appropriate combination of all these changes may explain the dramatic shifts observed in founder speciation. Mutability is a population-dependent phenomenon and can be mediated by transposition bursts under certain stressful conditions (see Fontdevila, 1988, for a revision). Besides, polygenic characters seem to be associated with mobile element architecture and mobility, as is documented by Ratner and Lyapunova (this volume). No detailed studies have been performed on the impact of founder events on mobile element transpositions, but this may be worth looking at.

If this proved to be true, transposition mutability would provide the needed supply of genetic variability advocated by population geneticists (Barton and Charlesworth, 1984) to explain many episodes of founder speciation without recurrence to special mechanisms proposed by founder evolutionists (e.g. transilience). Yet, in this case it is very doubtful that the new species architecture could be obtained without mutations that profoundly affect development, as is disclaimed by the proponent of the new founder theory of genome organization (Carson, 1982). In view of the outlined controversy on the founder theory substantiated by experimental and theoretical evidence, I rather stick to Carson's earlier view that "the basic characteristic of speciation, as opposed to subspecific differentiation, is thought to be a drastic shift in internal, interacting gene systems" (Carson, 1973). Much has to be learned until a detailed account of the mechanisms by which this speciation shift could be advanced, but it seems true that founder events, as described until now, may be a necessary, but not a sufficient, condition to speciation.

ACKNOWLEDGMENTS

This paper is made possible by the cooperative efforts of two research teams. One group, led by Dr. Osvaldo A.Reig (O.R.) consists of Dr. E.Hasson, Dr. J.Vilardi, Mr. J.Fanara and Ms C.Rodriguez and is associated with the Universidad de Buenos Aires, Departamento de Ciencias Biológicas, Grupo de Investigación en Biología Evolutiva, Argentina. The other group, led by Dr. A.Fontdevila (A.F.) consists of Dr. H.Naveira, Dr. F.Peris, Dr. C.Pla, Dr. A.Ruiz, Dr. A.Sánchez and Dr. M.Santos and is associated with the Universidad Autónoma de Barcelona, Departamento de Genética y Microbiología, Spain. However, the ideas and statements advanced in this paper are the sole responsability of the author. This research has been finantially supported on the Argentinian side by grants EX038 from the Universidad de Buenos Aires and grant 004/0422/87 from CONICET (Argentina) awarded to O.A. The Spanish side has funded this project by CICYT grants n°s 2920/76; 0910/81; 2825/83 and PB85/0071 awarded to A.F. during 10 years of research work.

Mr. Antonio Barbadilla and Ms. Julia Provecho typed and prepared the camera ready version of this manuscript and their cooperation was of utmost value for its completion.

REFERENCES

Allard RW, Khaler AL, Clegg MT (1977) Estimation of mating cycle components of selection in plants. In: Christansen FB and Fenchel T (eds) Measuring selection in natural populations, Lecture notes in biomathematics, vol 19. Springer Berlin Heidelberg New York, pp 1-19

Anderson WW, Levine L, Olvera O, Powell JR, de la Rosa ME, Salceda VM, Gaso MI, Guzmán J (1979) Evidence for selection by male mating success in natural populations of *Drosophila pseudoobscura*. Proc Nat Acad Sci USA 76: 1519-1523

Baker AJ, Moeed A (1987) Rapid genetic differentiation and founder effect in colonizing populations of common mynas *Acridotheres tristis*. Evolution 41: 525-538

Barker JSF, East PD, Phaff HJ, Miranda M (1984) The ecology of the yeast flora in necrotic *Opuntia* cacti and of associated *Drosophila* in Australia. Microb Ecol 10: 379-399

Barton NH, Charlesworth B (1984) Genetic revolutions, founder effects, and speciation. Ann Rev Ecol Syst 15: 133-164

Bryant EH, Van Dijk H, Van Delden W (1981) Genetic variability of the face fly, *Musca Autumnalis* de Geer, in relation to a population bottleneck. Evolution 35: 872-881

Bryant EH, Combs LM, McCommas SA (1986) The effect of an experimental bottleneck upon quantitative genetic variation in the housefly. Genetics 114: 1191-1211

Bryant E, Meffert LM (1988) Effect of an experimental bottleneck on morphological integration in the housefly. Evolution 42: 698-707

Carson HL (1958) The population genetics of *Drosophila robusta*. Advan Genet 9: 1-40

Carson HL (1959) Genetic conditions which promote or retard the formation of species. Cold Spr Harb Symp Quant Biol 24: 87-105

Carson HL (1965) Chromosomal morphism in geographically widespread species of *Drosophila*. In: Baker HG, Stebbins GL (eds) Genetics of colonizing species. Academic Press, New York, pp 503-531

Carson HL (1973) Reorganization of the gene pool during speciation. In: Morton NE (ed) Genetic structure of populations. Univ Press Hawaii

Carson HL (1982) Speciation as a major reorganization of polygenic balances. In: Barigozzi C (eds) Mechanisms of speciation. Liss, New York, pp 411-433

Carson HL (1985) Unification of speciation theory in plants and animals. Syst Bot 10: 380-390

Carson HL, Templeton AR (1984) Genetic revolutions in relation to speciation phenomena: The founding of new populations. Ann Rev Ecol Syst 15: 97-131

Carson HL (1986) Sexual selection and speciation. In: Karlin S, Nevo E (eds) Evolutionary processes and theory. Academic Press, pp 391-409

Clegg MT, Kahler AL, Allard RW (1978a) Estimation of life cycle components of selection in an experimental plant population. Genetics 89: 765-792

Clegg MT, Kahler AL, Allard RW (1978b) Genetic demography of plant populations. In: Brussard PF (ed) Ecological genetics: The interface. Springer, Berlin Heidelbreg New York, pp 173-178

Christiansen FB (1977) Population Genetics of *Zoarces viviparus* (L.) A review. In: Christiansen FB, Frydenberg O (eds) Measuring selection in natural populations. Springer, Berlin Heidelberg New York, pp 21-47

Crumpacker DW, Pyati J, Ehrman L (1977) Ecological genetics and chromosomal polymorphism in Colorado populations of *Drosophila pseudoobscura*. Evol Biol 10: 437-469

Dobzhansky T (1957) Genetics of natural populations. XXVI. Chromosomal variability in island and continental populations of *Drosophila willistoni* from Central America and West Indies. Evolution 11: 280-293

Dobzhansky T, Levene H (1948) Genetics of natural populations. XVII. Proof of operation of natural selection in wild populations of *Drosophila pseudoobscura*. Genetics 33: 537-547

Font Quer P (1973) Plantas Medicinales: El Dioscórides Renovado. Barcelona

Fontdevila A, Ruiz A, Alonso G, Ocaña J (1981) The evolutionary history of *Drosophila buzzatii*. I. Natural chromosomal polymorphism in colonized populations of the Old World. Evolution 35: 148-157

Fontdevila A (1982) Recent developments on the evolutionary history of the *Drosophila mulleri* complex in South America. In: Barker JSF, Starmer WT (eds) Ecological genetics and evolution. Academic Press Australia, pp 81-95

Fontdevila A, Ruiz A, Ocaña J, Alonso G (1982) The evolutionary history of *Drosophila buzzatii* II. How much has chromosomal polymorphism changed in colonization? Evolution 36: 843-851

Fontdevila A (1987) The unstable genome: An evolutionary approach. Genét Ibér 39: 315-349

Fontdevila A, Pla C, Hasson E, Wasserman M, Sánchez A, Naveira H, Ruiz A (1988) *Drosophila koepferae*: a new member of the *D. serido* superspecies taxon. Ann Ent Soc USA 81: 380-385

Goodnight CJ (1987) On the effect of founder events on epistatic genetic variance. Evolution 41: 80-91

Goodnight CJ (1988) Epistasis and the effect of founder events on the additive genetic variance. Evolution 42: 441-454

Hasson E (1988) Ecogenética evolutiva de *Drosophila buzzatii* y *Drosophila koepferae* (complejo mulleri, grupo repleta, Drosophilidae: Diptera) en las zonas áridas y semiáridas de Argentina. Ph.D.Thesis, Universidad de Buenos Aires, Argentina.

Hasson E, Vilardi JC, Fanara JJ, Rodriguez C, Reig OA, Fontdevila A (1989) The evolutionary history of *Drosophila buzzatii*. XVI. Fitness component analysis in an endemic natural population from Argentina. J. Evol Biol (submitted)

Kimura M (1971) Theoretical foundation of population genetics at the molecular level. Theor Popul Biol 2: 174-208

Lewontin R (1965) Selection for colonizing ability. In: Baker HG, Stebbins GL (eds) The genetics of colonizing species. Academic Press, New York London, pp 77-94

Lewontin RC (1974) The genetics of the evolutionary process. Columbia University Press

MacBean IY, McKenzie JA, Parsons PA (1971) A pair of closely linked genes controlling high scutellar chaeta number in *Drosophila*. Theor Appl Genet 41: 227-235

Mann J (1970) Cacti Naturalised in Australia and their Control. S G Reid Government Printer Brisbane

Nadeau JH, Baccus R (1981) Selection components of four allozymes in natural populations of *Peromyscus maniculatus*. Evolution 35: 11-20

Nadeu JH, Baccus R (1983) Gametic selection and hemoglobin polymorphism in *Peromyscus maniculatus*: a rejoinder. Evolution 37: 642-646

Peris F (1989) Adaptación trófica y evolución de *Drosophila* en zonas áridas. Ph. D. Thesis Universitat Autònoma de Barcelona

Ruiz A, Fontdevila A, Santos M, Seoane M, Torroja E (1986) The evolutionary history of *Drosophila buzzatii*. VIII. Evidence for endocyclic selection acting on the inversion polymorphism in a natural population. Evolution 40: 740-755

Ruiz A, Naveira H, Fontdevila A (1984) La historia evolutiva de *Drosophila buzzatii*. IV. Aspectos citogenéticos de su polimorfismo cromosómico. Genét Ibér 36: 13-35

Sánchez A (1986) Relaciones filogenéticas en los clusters *buzzatii* y *martensis* (grupo *repleta*) de *Drosophila*. Tesis doctoral Universitat Autónoma de Barcelona

St Louis VL, Barlow JC (1988) Genetic differentiation among ancestral and introduced populations of the eurasian tree sparrow *Passer montanus*. Evolution 42: 266-276

Thoday JM, Gibson JB, Spickett SG (1964) Regular responses to selection. 2. Recombination and accelerated response. Genet Res 5: 1-19

Vacek DC (1982) Interactions between microorganisms and cactophilic *Drosophila* in Australia. In: Barker JSF, Starmer WT (eds) Ecological genetics and evolution. Academic Press, Australia, pp 175-190

Wrigth S (1977) Evolution and the genetics of populations, vol III. Univ. Chicago Press.

Mating Probability, Body Size, and Inversion Polymorphism in a Colonizing Population of *Drosophila buzzatii*

A. Ruiz and M. Santos

Departamento de Genética y Microbiología, Universidad Autónoma de Barcelona, 08193 Bellaterra (Barcelona), Spain

> "*What we need is more knowledge about the ways in which populations, in fact, meet evolutionary challenges: What intensities of natural selection can they put up with, how far and how fast can they modify their phenotype (including their habitats)? Colonizing species are ones which we know to have been confronted by a challenge - that of their new location; and we often know, even quite precisely, how long they have been facing it.*"
>
> C.H. Waddington, 1965

INTRODUCTION

Drosophila buzzatii is a colonizing species mainly associated with the cactus genus *Opuntia*. Originally from South America, it has spread over the world within historical times following these cacti (Carson, 1965; Barker and Mulley, 1976; Fontdevila et al., 1981). It shows a moderately high inversion polymorphism in two of the four major autosomes encompassing about 15% of the total euchromatin (Carson and Wasserman, 1965; Fontdevila et al., 1981, 1982; Ruiz et al., 1984, 1986; Fontdevila, this volume). We are trying to answer a very simple question: how does natural selection work on this polymorphism in the wild? The answer, though, may be not so simple. Of course, indirect evidence from natural populations and experimental work carried out during the last forty years, shows that inversions are adaptive devices and that selection may be of considerable magnitude (Dobzhansky, 1970; Anderson et al., 1975; Spiess, 1977; Sperlich and Pfriem, 1986). Field studies of natural selection, on the other hand, are scarce and it is by no means clear which is the relative importance of the different selection components and which is the magnitude of the selection coefficients in nature. We think that inversions continue to be a very useful tool to learn about natural selection in this regard.

Several years ago, we undertook a study of selection components in a *D. buzzatii* natural population inhabiting an old *Opuntia ficus-indica* plantation near the fishing village of Carboneras (Almeria, southeastern Spain). This population has a relatively recent origin, probably being introduced not more than 250 years ago (Fontdevila et al., 1981; Fontdevila, this volume). The first results showed significant differences in larval viability among the second-chromosome karyotypes, but not among the fourth-chromosome karyotypes (Ruiz et al., 1986; Santos et al., 1989). Viability selection on the second chromosome was apparently directional, bringing about inversion frequency changes. This suggested the operation of selection in other parts of the life-cycle, given that inversion frequencies are quite stable through time in this population. Nevertheless, unambiguous evidence for selection acting on the adult phase was not found for either of the two chromosomes and, in particular, mating success -which is supposed to play a significant role in the maintenance of the chromosomal polymorphism in other species; e.g. Anderson et al. (1979), Santos et al. (1986)-, did not show a significant effect. This might be due, perhaps, to the limitations of the method used, namely the *indirect* comparison of inversion frequencies between consecutive life-cycle phases. This method cannot detect those patterns of selection which do not change gene frequencies, e.g. heterosis (see Ruiz et al., 1986, pp. 742-745). In addition, the power of the test employed to test for differential mating success among male karyotypes was very low, though not lower than that of similar tests found in the literature. With intermediate gene frequencies and sample sizes of the order of 500 chromosomes, even selection coefficients as high as 0.5 would be missed most of the time!

We present here a summary of the results obtained in recent field work performed in the same population of Carboneras and designed to overcome the limitations of the previous studies. Briefly, the approach involved two main improvements. First, the detection of sexual selection was based upon the *direct* comparison of karyotypic and inversion frequencies between mating and non-mating adults of both sexes. Second, body size was measured on the same adults that were being cytologically analyzed. The rationale for including this phenotypic trait in the study is the following: body size has been shown to affect mating success in many insects including *Drosophila* (Partridge and Farquhar, 1983; Thornhill and Alcock, 1983), and it has also been found to be correlated with karyotype in a few species (see references below). Since the correlation between mating success and the karyotype that we are trying to detect may not be large, the inclusion of a third character, such as body size, may greatly increase the possibility of detecting natural selection. At the same time, it may be a help in understanding the way selection operates. After all, natural selection acts primarily on the phenotype and only secondarily on the genotype, to the extent that it is correlated with the phenotype.

BODY SIZE AND MATING PROBABILITY

Empirical studies of sexual selection carried out in the laboratory with *Drosophila* species are abundant and have been quite successful in documenting those behavioural or morphological characterisitics that determine variation among individuals in mating success. To the contrary, the quantitative estimation of the intensity of sexual selection under natural conditions has been almost completely neglected. Only recently have there been some studies that analyzed whether sexual selection on morphological characters relating to body size, which is known to occur in laboratory situations (Ewing, 1961, 1964; Monclús and Prevosti, 1971; Partridge and Farquhar, 1983), is also taking place in nature (Partridge et al., 1987; Santos et al., 1988; Taylor and Kekić, 1988).

A field study designed to test for the occurrence of sexual selection on body size in *D. buzzatii* was conducted in June 1987. Every evening from 19:00 to 21:00 hours for five consecutive days, the flies that were attracted to a number of rotting *Opuntia* cladodes were observed and samples of mating and non-mating adults were gently collected with an insect aspirator, from the surface of each rot. A comprehensive account of the ecology and mating behaviour of *D. buzzatii* and details about the sampling procedure are given elsewhere (Ruiz et al., 1986; Santos et al., 1988, 1989). The study was repeated, using a similar sampling scheme, during six days in June 1988. On both occasions, the thorax length of each wild fly was measured as an indicator of its body size. Table 1 gives the mean thorax length of the mating and non-mating flies for the

Table 1: Mean thorax size ($\overline{X}$) in mm and standard deviation (SD) of random samples of wild mating and non-mating *Drosophila buzzatii* adults caught in two successive years in the population of Carboneras (Spain).

Date	Sex		Non-mating	Mating	Standardized selection differential i	t†
June 1987	Males	$\overline{X}$	0.963	0.984	0.34	4.54∗ ∗ ∗
		SD	0.064	0.057		
		N	299	352		
	Females	$\overline{X}$	1.037	1.048	0.16	2.19 ∗
		SD	0.071	0.057		
		N	287	321		
June 1988	Males	$\overline{X}$	0.983	1.000	0.25	3.22 ∗∗
		SD	0.068	0.061		
		N	303	298		
	Females	$\overline{X}$	1.049	1.064	0.24	3.02 ∗∗
		SD	0.064	0.060		
		N	300	298		

† A modified t-test was applied to the female samples collected in June 1987 due to the significantly higher variance of the character in the non-mating females.
$*P < 0.05; **P < 0.01; ***P < 0.001$.

two years, together with estimates of the standardized selection differentials and their statistical significance (Endler, 1986). The selection intensities were calculated assuming that the non-mating flies accurately represented the population before selection. All the selection differentials are statistically significant, with the same average value of 0.25 for the two sexes in each year. The obvious conclusion is that directional selection for mating success on body size is taking place for both males and females in *D. buzzatii*. However, solely from the information given in Table 1, the interpretation of our field study may not be as straightforward. *Drosophila* natural populations have overlapping generations and the flies found at any given place and time are not a cohort but a mixture of adults from various age classes. Since body size is positively correlated with longevity in *D. melanogaster* (Partridge and Farquhar, 1983), differences in size between mating and non-mating flies could be explained, at least in part, if variance in adult survivorship related with body size occurs in *D. buzzatii.* We addressed this point in the June 1988 field study, and the preliminary results suggest that both longevity and mating success are positively correlated with thorax length.

Our findings of statistically significant intensities of selection in both males and females contrast with the results obtained by Partridge et al. (1987), which show that directional sexual selection exists for males ($i \simeq 0.75$) but not for females in *D. melanogaster.* Taylor and Kekić (1988) also detected a statistically significant intensity of sexual selection for *D. melanogaster* males with a value of $i = 0.48$, although they did not report data for females. Taking into account the well established principle that males have a higher fitness variance for mating success than do females (Bateman, 1948; Wade, 1979; Wade and Arnold, 1980), differences in sexual selection intensities are expected between sexes, with males experiencing stronger sexual selection. This situation would arise if there are differences in the time each sex invests in activities related to mating, and we would therefore find females to show less variance and males to have a greater variance in polygynous species (see Sutherland, 1985, 1987, for a discussion of Bateman's results). Mating systems of cactophilic *Drosophila* species have the striking characteristic that female remating is rather frequent, with a mean number of mates per individual female as high as 3.89 during a four hour period in the Sonoran Desert species, *D. nigrospiracula* (Markow, 1982). This contrasts markedly with the mating behaviour of species classically used in laboratory studies, such as *D. melanogaster, D. pseudoobscura,* and *D. subobscura,* where females eventually remate after a number of days (cf, Pruzan, 1976; Loukas et al., 1981; Markow, 1982, 1985; Santos et al., 1986). Thus, variance for mating success might be of the same order of magnitude for both sexes in some cactophilic species, including *D. buzzatii* (unpublished laboratory work shows that female remating is also frequent in this species), which could explain our findings of similar directional selection for body size in both males and females. In addition, female mating success is probably an important fitness component in *D. buzzatii*, since the number of offspring produced by a female depends on the number of mates (Barbadilla et al., submitted).

KARYOTYPE FREQUENCIES ARE CORRELATED WITH BODY SIZE

The Carboneras population is polymorphic for four arrangements on the second chromosome: *2 standard (st)*, *2j*, *2jz*^{3} and *2jq*^{7}; and two arrangements on the fourth chromosome, *4st* and *4s* (Fontdevila et al., 1981; Ruiz et al., 1986). Hence, ten different second-chromosome karyotypes and three fourth-chromosome karyotypes are possible in this population. Given the low frequencies of arrangements *2jz*^{3} and *2jq*^{7} (usually < 10%), some of the ten second-chromosome karyotypes may be scarce or even completely absent in small samples. For this reason, in this account, and for the sake of simplicity, those two arrangements have been pooled with arrangement *2j* into a single class (that we will denote as *2j•*). All three arrangements, *2j*, *2jz*^{3} and *2jq*^{7}, share inversion *j* and are thus derived, compared to the standard chromosome which is the species' ancestral arrangement (Ruiz et al., 1982). Therefore, their pooling also seems justified on phylogenetic grounds.

Only the wild flies collected in June 1987 were cytologically analyzed. Their karyotypes were inferred from those of their progeny when crossed individually to virgin adults of a laboratory stock homozygous for the *2j* and *4st* chromosome arrangements (Ruiz et al., in preparation). In total, we obtained information on the body size and the second-chromosome karyotype of 565 males and 515 females (563 males and 516 females for the fourth chromosome). In order to test for an association between body size and karyotype, the two samples of mating and non-mating males were first divided into three classes according to body size: small (thorax length < 0.950 mm), medium (0.950-1.000 mm), and large (> 1.000 mm). Mating and non-mating females were similarly classified into three groups: small (thorax length < 1.025 mm), medium (1.025-1.075 mm), and large (> 1.075 mm). Then, for each sample, the frequencies of the various karyotypes at each chromosome were compared among body size classes by means of a 3 x 3 contingency table (Sokal and Rohlf, 1981). This table was partitioned into two orthogonal comparisons which are of interest: one testing the difference between homokaryotypes and heterokaryotypes and another one comparing the two homokaryotypes. In addition, inversion frequencies were also compared among body size classes by means of a 2 x 3 contingency table.

The second-chromosome karyotype frequencies in the various samples

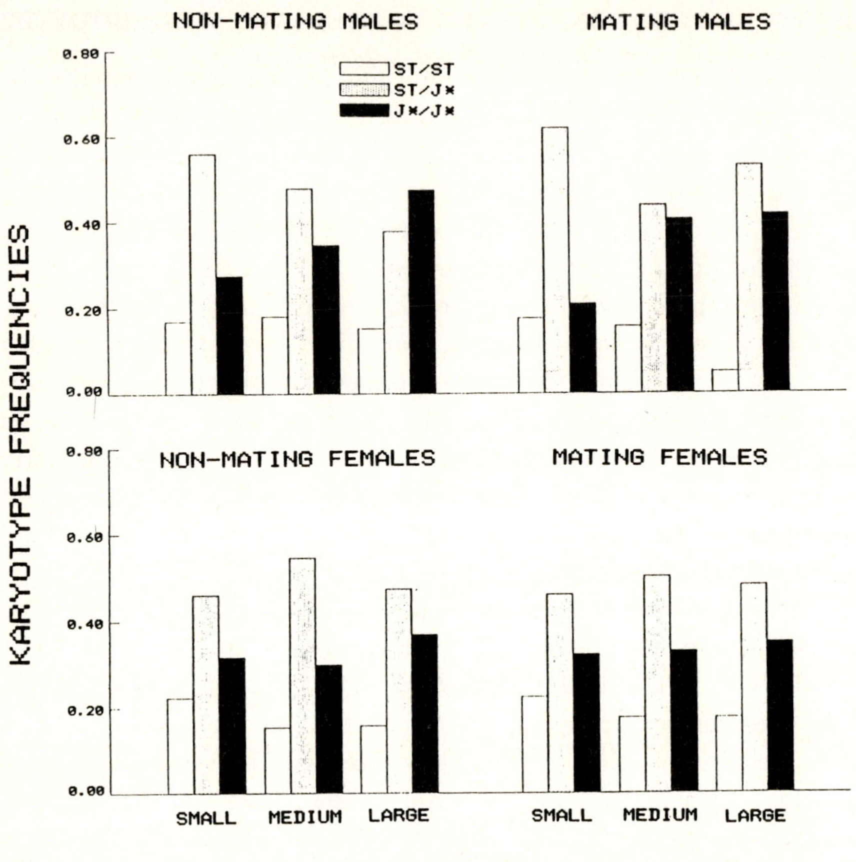

Figure 1: Second-chromosome karyotype frequencies in samples of mating and non-mating *D. buzzatii* adults, classified according to body size.

of wild individuals, classified according to body size, are shown in Figure 1. In both samples of males, the frequency of the *2j•/ j•* karyotype was higher (by about 20%) in large males than in small males, whereas that of *2st/st* and *2st/ j•* showed the opposite trend. The G-value for the differences among the three karyotypes was highly significant in the sample of mating males, but non-significant in the sample of non-mating males (Table 2). The total G-value for the two samples was also highly significant. In the sample of mating males, the apportionment of the G-value among karyotypes showed that the difference between the two homokaryotypes was making the greater contribution, yet the difference between

Table 2: Values of G-statistics for the comparison of karyotype and arrangement frequencies among body size classes in different samples of *Drosophila buzzatii* males.

	Non-mating		Mating		Total	
	df	G	df	G	df	G
Second chromosome						
Karyotypes	4	5.84	4	16.79**	8	22.63**
Hom. vs Het.	2	4.17	2	6.15*	4	10.32*
Between Hom.	2	1.67	2	10.64**	4	12.31*
Arrangements	2	3.20	2	8.92*	4	12.12*
Fourth chromosome						
Karyotypes	4	7.30	4	6.12	8	13.42a
Hom. vs Het.	2	7.07*	2	4.32	4	11.39*
Between Hom.	2	0.23	2	1.80	4	2.03
Arrangements	2	3.14	2	5.11a	4	8.25a

a $0.05 < P < 0.10$; $*P < 0.05$; $**P < 0.01$.

homokaryotypes and heterokaryotypes was also significant. The comparison of the inversion frequencies among body size classes also yielded significant results in males (Table 2). In both samples, the *2j*• arrangement increased and the *2st* decreased in frequency with increasing body size. For females, the pattern observed was qualitatively identical to that we just described in males (Figure 1), yet none of the comparisons of the karyotype and inversion frequencies gave significant results.

The results obtained for the fourth chromosome are illustrated in Figure 2. Again, the pattern was qualitatively identical in the two samples of males as well as in those of females. In all four samples, the frequency of the *4st/s* heterokaryotype was higher (by about 10-18%) in large flies than in small flies, while it was intermediate in medium-sized flies. This coincidence in four independent samples has an associated probability lower than 0.005 of being a chance event. Meanwhile, the frequency of the *4st/st* homokaryotype tended to decrease with increasing size. The *4s/s* homokaryotype was scarce, and a regular trend was not apparent here. As in the case of the second-chromosome, contingency tables yielded significant results only for males (Table 2). However, in this case, the only significant comparison was between heterokaryotypes

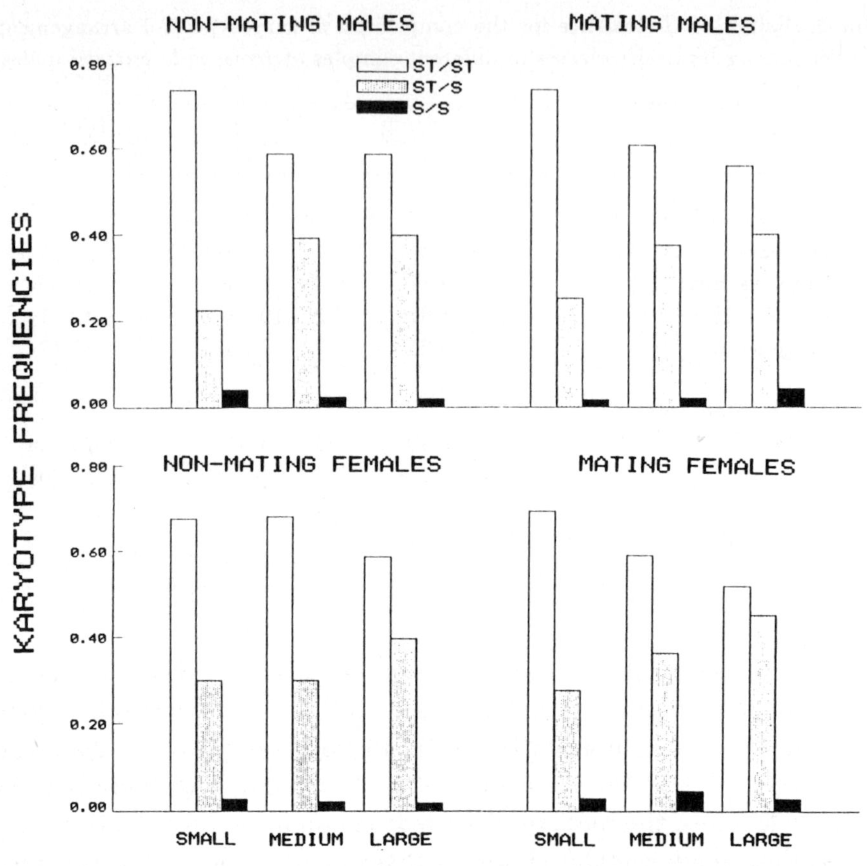

Figure 2: Fourth-chromosome karyotype frequencies in samples of mating and non-mating *D. buzzatii* adults, classified according to body size.

and homokaryotypes. No significant differences in inversion frequencies among body size classes were obseved, but, in all samples, the frequency of the *4s* inversion in large flies was slightly higher than in small flies.

In summary, both polymorphic chromosomes showed a correlation with body size which was qualitatively similar in the two sexes but more pronounced in males than in females. Each chromosome, however, exhibited a different pattern. For the second chromosome, a large size was associated with a higher frequency of the *2j•/j•* homokaryotype, resulting in a significant increase of the *2j•* arrangement with increasing size. On the other hand, for the fourth chromosome a larger size was associated with a higher frequency of the *4st/s* heterokaryotype, and a slightly but non-significantly higher frequency of the *4s* inversion. Similar observations have been made previously in natural populations of other species (see Table 3 for a review). In all these cases but one, only males were studied. In the seaweed fly, *Coelopa frigida*, the single exception up to now, the correlation found was similar in males and females, but the effect was larger in males (3 out of 4 samples significant) than in females (1 out of 3), just as in *D. buzzatii.*

The outcome of a phenotypic correlation between body size and karyotype in the field can be explained in at least three different ways. Two of them are illustrated, following Robertson (1955), in Figure 3. Firstly, it might be that this correlation is purely environmental in origin. If both body size and karyotype are correlated with longevity, due to separate causes, they would also show an association with each other in the adult population, which is a mixture of variously aged individuals (Figure 3a). Secondly, and perhaps a more likely possibility, is that the correlation has a genetic cause; i.e. the various karyotypes directly affect the body size of their carriers. This might occur through a position effect of the inversions or, most likely, if the chromosome arrangements are predominantly associated with different alleles at particular loci, influencing the development of the imago (Figure 3b, c). A correlation betweeen body size and karyotype would also arise indirectly if each karyotype selects a particular niche, where its development may produce a characteristic adult size. This third explanation (not shown in Figure 3) is, from the point of view of future adult selection, equivalent to the previous genetic one.

Former studies in other species have not always gone so far as to distinguish among these alternatives. The field observations in *D. subosbcura* (see Table 3) have been corroborated by selection experiments (Prevosti, 1960, 1967), thus providing evidence for a genetic correlation due to the genic content of inversions in this species. In contrast, the grasshopper, *Keyacris scurra* (formerly *Moraba scurra*), has apparently not been amenable to laboratory studies, such as those necessary to discriminate among the alternative explanations, accounting for the extensive observations carried out in natural populations of this species (see Table 3). In the seaweed fly, *Coelopa frigida*, some of the samples, showing a sig-

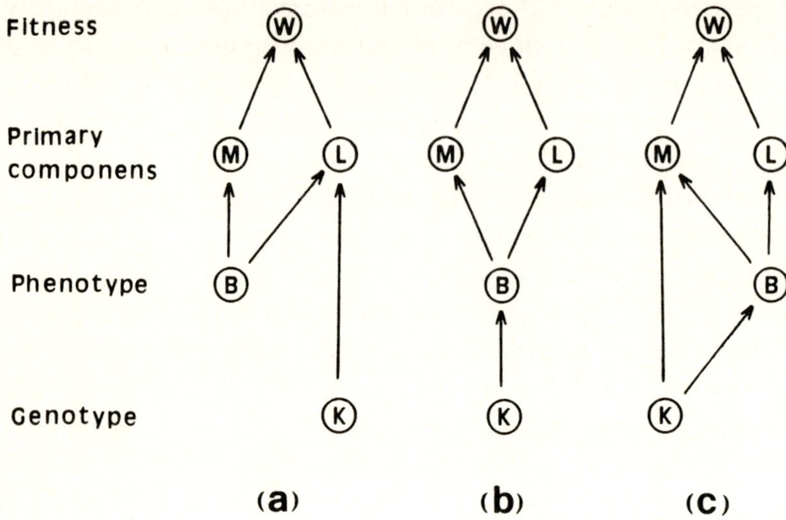

Figure 3. Schematic representation of various possible relationships among total fitness (W), mating success (M), longevity (L), body size (B) and karyotype (K). Upward arrows indicate genetic correlations. (a) K is genetically correlated with L and environmentally correlated with B; (b) B is the only relevant phenotypic trait genetically correlated with K; (c) K is genetically correlated with other traits affecting M besides B.

nificant effect of karyotype on body size, were collected as larvae in the field, thus ruling out a correlation with longevity but not an indirect correlation due to habitat selection. In *D. buzzatii*, neither of the various explanations can be ruled out definitively at the moment. However, preliminary tests carried out with laboratory stocks derived from the Argentinian population of Arroyo Escobar (see Fontdevila et al., 1982, for a description of this population), have indicated that flies carrying the *2j* and *2jz*^{3} arrangements are, on the average, larger than flies carrying the *2st* arrangement (E. Hasson, personal communication).

KARYOTYPE AND INVERSION FREQUENCY CHANGES WITH MATING

We have already seen that larger flies have a higher mating probability in the population of Carboneras. A correlation between body size and karyotype for both the second and fourth chromosomes was also detected in the June 1987 samples. Let us assume, as a working hypothesis,

Table 3: Correlation between body size and karyotype in natural populations.

Species	Population	Trait	Polymorphism	Correlation	Ref.
Keyacris scurra (males)	Several populations in a large area of at least 11,000 Km^2 in Australia	Live weight	2 pericentric inversions on autosomes CD and EF	Decreasing effect of inversions and increasing effect of standard arrangements (heterokaryotypes intermediate)	(1-3)
Coelopa frigida (males and females)	3 natural populations from England and Wales separated by 850 Km of coastline	Wing length	The α/β inversion on chromosome I	$\alpha\alpha > \alpha\beta > \beta\beta$ in both sexes but larger differences in males	(4)
Drosophila subobscura (males)	1 natural population from Barcelona (Spain)	Wing length	Many inversions in all chromosome pairs	Body size correlated with inversion heterozygosity	(5)
Drosophila subobscura (males)	1 natural population from Mt. Parnes (Greece)	Wing length	Several inversions on chromosomes A and U	$A_{ST} > A_1 > A_2$ and the heterokaryotypes for the U chromosome larger than the average homokaryotype	(6-7)
Drosophila melanogaster (males)	1 natural population from Missouri (USA)	Wing/thorax index	Four inversions in chromosome arms 2L, 2R, 3L and 3R	Standard chromosomes show lower wing-loading index than inversions, especially in the case of 2R and 3R	(8)

(1) White and Andrew, 1960; (2) White and Andrew, 1962; (3) White et al., 1963; (4) Butlin et al., 1982; (5) Prevosti, 1966; (6) Krimbas, 1967; (7) Krimbas and Loukas, 1980; (8) Stalker, 1980

that this correlation is genetic and not purely environmental as discussed above. Assume further, that body size is the only relevant phenotypic trait affected by karyotype as shown in Figure 3b. This is not necessarily true for both chromosomes; one or the two of them could also affect other phenotypic trait, e.g. those involved in courtship behaviour, with direct consequences upon mating success (Figure 3c). The former hypothesis, however, is more simple and has predictable consequences. If it is correct, it follows that karyotype and inversion frequencies would be expected to experience certain changes with mating due to their correlation with body size. The magnitude of these changes, though, is probably negligible in females, for selection intensity was relatively small in this sex, and the association between body size and karyotype was statistically non-significant. Males, on the other hand, showed a higher selection intensity, but the correlation between body size and karyotype was still quite small. Thus, the expected frecuency changes are likely not very large even in this case.

A summary of the observed karyotype and inversion frequencies in mating and non-mating flies is given in Table 4. Sample sizes in this table are not identical to those in the last section, for we obtained additional cytological information from nearly one hundred adults not scored for thorax length. It can be seen that the *2j•* arrangement increased in mating males compared to non-mating males by about 5%, just the result we would expect given the association found between body size and second-chromosome karyotype in males. The *2j•* arrangement also increased in females, but only by 1.6%, as expected from previous information. The frequency change was statistically non-significant in both cases. However, when the inversion frequencies were compared between sexes, a significant difference was found in the mating flies but not in the non-mating flies (Table 4). This would suggest that the effect is indeed real.

In the fourth chromosome, the frequency of the *4st/s* heterokaryotype increased by 2% and 5.3% in males and females, respectively. Once again, these changes, yet non-significant, went in the expected direction, since the *4st/s* heterokaryotype was associated with large body size. In addition, the values of the fixation index showed that the frequency of the *4st/s* heterokaryotype was over that expected by the binomial square

Table 4: Karyotype and arrangement frequencies in the samples of mating and non-mating *Drosophila buzzatii* adults collected in June 1987. The values of the fixation index (F) for the fourth chromosome are given. N = number of individuals.

	Second chromosome Arrangement			*Fourth chromosome* Karyotype				
Sample	*st*	*j•*	2N	*st/st*	*st/s*	*s/s*	N	*F*
Non-mating								
Males	0.418	0.582	552	0.627	0.341	0.033	276	-0.0529
Females	0.441	0.559	506	0.647	0.329	0.024	255	-0.0780
Heterogeneity	df = 1, G = 0.53			df = 2, G = 0.52				
Mating								
Males	0.368	0.632	660	0.605	0.361	0.033	332	-0.0748
Females	0.425	0.575	630	0.583	0.382	0.035	314	-0.0920
Heterogeneity	df = 1, G = 4.41*			df = 2, G = 0.34				
Pooled								
Non-mating	0.429	0.571	1058	0.637	0.335	0.028	531	-0.0642
Mating	0.396	0.604	1290	0.594	0.372	0.034	646	-0.0832*

$*P < 0.05$

rule in all samples, and that the excess increased regularly with mating. Comparisons with the Hardy-Weinberg expectations were tested by the formula $\chi^2 = \mathrm{N}\ F^2$, where N is the number of individuals in the sample (Li and Horvitz, 1953). The departure from the expected proportions was non-significant in the pooled non-mating sample, but it was significant in the pooled mating sample (Table 4). Therefore, the overall pattern seems qualitatively consistent with our hypothesis. This does not mean that this hypothesis is necesarily correct, but rather that it cannot be ruled out. In any case, a deeper and quantitative analysis, including all the second chromosome arrangements, would be required before drawing definitive conclusions.

CONCLUDING REMARKS

The detection and analysis of selection in the adult phase of the life-cycle is not an easy task in *Drosophila*. The data presented in this perforce brief account show that there is a positive correlation between mating probability and body size in the *D. buzzatii* population of Carboneras. This correlation, which was significant in the two sexes, might be due to larger flies having an advantage in mating success, longevity and/or any combination of these two components of fitness. It is evident that additional information is necessary to further elucidate this point.

Our results also detected a correlation between body size and karyotype. This correlation can be explained in different ways, but with future laboratory work we should easily be able to choose among the various alternatives. The important point, at this moment, is that whatever the explanation for the correlation is, it always implies the operation of adult selection on the inversion polymorphism of *D. buzzatii* (Figure 3). This is a remarkable conclusion which embraces a hidden lesson. If we had only looked at the karyotype of the flies, little or no evidence of the operation of selection would have been found, since the comparison of karyotype and inversion frequencies between mating and non-mating flies always gave non-significant results. This points to the low power of the usual chi-square tests which, to detect small and moderately large selection coefficients, require samples sizes often too large to be practical. This is one, maybe the most important one, of the many reasons for the failure in detecting natural selection in the wild (Endler, 1986). The complementary strategy of looking a few relevant phenotypic characters and work out the partial correlations, promises to be fruitful, as the example presented here shows.

ACKNOWLEDGEMENTS

This address was based on data gathered by a team of five other people besides the authors, namely: Antonio Barbadilla, Jorge E. Quezada-Díaz, Esteban Hasson, Francesc Peris, and Antonio Fontdevila. Their enthusiastic and thorough job is deeply acknowledged. In addition, we are much indebted to Mr. A. Barbadilla and Prof. A. Fontdevila for estimulating discussion during the course of the experiment. However, the ideas and concepts put forth in this paper are the sole responsability of the authors. The paper has been made much more readable by the careful work of Linda MacNamee, whose expertise and knowledge of English is well above our capacities. This research was supported by grant # PB85-0071 to A. Fontdevila from the Comisión Interministerial de Ciencia y Tecnología (CICYT), Spain.

REFERENCES

Anderson WW, Dobzhansky Th, Pavlovsky O, Powell J and Yardley D (1975) Genetics of natural populations. XLII. Three decades of genetic change in *Drosophila pseudoobscura*. Evolution 29: 24-36

Anderson WW, Levine L, Olvera O, Powell JR, de la Rosa ME, Salceda VM, Gaso MI and Guzmán J (1979) Evidence for selection by male mating success in natural populations of *Drosophila pseudoobscura*. Proc Nat Acad Sci USA 76: 1519-1523

Barker JSF and Mulley JC (1976) Isozyme variation in natural populations of *Drosophila buzzatii*. Evolution 30: 213-233

Bateman AJ (1948) Intra-sexual selection in *Drosophila*. Heredity 2: 349-368

Butlin RK, Read IL and Day TH (1982) The effects of a chromosomal inversion on adult size and male mating success in the seaweed fly, *Coelopa frigida*. Heredity 49: 51-62

Carson HL (1965) Chromosomal morphism in geographically widespread species of *Drosophila*, pp. 503-531. In: Baker HG and Stebbins GL (eds). The Genetics of Colonizing Species Academic Press New York

Carson HL and Wasserman M (1965) A widespread chromosomal polymorphism in a widespread species *Drosophila buzzatii*. Amer Natur 99: 111-115

Dobzhansky Th (1970) Genetics of the Evolutionary Process. Columbia Univ Press New York

Endler JA (1986) Natural Selection in the Wild. Princeton Univ Press Princeton New Jersey

Ewing AW (1961) Body size and courtship behaviour in *Drosophila melanogaster*. Anim Behav 9: 93-99

Ewing AW (1964) The influence of wing area on the courtship behaviour in *Drosophila melanogaster*. Anim Behav 12: 316-320

Fontdevila A, Ruiz A, Alonso G and Ocaña J (1981) Evolutionary history of *Drosophila buzzatii*. I. Natural chromosomal polymorphism in colonized populations of the Old World. Evolution 35: 148-157

Fontdevila A, Ruiz A, Ocaña J and Alonso G (1982) Evolutionary history of *Drosophila buzzatii*. II. How much has chromosomal polymorphism changed in colonization. Evolution 36: 843-851

Krimbas CB (1967) The genetics of *Drosophila subobscura* populations. III. Inversion polymorphism and climatic factors. Molec Gen Genetics 99: 133-150

Krimbas CB and Loukas M (1980) The inversion polymorphism of *Drosophila subobscura*. Evol Biol 12: 163-234

Li CC and Horvitz DG (1953) Some methods of estimating the inbreeding coefficient. Am J Hum Genet 5: 107-117

Loukas M, Vergini Y and Krimbas CB (1981) The genetics of *Drosophila subobscura* populations. XVIII. Multiple insemination and sperm displacement in *Drosophila subobscura*. Genetica 57: 29-37

Markow TA (1982) Mating systems of cactophilic *Drosophila* pp 273-287. In: Barker JSF and Starmer WT (eds) Ecological Genetics and Evolution: The Cactus-Yeast-*Drosophila* Model System Academic Press

Markow TA (1985) A comparative investigation of the mating system of *Drosophila hydei*. Anim Behav 33: 775-781

Monclús M and Prevosti A (1971) The relationship between mating speed and wing length in *Drosophila subobscura*. Evolution 25: 214-217

Partridge L and Farquhar M (1983) Lifetime mating success of males fruitflies (*Drosophila melanogaster*) is related to their size. Anim Behav 31: 871-877

Partridge L, Hoffman A and Jones JS (1987) Male size and mating success in *Drosophila melanogaster* and *Drosophila pseudoobscura* under field conditions. Anim Behav 35: 468-476

Prevosti A (1960) Cambios en la heterocigosis por inversión cromosómica al variar por selección la longitud del ala en *Drosophila subobscura*. Genét Ibér 12: 27-41

Prevosti A (1966) Inversion heterozygosity and size in a natural population of *Drosophila subobscura* pp 49-54. Symposium on the Mutational Process: Mutation in Population Prague August 9-11 1965

Prevosti A (1967) Inversion heterozygosity and selection for wing length in *Drosophila subobscura*. Gent Res Camb 10: 81-93

Pruzan A (1976) Effects of age, rearing and mating experiences on frequency dependent sexual selection in *Drosophila pseudoobscura*. Evolution 30: 130-145

Robertson A (1955) Selection in animals: synthesis. Cold Spring Harbor Symp Quant Biol 20: 225-229

Ruiz A, Fontdevila A, Santos M, Seoane M and Torroja E (1986) The evolutionary history of *Drosophila buzzatii*. VIII. Evidence for endocyclic selection acting on the inversion polymorphism in a natural population. Evolution 40: 740-755

Ruiz A, Fontdevila A and Wasserman M (1982) The evolutionary history of *Drosophila buzzatii*. III. Cytogenetic relationships between two sibling species of the *buzzatii* cluster. Genetics 101: 503-518

Ruiz A, Naveira H and Fontdevila A (1984) La historia evolutiva de *Drosophila buzzatii*. IV. Aspectos citogenéticos de su polimorfismo cromosómico. Genét Ibér 36: 13-35

Santos M, Ruiz A, Barbadilla A, Quezada-Díaz JE, Hasson E and Fontdevila A (1988) The evolutionary history of *Drosophila buzzatii*. XIV. Larger flies mate more often in nature. Heredity 61: 255-262

Santos M, Ruiz A and Fontdevila A (1989) The evolutionary history of *Drosophila buzzatii*. XIII. Random differentiation as a partial explanation of the observed chromosomal variation in a structured natural population. Amer Natur 133: 183-197

Santos M, Tarrio R, Zapata C and Alvarez G (1986) Sexual selection on chromosomal polymorphism in *Drosophila subobscura*. Heredity 57: 161-169

Sokal RR and Rohlf FJ (1981) Biometry, 2nd ed WH Freeman New York

Sperlich D and Pfriem P (1986) Chromosomal polymorphism in natural and experimental populations pp 257-309. In: Ashburner M, Carson HL and Thompson Jr JN (eds). The Genetics and Biology of *Drosophila* Vol 3e Academic Press London

Spiess EB (1977) Genes in Populations. John Wiley & Sons, New York

Stalker HD (1980) Chromosome studies in wild populations of *Drosophila melanogaster*. II. Relationship of inversion frequencies to latitude, season, wing-loading and flight activity. Genetics 95: 211-223

Sutherland WJ (1985) Chance can produce a sex difference in variance in mating success and account for Bateman's data. Anim Behav 34: 1349-1352

Sutherland WJ (1987) Measures of sexual selection. Oxf Surv Evol Biol 2: 90-101

Taylor CE and Kekić (1988) Sexual selection in a natural population of *Drosophila melanogaster*. Evolution 42: 197-199

Thornhill R and Alcock J (1983) The Evolution of Insect Mating Systems. Harvard Univ Press Cambridge

Waddington CH (1965) Introduction to the Symposium pp 1-6. In Baker HG and Stebbins GL (eds). The Genetics of Colonizing Species. Academic Press New York

Wade MJ (1979) Sexual selection and variance in reproductive success. Amer Natur 114: 742-764

Wade MJ and Arnold SJ (1980) The intensity of sexual selection in relation to male sexual behavior, female choice, and sperm precedence. Anim Behav 28: 446-461

White MJD and Andrew LE (1960) Cytogenetics of the grasshopper *Moraba scurra*. V. Biometric effects of chromosomal inversions. Evolution 14: 284-292

White MJD and Andrew LE (1962) Effects of chromosomal inversions on size and relative viability in the grasshopper *Moraba scurra* pp 94-101. In: Leeper GW (ed) The Evolution of Living Organisms Melbourne Univ Press

White MJD, Lewontin RC and Andrew LE (1963) Cytogenetics of the grasshopper *Moraba scurra*. VII. Geographic variation of adaptive properties of inversions. Evolution 17: 147-162

Colonization and Establishment of the Paleartic Species *Drosophila Subobscura* in North and South America

A. Prevosti, L. Serra, M. Aguadé, G. Ribo, F. Mestres, J. Balañá, and M. Monclus

Department of Genetics, Faculty of Biology, University of Barcelona,
Diagonal 645, 08028 Barcelona, Spain

Until 1978 ***Drosophila subobscura*** was a Palearctic species distributed all over Europe (except in Central and Northern Scandinavia), the Macaronesian Islands, North Africa and some parts of Western Asia. In most of this area it is a common species with rather dense populations. This species was detected for the first time in Chile in February 1978, in Puerto Montt in the South of the country (Brncic et al. 1981). Subsequently it has spread very quickly in Chile and in 1981 was present from La Serena (29º55' LS) to Punta Arenas (53º10' LS) (Budnik and Brncic 1982). In November 1981 we did find the species in large numbers in San Carlos de Bariloche (Argentina), east of the Andes (Prevosti et al. 1983). Later on, in November 1986, it was collected east of the Andes in Argentina from San Juan (31º33' LS) to Esquel (42º55' LS) (Table 1). A small and isolated collection by López (1985) in Mar del Plata on the Atlantic Coast, about 400 km south of Buenos Aires, is the only finding of the species in other parts of Argentina. The same author (personal communication) did not find the species in other parts of the Buenos Aires province.

Drosophila subobscura was detected in North America in the summer of 1982 by A. Beckenbach, in Port Towsend (Wa, USA) (Beckenbach and Prevosti 1986). In the same season he found the species in several parts of British Columbia near Vancouver (Canada) and in the states of Washington and Oregon (USA). Later on, in 1983 and 1984, it was collected in California, as far South as Ojai (34º28' LN), about 100 km northwest of Los Angeles (Prevosti et al. 1987). In collections carried out in summer 1986 it was found in large numbers in the states of Washington and Oregon (Table 2) and in small numbers in several parts of the Sierra Nevada in California and in Genoa (Nevada) (Table 3).

CHARACTERISTICS OF THE COLONIZED AREAS

The colonized areas occupy symmetrical geographical locations in the Northern and the Southern hemispheres, with climatic conditions that parallel those of the native area of the species. The only regions with Mediterranean climatic conditions in America are in Central Chile

Table 1. Collections in Argentina. November 1986

	S. Juan 31º 33' LS				Mendoza 32º 48' LS				S.Carlos de Bariloche 41º 11' LS				Esquel 42º 55' LS			
	♂	♀	Total	%	♂	♀	Total	%	♂	♀	Total	%	♂	♀	Total	%
D. subobscura	1	2	3	2.4	20	13	33	36.7	247	318	565	90.1	27	41	68	20.7
D. melanogaster	7				10				-				-			
		(1) 38	60	48.8		16	33	36.7		-	-	-		-	-	-
D. simulans	15				7				-				-			
D. immigrans	13	24	37	30.1	5	12	17	18.9	-	-	-	-	-	-	-	-
D. hydei	9	14	23	18.7	-	-	-	-	-	-	-	-	-	-	-	-
repleta group	-	-	-	-	1	2	3	3.3	-	-	-	-	-	-	-	-
D. pavani	-	-	-	-	1	3	4	4.4	-	-	-	-		-	-	-
Scaptomyza sp.	-	-	-	-	-	-	-	-	27	35	62	9.9	124	137	261	79.3
Total	45	78	123		44	46	90		274	353	627		151	178	329	

(1) Figures in this row correspond to individuals of both *D. melanogaster* and *D. simulans*.

Table 2. Collections in Western North America, July1986

	Arlington, Wa 48º 12' LN				Centralia, Wa 46º 43' LN				Woodburn, Or 45º 09' LN				Medford, Or 42º 20' LN				Davis, Ca 38º 33' LN			
	♂	♀	Total	%	♂	♀	Total	%	♂	♀	Total	%	♂	♀	Total	%	♂	♀	Total	%
D.pseudoobscura	23				80				158				518	394	911	45.9	20	31	51	17.3
		(1) 95	201	1.6		473	964	36.6		136	314	12.0								
D. athabasca	83				411				20				-	-	-	-	-	-	-	-
D. subobscura	942	670	1611	85.0	697	857	1554	59.1	1193	758	1951	74.7	275	243	518	26.1	3	3	6	2.1
D. ambigua	2	-	2	.2	-	-	-	-	-	-	-	-	-	-	-	-	-	-	-	-
D. melanogaster	4	4	8	.4	5	7	12	.5	148				214	314	528	26.6	65			
										(2) 99	248	9.5						110	202	68.7
D. simulans	-	-	-	-	-	-	-	-	1				-	-	-	-	27			
D. immigrans	18	19	37	1.9	1	3	4	.1	38	26	64	2.4	2	-	2	.1	1	3	4	1.4
D.pinnicola	2	1	4	.2	1	-	1	.04	-	-	-	-	3	1	4	.2	-	-	-	-
D. hydei	3	1	4	.2	-	-	-	-	1	-	1	.04	4	2	6	.3	-	-	-	-
repleta group	-	-	-	-	-	-	-	-	-	-	-	-	-	-	-	-	14	17	31	1.5
D. transversa	2	5	7	.4	16	39	55	2.1	9	21	30	1.1	-	-	-	-	-	-	-	-
D. subquinaria	-	2	2	.1	5	10	15	.6	1	-	1	.04	2	1	3	.1	-	-	-	-
D. virilis	1	2	3	.2	20	5	25	.9	-	-	-	-	1	1	2	.1	-	-	-	-
D. busckii	1	3	4	.2	-	-	-	-	1	2	3	.1	5	2	7	.3	-	-	-	-
D. funebris	-	-	-	-	-	-	-	-	1	-	-	.04	-	-	-	-	-	-	-	-
Sc. pallida	-	2	2	.1	-	-	-	-	1	-	1	.04	1	1	2	.1	-	-	-	-
Sc. sp.	5	5	10	.5	1	-	1	.04	-	-	-	-	1	1	2	.1	-	-	-	-
Total	1085	810	1895		1237	1394	2631		1571	1042	2613		1025	958	1983		130	164	294	

(1) Figures in this row correspond to individuals of both *D. pseudoobscura* and *D. athabasca*.

(2) Figures in this row correspond to individuals of both *D. melanogaster* and *D. simulans*.

Table 3. Collections in Sierra Nevada. August 1986

	Groveland, Ca el. 867 m				Mather, Ca el. 1378 m				Genoa, Ne el. 1463 m				Truckee, Ca el. 1774 m			
	♂	♀	Total	%	♂	♀	Total	%	♂	♀	Total	%	♂	♀	Total	%
D. pseudoobscura	89				90				51				130			
D. miranda (1)	-	265	602	75.7	1	262	569	92.4	1	29	61	79.2	-	105	237	60.8
D. azteca	248				216				-				2			
D. subobscura	9	43	52	6.5	3	25	28	4.5	-	1	1	1.3	1	7	8	2.1
D. melanogaster	7				2				8	5	13	16.9	-	-	-	-
		(2) 78	137	17.2		3	6	1.0								
D. simulans	52				1				-	-	-	-	-	-	-	-
D. immigrans	1	3	4	.5	-	1	1	.2	-	-	-	-	-	-	-	-
D. occidentalis	-	-	-	-	9	3	12	1.9	1	1	2	2.6	59	69	128	32.8
D. pinnicola	-	-	-	-	-	-	-	-	-	-	-	-	7	5	12	3.1
Sc.sp.	-	-	-	-	-	-	-	-	-	-	-	-	4	1	5	1.3
Total	406	389	795		322	294	515		41	36	77		203	187	390	

(1) Figures in this row correspond to individuals of *D. pseudoobscura* , *D. miranda* and *D. athabasca* (due to the difficulty of separating females of these three species), except males which correspond only to *D. miranda*.
(2) Figures in this row correspond to individuals of both *D. melanogaster* and *D. simulans*.

and in California (Figure 1). Southward in Chile and Northward in North America, the Mediterranean climate merges into a Western maritime climate; Northward in Chile and Southward in North America it merges into semiarid and arid conditions. This latitudinal distribution of the climates is similar to that in the original area of distribution of ***D.subobscura*** in the Old World. This similarity of climatic conditions is also reflected in a greater resemblance of the vegetation and of the cultivated plants of the colonized areas to the original area of the species than in other American regions.

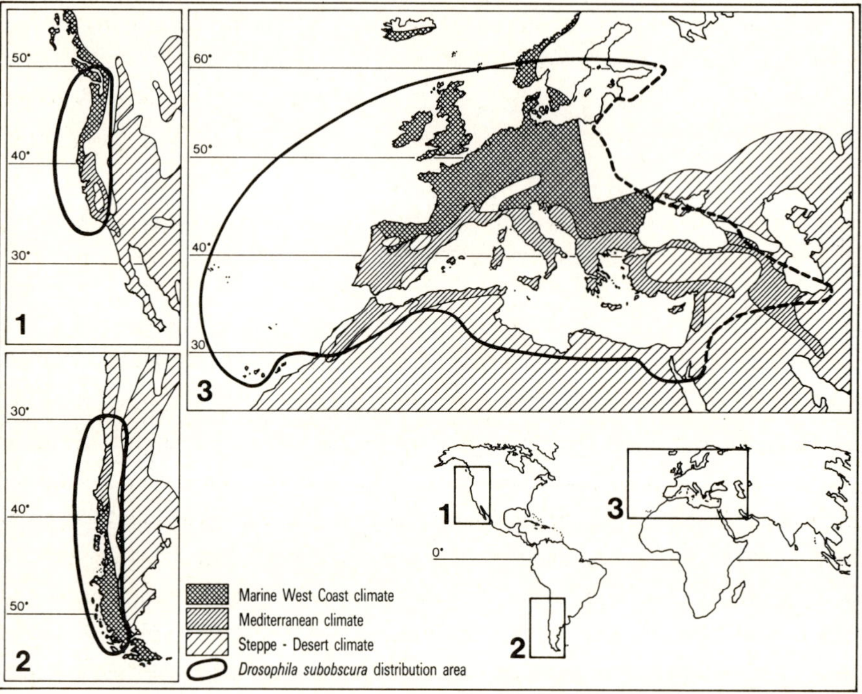

Fig.1. Climatic characteristics of the distribution areas of *D.subobscura*. Dashed line outlining the distribution perimeter in the Old World represents uncertainty concerning the eastern boundary of the distribution.

The colonization of America by ***D.subobscura*** is in this sense different from other colonizations by ***Drosophila*** species. These species often invade areas with different climatic conditions. This is the case of cosmopolitan species, such as ***D.melanogaster***, which show a great adaptability to new conditions. At least up to now, this does not seem to be the case of ***D.subobscura***.

An important difference between both colonized areas is the presence in North America of taxonomically and probably ecologically very similar species of ***Drosophila*** also included in the Obscura group. One of these species is ***D.pseudoobscura***, well known as a favourite material for evolutionary studies of American population geneticists. In South America no ***Drosophila*** species thus related to ***D.subobscura*** is present. Therefore, it seems admissible as an educated guess to suppose that ***D.subobscura*** has found an open niche facilitating its expansion in South America, whereas in North America it has perhaps found competitors in ***D.pseudoobscura*** or some of its close relatives ***D.persimilis***, ***D.miranda***, ***D.azteca*** or ***D.athabasca***.

Up to now some differences in the temperature requirements of ***D.subobscura*** and ***D.pseudoobscura*** have been detected. The Old World species is adapted to cooler conditions. Its optimum breeding temperature in the laboratory is about 18ºC. The American species is usually bred at 25º C, but at this temperature ***D.subosbscura*** cannot be bred. The populations of ***D.subobscura*** are large in early spring in California. However, the populations of ***D.pseudoobscura*** grow later in the spring. Also, in summer (Table 2) the collections of ***D. subobscura*** are larger in the North and this is not true of the collections of ***D.pseudoobscura*** .

THE EXPANSION OF THE COLONIZATION

It is safe to assume that the colonization of America by ***D.subobscura*** was detected at its beginnings, D. Brncic and his collaborators of the University of Chile in Santiago had collected Drosophila every year in Puerto Montt for more than two decades. Before 1978 they had never collected ***D.subobscura*** there nor in other sites in Chile, where they carried out extensive collections at various times. The area colonized in North America includes places where research in natural populations of ***D.pseudoobscura*** had been active for many years, especially by Dobzhansky and coworkers. Distinguishing the females of ***D.pseudoobscura*** and ***D.subobscura*** by morphological criteria is difficult, but the males can be readily identified, and observation of the polytene chromosomes makes differentiation of the two species an easy task. Had ***D.subobscura*** been present in Western North America before 1975 Dobzhansky would certainly have detected it among the thousands of individuals from scores of places that he examined for chromosomal polymorphism.

The colonizers which arrived in North and South America are representative of the same population, as has been ascertained through the study of the chromosomal inversion polymorphism (Beckenbach and Prevosti 1986; Prevosti et al. 1988). ***D.subobscura*** is highly polymorphic for its chromosomal arrangements. The elements of its kariotype are 5 rods and one dot. The five rods are named A (X chromosome), J, U, E and O and all are highly polymorphic. The inversions are indicated by numbers . Thus, O_5 means an O chromosome arrangement differing from the O standard (O_{St}) arrangement in the presence of the 5 inversion. O_{3+4+7} means an arrangement differing from O_{St} in the inversions 3, 4 and 7, and so on. Over 80 different chromosomal arrangements have been described in ***D.subobscura***, and the 19 arrangements which arrived in America are exactly the same in both colonizations. This very coincidence makes it highly improbable that each set of colonizers proceeds from a different population. The presence of the O_5 arrangement makes it difficult to ascertain the geographical origin of the colonizers. This arrangement has been found only in some populations from the northwest of Europe and from the Eastern Mediterranean area. However, it has never been detected in the western and central Mediterranean regions. Excluding O_5, all arrangements present in America point to the West of the Mediterranean as the colonizers original area.

The analysis of the allelism of recessive lethals in the North American population of Gilroy (Mestres et al., in preparation) showed that all the O_5 chromosomes present in the analysed sample, were carriers of the same lethal. O_5 chromosomes from Puerto Montt (Chile) were then analysed and also all carried the same lethal present in Gilroy. This is a confirmation of the common origin of both colonizations, since some O_5 chromosomes collected in Europe have also been analyzed and do not carry the lethal. These results indicate that only one O_5 chromosome arrived with the colonizers in America and that it brought this lethal. The finding of this lethal in the Old World would be a strong indication of the origin of the colonizers, although the probability of this finding seems to be very small indeed. The analysis of the allozyme systems (Prevosti et al. 1982) and the mitochondrial DNA (Latorre et al. 1986) gives no clue about the origin of the colonizers either.

THE FOUNDER EFFECT

Although the origin of the colonizers is not known, the data on chromosomal polymorphism indicate that a significant fraction of the gene pool of ***D.subobscura*** is present in America. Many of the chromosomal arrangements found in this continent are among the most frequent and widespread in the Old World. Assuming a western Mediterranean origin, we calculated the probability of taking a random sample of flies carrying the arrangements present in America (with the exception of O_5) from some populations of this area, taking into account the frequencies of the chromosomal arrangements in these possible original populations. The size of the sample with the maximum probability of carrying the arrangements present in America oscillates between 10 to 15 individuals (20 to 30 chromosomes). Obviously this estimate is only indicative. It is only possible to infer the minimum number of colonizers with certainty. This is 3 individuals, since 6 different arrangements of the O chromosome and 5 of the E chrromosome arrived at the New World.

The data on the allozyme systems (Table 4) also point to a rather mild founder effect. A reduction in the number of alleles is observed, but in general only the rare alleles are not present. There are some exceptions in data not yet published, i.e. in the Idh system, showing an increase in the frequency of a rare allele in America, which is also explained by the founder effect.

Table 4. Allelic frequencies in European and American populations of *Drosophila subobscura*. Data from Barcelona by Loukas et al. (1979).

locus	Barcelona	Gilroy	Davis	Eureka	Cntlia.	Arlton.
Acph						
025	0.022	--	--	--	--	--
054	0.064	--	0.055	0.034	0.030	0.025
100	0.866	1.000	0.945	0.966	0.970	0.975
188	0.043	--	--	--	--	--
200	0.005	--	--	--	--	--
n=	*187*	*38*	*181*	*147*	*164*	*162*
Est-5						
078	0.016	--	--	--	--	--
086	0.069	--	--	--	--	--
090	0.519	0.404	0.467	0.331	0.220	0.296
100	0.390	0.596	0.533	0.669	0.780	0.704
106	0.005	--	--	--	--	--
n=	*187*	*109*	*195*	*166*	*164*	*162*
Idh						
100	0.989	--	--	--	0.833	0.786
117	0.011	--	--	--	0.167	0.214
n=	*187*				*150*	*154*
Lap						
069	0.005	--	--	--	--	--
086	0.048	--	--	--	--	--
100	0.690	0.595	0.525	0.503	0.515	0.482
106	0.118	0.347	0.328	0.347	0.301	0.348
111	0.107	0.058	0.147	0.150	0.184	0.171
118	0.027	--	--	--	--	--
125	0.005	--	--	--	--	--
n=	*187*	*121*	*198*	*167*	*163*	*164*
Pept-1						
040	0.642	0.882	0.697	0.684	0.571	0.635
100	0.347	0.118	0.303	0.316	0.429	0.365
160	0.011	--	--	--	--	--
n=	*187*	*34*	*198*	*171*	*163*	*167*
Pgm						
042	0.005	--	--	--	--	--
071	0.016	--	--	--	--	--
100	0.920	--	1.000	1.000	1.000	1.000
132	0.048	--	--	--	--	--
158	0.011	--	--	--	--	--
n=	*187*		*146*	*352*	*154*	*160*

The frequencies of lethals and their allelism also support the idea that the founder effect was probably mild. In one European population, from Bordils (Gerona province) in the Northeast of Spain, an analysis of the frequency of recessive lethals of the O chromosome has been carried out. There were 33 different lethals among the 38 detected, corresponding to a frequency of .290 ± .040. Carrying out the same analysis in Gilroy, among 16 lethals detected 11 were different (.144 ± .033). Whilst 5 lethals were twice detected in Bordils, in Gilroy one lethal was found 4 times and another 3 (Mestres et al. in preparation).

The founder effect is also manifest in the associations between allozyme alleles and chromosomal arrangements. Three associations found in Western European populations, Lap 1.00 and Pept-1 0.40 with O_{3+4} and Est-5 1.00 with O_{St} were also found, although strengthened, in Chile.

To measure the strength of the association we use the index of Nei and Li (1980). As an example, Table 5 shows data corresponding to the Lap system. The frequencies of Lap 1.00 within the chromosomes with an arrangement of the O_{3+4} phylad are given in the first column. In the second column the frequencies of the same alleles within O_{St} are recorded. The differences between both frequencies are the values of the indexes of Nei and Li and are given in the third column. As shown in this table, the values of this index are higher in the Chilean populations. The strengthening of the associations in America can be easily explained by the founder effect.

Some systems, like Est-7 and Acph, only show associations in some European populations, because they are monomorphic in Chile. New associations have also appeared in America. The values of the indices of Nei and Li given in Table 5 indicate a clear association between J_{St} and Est-3 1.00 in the three Chilean populations analysed. Moreover, the chi-squares calculated to test the associations are highly significant in the three cases. The new associations can also be easily explained by the founder effect.

NO FOUNDER EFFECTS OCURRED DURING THE COLONIZATION PROCESS

Table 6 shows the frequencies of the chromosomal arrangements in the populations of South America and North America. The same arrangements are present in both areas with the exception of E_{17}. This arrangement was detected in 1981 in two Chilean populations. It was new and had never been found in the Old World. Probably it appeared in Chile after the beginning of the colonization, but it has not been found in the collections carried out in 1986 in the same places as in 1981. It has probably been lost, and had no significance for the colonization. The other arrangements have frequencies which are not very different in both colonized areas.

The frequencies of the allozyme alleles, as well as the associations of these alleles with the chromosomal arrangements also show small differences between South and North American populations.

Another stricking similarity between both colonizations is the presence of the same recessive lethal gene associated with the O_5 inversion in both areas, as indicated above.

The great similarities of the gene pools of the colonizers of both American areas strongly support the hypothesis of its common origin. But it is also highly improbable that they arrived independently from the Old World to America. Two independent samples from the same original population have a very small probability of carrying gene pools as similar as those found in both American areas. Thus, it seems unavoidable to suppose that the colonization has been sequential. One area might have been colonized first and flies coming from it could have arrived later to the second area. On this assumption we also have to admit that the number of founders of the second colonization should be high enough to carry the whole gene pool of the colonizers of the first area. Actually, no indications of the founder effect are observed when comparing the genetic characteristics of the South American and the North American populations.

The consideration of the genetic differences between the populations of the same colonized area gives no indication of founder effects either. As supported by the data on the progress of the colonization in Chile (Brncic et al. 1981), the expansion of the colonization has been very rapid and massive, not allowing founder effects to occur. Thus, it seems safe to conclude that during the process of colonization no new founder cases occurred. Although some differences betwen populations of the same area have already been detected, especially in chromosomal polymorphism, as we will now see these differences cannot be accounted for by random factors such as the founder effect.

CLINES IN THE CHROMOSOMAL POLYMORPHISM

The most striking results obtained up to now in the study of the colonization are the clines which appeared in the frequencies of the chromosomal arrangements (Prevosti et al. 1985, 1988).

The frequencies of the chromosomal arrangements in the populations from South and North America are reported in Table 6. The coefficients of correlation between these frequencies and the latitude of the places where the populations were collected are given in Table 7. This table also includes the correlations for a set of populations sampled in Western Europe along a latitudinal transect. Most of these correlations have very high values and are highly significant in the Old World, and some of them are also significant in America. Moreover a high coincidence is observed in the signs of the correlations in both continents, and in America no correlations having a different sign in Europe are significant. The correlation of the frequencies of O_5 with latitude has not been calculated in Europe because this arrangement has been found there only in some northern populations. Thus the positive correlation with latitude of the frequencies of this arrangement in North and South America is in agreement with its distribution in Western Europe. All these results dramatically support the proposition that climatic factors changing

Table 5. Associations between allozyme alleles and chromosomal arrangements (Index of Nei and Li, 1980)

	LAP			EST-3		
	Frequency of 1.00 in O_{3+4}	Frequency of 1.00 in O_{St}	Index	Frequency of 1.00 in J_{St}	Frequency of 1.00 in J_1	Index
	x	y	d = x - y	x	y	d = x - y
Chile						
Santiago	.810	.235	.575	.667	.288	.379
Chillán	.777	.045	.733	.824	.345	.479
Laja	.743	.053	.690	.821	.394	.427
Valdivia	.798	.000	.798	-	-	-
Old World						
Edinbourgh	.828	.428	.400	-	-	-
Louvain	.795	.500	.295	.463	.643	-.180
Villars	.795	.447	.348	.482	.487	-.005
Montpellier	.821	.458	.363	-	-	-
La Magdalena	.783	.500	.283	-	-	-
Picos Europa	.694	.538	.156	-	-	-
Uztegui	.813	.375	.438	-	-	-
Barcelona	.883	.282	.601	.333	.456	-.123
Navacerrada	-	-	-	.459	.404	.055
Valencia	.810	.378	.432	-	-	-
Córdoba	.786	.600	.186	.733	.506	.227
Ronda	.760	.500	.260	-	-	-
Angra (Azores)	.951	.400	.551	.414	.394	.020
Faial (Azores)	.798	.361	.437	.500	.473	.027
Madeira	-	-	-	.500	.400	.100

Table 6. Chromosomal arrangement frequencies in Chile

Chromosomal arrangements	Viña del Mar 33º02´S n=38	Santiago 33º30´S n=192	Chillán 36º37´S n=218	Laja 37º10´S n=301	Valdivia 39º46´S n=186	Bariloche 41º11´S n=133	Puerto Montt 41º28´S n=146	Castro 42º30´S n=199	Coyhaique 45º35´S n=213
A_{st}	44,4	39,7	46,0	51,7	50,0	50,0	41,1	43,1	55,9
A_2	55,6	60,3	54,2	48,3	50,0	50,0	58,9	56,9	44,1
J_{st}	23,7	21,7	34,3	29,4	20,4	40,6	30,1	34,2	29,3
J_1	76,3	78,3	65,7	70,6	79,6	59,4	69,9	65,8	70,7
U_{st}	43,6	46,4	52,1	50,8	52,2	56,6	47,3	49,6	58,2
U_{1+2}	35,9	37,6	30,5	30,0	31,5	28,6	32,2	36,9	23,0
U_{1+2+8}	20,5	16,0	17,4	19,3	16,3	15,8	20,5	13,4	18,9
E_{st}	47,4	50,0	61,5	53,8	62,9	66,9	56,8	64,4	65,9
E_{1+2}			0,5	1,0	0,5				
E_{1+2+9}	23,7	14,4	7,3	18,6	11,3	3,8	8,9	11,0	10,6
$E_{1+2+9+3}$	23,7	23,4	18,3	13,0	14,5	21,8	21,9	20,3	16,3
$E_{1+2+9+12}$	5,3	11,5	11,0	13,6	10,8	7,5	12,3	4,2	7,3
E_{17}		0,5	1,4						
O_{st}	26,3	24,0	27,5	20,8	15,3	24,1	24,7	22,7	34,1
O_5		2,6	5,5	5,1	9,5	9,0	13,0	10,1	5,7
O_7		0,5		0,3					
O_{3+4}	2,6	2,1	5,0	4,4	5,3	3,0	2,1	1,7	1,6
O_{3+4+2}	34,2	35,9	32,6	33,8	38,4	29,3	31,5	30,3	24,4
O_{3+4+7}	15,8	15,6	10,6	13,3	7,4	10,5	13,0	16,8	3,3
O_{3++4+8}	21,1	19,3	18,8	22,2	24,2	24,1	15,8	18,5	30,9

The E_{17} inversion has never been found in the Old World. O_7 is the result of recombination between O_{3+4+7} and O_{st}. Altough O_7 does not overlap with O_{3+4}, a strong linkage disequilibrium mantains the arrangement O_{3+4+7}.

Table 6 continuation. Chromosomal arrangement frequencies in North America.

Chromosomal arrangements	Arlington 48°12′N n=178	Centralia 46°43′N n=170	Woodburn 45°09′N n=70	Medford 42°20′N n=94	Eureka 40°49′N n=128	Davis 38°33′N n=230	Gilroy 37°00′N n=142
A_{st}	53,1	56,1	57,9	59,5	53,7	53,7	52,1
A_2	46,9	43,9	42,1	40,5	46,3	46,3	47,9
J_{st}	36,0	36,7	30,0	44,2	41,9	41,1	36,9
J_1	64,0	63,3	70,0	55,8	58,1	58,9	63,1
U_{st}	48,3	41,2	46,5	41,5	43,7	33,3	28,0
U_{1+2}	33,7	39,4	32,4	31,9	36,1	36,0	38,5
U_{1+2+8}	18,0	19,4	21,1	26,6	20,2	30,7	33,6
E_{st}	66,8	74,1	63,4	66,0	62,1	51,7	51,4
E_{1+2}	0,6	1,8	5,6	2,1	1,2	2,2	2,1
E_{1+2+9}	11,2	8,2	9,9	16,0	18,6	25,0	22,5
$E_{1+2+9+3}$	15,2	12,9	11,3	8,5	9,3	9,9	14,1
$E_{1+2+9+12}$	6,2	2,9	9,9	7,4	8,7	11,2	9,9
O_{st}	31,5	30,0	24,3	14,9	26,9	17,6	14,9
O_5	5,6	10,6	8,6	6,4	7,7	1,8	0,7
O_7							0,7
O_{3+4}	2,8	3,5	4,3	2,1	3,2	3,1	7,8
O_{3+4+2}	28,1	28,8	32,9	45,7	34,0	37,0	39,7
O_{3+4+7}	8,4	2,3	17,1	7,4	9,6	21,1	19,9
O_{3+4+8}	23,6	24,7	12,9	23,4	18,6	19,4	16,3

O_7 is the result of recombination between O_{3+4+7} and Ost. Although O_7 does not overlap with O_{3+4}, a strong linkage disequilibrium mantains the arrangement O_{3+4+7}

gradually with latitude are the cause of the clines in the chromosomal arrangements.

Seven of the populations analysed in Chile in 1981 were sampled again in 1986, in order to investigate how the colonization was progressing. The values of the correlations of the frequencies of the chromosomal arrangements with latitude in these populations are also shown in Table 7. A general increase in the values of these correlation coefficients is observed in

Table 7. Correlation coefficients between arrangement frequencies and latitude.

	Old World	North America	South America 1981 (9 populations)	South America 1986 (7 populations)
A_{st}	.880 ***	.347	.492	.496
A_2	-.864 ***	-.345	-.492	-.496
J_{st}	.972 ***	-.433	.457	.547
J_1	-.973 ***	.433	-.457	-.547
U_{st}	.974 ***	.850 *	.727 **	.769 *
U_{1+2}	-.387	-.283	-.626 *	.103
U_{1+2+8}	-.944 ***	-.881 **	-.197	-.828 *
E_{st}	.973 ***	.868 **	.838 ***	.880 **
E_{1+2+9}	-.575	-.930 ***	-.612 *	-.839 *
$E_{1+2+9+3}$	-.834 ***	.348	-.306	-.537
$E_{1+2+9+12}$	-.968 ***	-.696	-.226	-.552
O_{st}	.870 ***	.793 *	.219	.730
O_5(1)		.770 *(2)	.730 **(2)	.777 *(2)
O_{3+4}	-.870 ***	- .460	-.325	.631
O_{3+4+2}	-.267	-.690	-.718 *	-.328
O_{3+4+7}	-.577 *	-.674	-.618 *	-.930 ***
O_{3+4+8}	-.804 ***	.385	.412	.044

(1) The correlation of latitude with O_5 frequencies has not been calculated for the Old World because this arrangement is only present in two populations used to calculate the correlation. (2) This positive correlation agrees with the distribution of O_5 in the Old World, since it is characteristic of northern populations.

relation to the values of 1981, although there are some exceptions. In fact, the correlation with U_{1+2} has lost its statistical significance and has changed the sign. It is interesting to note that this correlation is one of the few which are not significant in Europe. Actually, U_{1+2} has very low frequencies or is not present in the North of Europe where U_{St} is the dominant arrangement. Southwards U_{St} decreases and U_{1+2} becomes more abundant. However, in the Mediterranean countries U_{1+2} is replaced by more complex arrangements, such as U_{1+2+8} in the Western Mediterranean areas, and U_{1+2+6} and U_{1+2+7} in other parts of the Mediterranean. This distribution of U_{1+2}, which has its highest frequency in intermediate latitudes, is the reason for the lack of significant correlation of this arrangement with latitude. One of the arrangements showing a higher increase in its correlation with latitude is U_{1+2+8}, which became significant in

1986. This indicates that the frequency of U_{1+2+8} has increased in the North of Chile at the expense of that of U_{1+2}, and thus the pattern of distribution of the chromosome U arrangements in Chile is becoming more similar to that of the European populations.

The value of the O_{3+4+2} correlation also dropped considerably in 1986, losing its statistical significance. As with U_{1+2} the correlation with latitude of the frequencies of O_{3+4+2} is not statistically significant in Europe, where the frequencies of this arrangement are usually very low and show a rather irregular geographical variation. The correlation of O_5 with latitude increased in 1986, becoming statistically significant. The correlation coefficient with O_{St} has also increased considerably and is almost significant at the .05 level. A decrease in the correlation coefficient is observed for O_{3+4+8}, although its sign has changed, becoming now negative as in Western Europe. Thus, it now resembles the European correlation more than in 1981. Only the coefficient of O_{3+4} departs from the European correlation more than in 1981. However, this correlation is not very indicative in America, because it shows very low frequencies, in general lower than .05.

In the chromosomes A, J and E all the correlations show a more or less accentuated tendency to increase in relation to 1981.

CONCLUSIONS

The double colonization of America by ***D.subobscura*** offers a unique opportunity for evolutionary studies. It is like a large scale natural experiment with two replicates.

Unfortunately, we do not know the number of founders which arrived in America. Actually, we did several estimates based on different hypotheses and all the results agree in giving as most probable a number of founders between 10 and 15 individuals. What is certain is that whatever this number was, because of the rapid expansion of the initial population the founder effect was mild. Consequently, a significant fraction of the gene pool of ***D.subobscura*** is present in the American populations.

The almost complete identity of the gene pool of the populations of both American colonizations, dramatically confirmed by the association of all O_5 chromosomes present in America with the same recessive lethal gene, supports the idea that these colonizations were sequential, with no founder effect at all in the passage from the first to the second colonized area. Soon after the beginning of the colonization a great flush of the populations of ***D.subobscura*** occurred, which led to a rapid expansion of the colonized areas without any further founder effects. Up to now no crash has followed this initial flush. Ten years after the beginning of the colonization ***D.subobscura*** seems to be firmly established in both colonized areas; in North America, in spite of the presence of authoctonous closely related species such as ***D.pseudoobscura*** and its relatives. In South America as well as in North America, depending on the site and the season of the year in which the collections are carried out, ***D.subobscura*** is now the dominant species.

Adaptive processes are occurring within the colonized areas, promoting genetic differences between the populations. The latitudinal clines established in the frequencies of the chromosomal arragements are the most significant observations supporting the presence of these adaptive processes. These clines parallel those established in the Old World. The same arrangements decrease or increase in frequency in both American areas and in the Palearctic Region. The adaptive differentiation of the populations was initially very rapid. At least in Chile the colonization was detected early at its beginnings. Three years later, in 1981, the clines were already clearly detected. The analysis of collections carried out later, in 1986, shows an increase of the similarity to the pattern of the European clines. The adaptive process continued, but at a lower pace. The dynamical properties of the genetic variability of ***D.subobscura*** in relation to its adaptation to climatic factors are manifest in these results.

ACKNOWLEDGEMENTS

This work was supported by Grant 2844 from the Comisión Asesora para la Investigación Científica y Técnica, Spain, and Grant CCB-8504013 from the United States-Spain Joint Committee for Scientific and Technological Cooperation.

BIBLIOGRAPHY

Beckenbach A and Prevosti A (1986). Colonization of North America by the European species ***D.subobscura*** and ***D.ambigua*** . The American Midland Naturalist 115: 10-18.

Brncic D, Prevosti A, Budnik M, Monclús M and Ocaña J (1981). Colonization of ***D.subobscura*** in Chile. I. First population and cytogenetic studies. Genetica 56: 3-9.

Budnik M and Brncic D (1982). Colonización de ***Drosophila subobscura*** Collin en Chile. In R. Cruz-Coke and D. Brncic (eds.) Genetica. V Congreso Latinoamericano de Genética. Asociación Latinoamericana de Genética y Sociedad de Genética de Chile, Santiago.

Latorre A, Moya A and Ayala F J (1986). Evolution of mitochondrial DNA in ***D.subobscura***. Proc Natl Acad Sci USA 83: 8649-8653.

López M (1985). ***Drosophila subobscura*** has been found in the Atlantic Coast of Argentine. D.I.S. 61: 113.

Mestres F, Pegueroles G, Prevosti A and Serra L. Colonization of America by ***Drosophila subobscura*** : Lethal genes and the problem of the O_5 inversion. (in preparation).

Nei M and Li W H (1980). Non random association between electromorphs and inversion chromosomes in finite populations. Gen Res 33: 65-85.

Prevosti A, Ribó G, García MP, Sagarra E, Aguadé M, Serra L and Monclús M (1982). Los polimorfismos cromosómico y aloenzimático en las poblaciones de ***D.subobscura*** colonizadoras de Chile. Actas V Congr Latinoam Genética: 189-197.

Prevosti A, Ribó G, Serra L, Aguadé M, Balañá J, Monclús M and Mestres F (1988). Colonization of America by ***D.subobscura***: Experiment in natural populations that supports the adaptive role of chromosomal inversion polymorphism. Proc Natl Acad Sci USA 85: 5597-5600.

Prevosti A, Serra L and Monclús M (1983). ***Drosophila subobscura*** has been found in Argentina. D.I.S. 59: 103.

Prevosti A, Serra L, Monclús M, Mestres F, Latorre A, Ribó G and Aguadé M (1987). Colonización de América por ***Drosophila subobscura*** . Evolución Biológica (Bogotá) 1: 1-24.

Prevosti A, Serra L, Ribó G, Aguadé M, Sagarra E, Monclús M and García MP (1985). The colonization of ***D.subobscura*** in Chile. II. Clines in the chromosomal arrangements. Evolution 39: 838-844.

Short Range Genetic Variations and Alcoholic Resources in *Drosophila melanogaster*

J.R. David[1], A. Alonso-Moraga[2], P. Capy[1], A. Muñoz-Serrano[2], and J. Vouidibio[3]

[1] Laboratoire de Biologie et Génétique Evolutives, C.N.R.S., 91198 Gif-sur-Yvette, France
[2] Department of Genetics, University of Córdoba, 14071 Córdoba, Spain
[3] Département de Biologie Cellulaire, Faculté des Sciences, B.P. 69, Brazzaville, R.P. Congo

Abstract

Short range genetic variations, sometimes occurring over less than a kilometer, were found at the Adh locus in natural populations from southern Spain and the Congo. Such variations seem to imply strong selective pressure exerted by environmental alcohol in cellars or in breweries. Field populations are apparently selected in an opposite direction, and the occurrence of opposite adaptive peaks seems to explain the genetic structure of samples collected. Three other enzyme loci exhibited correlated variations with Adh, i.e. Gpdh and Est-6 in Spain and Gpdh and G6pd in the Congo. By contrast, French populations are genetically homogeneous over a large area and could be considered to be a single panmictic unit. However, a release-recapture experiment of genetically marked flies suggested that the larval breeding sites in this country also correspond to small, dispersed and ephemeral patches.

Introduction

Analysing and interpreting the spatial structure of natural populations is a key issue in ecological genetics and evolutionary biology. In this respect, *Drosophila melanogaster* is a remarkable model illustrating both the evolution of ideas and the discovery of new data. For a long time it has been assumed, mainly a priori, that this domestic species was permanently transportated around the world by the activities of modern man, and that the resulting gene flow would prevent any geographic differentiation. Then, when genetic comparisons of distant populations became available, it turned out that D. melanogaster exhibited very large divergences between geographic races, these resulting from founder effects, historical events but also from ecological selection and climatic adaptations (see Lemeunier at al, 1986 and David and Capy, 1988, for reviews). Numerous traits are arranged according to long range latitudinal clines, often parallel on different continents, and this strongly suggests the adaptive significance of these clines (Endler, 1986). In contrast, medium (a few hundred kilometers) and short range (a few kilometers) variations have long remained poorly documented. This has been taken as evidence of a fairly high dispersal capacity of adult flies : only strong selective pressures may counterbalance the homogenizing influence of gene flow (Slatkin, 1987). In the case of D. melanogaster, the best evidence of some recent worldwide gene flow is the probable spreading of the P transposable element in all natural populations (Anxolabéhère et al. 1988).

An original feature in D. melanogaster ecology is the capacity to use artificial, man-made alcoholic resources, especially in wine production areas of temperate countries. This capacity is related to alcohol dehydrogenase (Adh) locus polymorphism and long range latitudinal clines are well documented both for Adh gene frequencies and ethanol tolerance (Van Delden, 1982 ; David and Capy, 1988). In wine production areas, alcoholic resources are available but generally for a short duration in the year, so natural populations are exposed to different selective pressures according to season and local conditions. Such a situation, i.e. different resources among nearby habitats, has provided good opportunity for investigating short range adaptive genetic variation. However, the results are variable according to the country under study, in spite of their similar environments (Gibson and Wilks, 1988).

Investigations in Australia failed to show clear cut differences in _Adh_ gene frequencies between cellar and field populations, although some variations in ethanol tolerance were observed (McKenzie and Parsons, 1974 ; McKenzie and McKechnie, 1981 ; Oakeshott and Gibson, 1984). A similar observation was made in California (Marks _et al_, 1980). In contrast, Spanish populations exhibit differences according to their habitat, i.e. more _Adh-F_ alleles in cellars (Briscoe _et al_, 1975 ; Alonso-Moraga _et al_, 1985). A similar observation was made in Canada (Hickey and McLean, 1980). In France, there are apparently no variations in _Adh-F_ frequency between wine production areas and other places (Charles-Palabost _et al_, 1985), although a higher ethanol tolerance is suspected in cellar populations during vintage time (Capy _et al_, 1987a, and unpublished results). A similar observation has been recently made in Hungary (Pecsenye, 1989).

In this paper, various results will be described, which either demonstrate significant variations over very short distances in natural populations, or represent evidence of different breeding sites of apparently sympatric demes. We shall first consider data collected in southern Spain, in and around wine cellars ; secondly, observations made in a completely different ecological situation, i.e. the urban environment of an Afroequatorial city ; thirdly, how the release and recapture of _Drosophila_ adults in France suggests that natural populations may indeed be subdivided into small independent entities.

Populations in southern Spain

During the last 6 years, numerous samples of _D. melanogaster_ adults have been collected in the Cordoba region. The whole area prospected covers about 40,000 square km, but the origin of most samples is within a smaller vicinity, i.e. a radius of about 30 km (Alonso-Moraga _et al_, 1985 ; Muñoz-Serrano _et al_, 1985 ; Alonso-Moraga et Muñoz-Serrano, 1986).

An overall survey enables the distinction of two types of samples, those collected in wine cellars and those collected outdoors, in field conditions. As in other parts of the world, cellar populations are characterized by a demographic outburst in Autumn, which is mediated by the availability of large amounts of grape residues with high ethanol content. However, permanent, small-sized populations may be found in cellars all year round, breeding on wine exuding from barrels or on the surface of wine jars. Outdoors, field populations are attracted to fermenting baits and can also be found all year round except during the very hot Summer months. Identified breeding sites include oranges, figs, prickly pears and garbage. The dynamic relationships between cellar and field populations are not well understood and remain the subject of active investigation. In several cases cellar and field samples were collected a few hundred meters apart but still exhibited clear genetic differences (Alonso-Moraga *et al*, 1988). The results presented here may thus be considered as corresponding to short range variations, i.e. the distance which is easily covered by a fly during its life time.

Three allozyme loci, *Adh*, *Gpdh* (glycerophosphate dehydrogenase) and *Est-6* (esterase-6) were analysed in wild-caught flies. Each locus exhibits 2 alleles, referred to as *F* and *S* according to electrophoretic mobility. In the case of *Est-6* a few rare alleles were found but ignored. The genotypic frequencies were analysed by calculating the fixation index Fis, and the between-sample variability with the index Fst (Wright, 1951). Average values for field and cellar samples are given in Table 1.

For the *Adh* gene, a clear difference exists between the two habitats. The *F* allele, which results in more enzyme activity, and is generally favored when experimental cultures are treated with ethanol (see Van Delden, 1982), is more frequent in cellars. The distributions of these two kinds of samples are shown in Figure 1. The between sample variability, estimated by Fst is also greater in field

Table 1. Comparison of allelic frequencies at 3 allozyme loci and of fixation indices in samples collected in field or cellar habitats in Cordoba vicinity. n : number of different samples ; comparisons made with Student's t parameter ; level of significance : ** $p<0.01$. Fis is the within sample fixation index (departure from H.W. equilibrium). Fst is the between sample fixation index. Fit is calculated according to the formula (1-Fit) = (1-Fis) (1-Fst).

locus	habitat	n	percent F allele	Fis	Fst	Fit
Adh	field	23	54.70±4.67	0.497±0.047	0.203	0.600
	cellar	29	81.84±1.81	0.100±0.037	0.064	0.158
	comparison		5.75**	6.75**		
Gpdh	field	23	70.13±2.76	0.218±0.041	0.083	0.299
	cellar	29	53.27±1.63	0.077±0.022	0.031	0.091
	comparison		3.91**	3.20**		
Est-6	field	17	46.49±2.71	0.210±0.034	0.050	0.249
	cellar	22	36.28±1.45	0.190±0.041	0.020	0.207
	comparison		3.52**	0.38		

populations. The most surprising result concern Fis, which measures the departure from a Hardy-Weinberg equilibrium in each sample. For the cellar populations, the average is not significantly different from zero. By contrast, field samples exhibit a high, positive mean, close to 0.50, indicating a severe deficit of heterozygous flies.

The two other studied loci show similar, although less extreme features. The frequency of Gpdh-F is significantly lower in cellars than in fields, while Est-6-F is more frequent in fields than in cellars. A higher fixation index Fis is found in field samples for Gpdh. For Est-6, no difference exists between the two types of habitats but the heterozygote deficit is significant in both cases. For the two loci, the variability among samples (Fst) is greater for field collected flies.

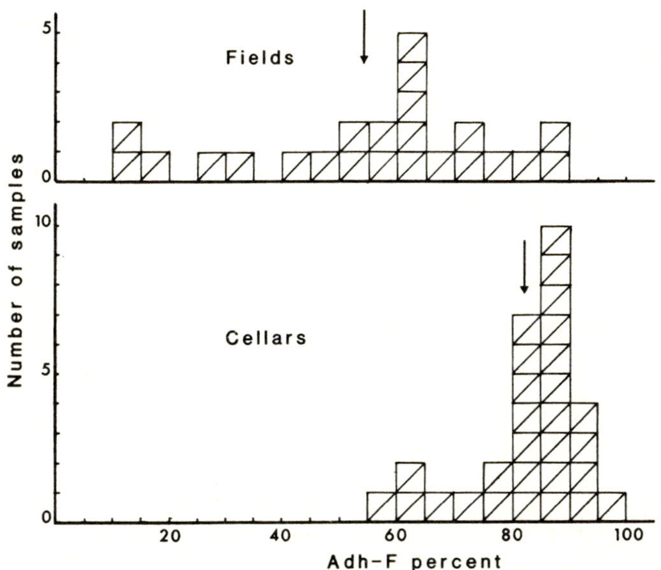

Figure 1 Adh-F frequencies in various D. melanogaster samples collected in field or cellar conditions in southern Spain. Arrows point to the mean of each sample group.

The fact, already pointed out, that significant differences may be found between samples collected at the same time over very short distances, strongly suggests that the differences are maintained by strong and divergent selection. The only alternative explanation would be a long standing interruption of gene flow, an ecologically unrealistic hypothesis. A most surprising observation is a consistent deficit in heterozygotes, which is found, with different modalities, for the three loci.

Seven different mechanisms, which may explain a deficit in heterozygotes, are listed below :

1. High frequency of null alleles ;
2. Strong inbreeding ;
3. Mixing of genetically different populations (Wahlund effect) ;
4. Selection favoring the homozygotes ;
5. Selection against heterozygotes ;

6. Genetic homogamy ;
7. Resource-habitat selection by different genotypes.

It is hoped that future investigations will test these possibilities. For the moment we propose that several of the above mechanism may be simultaneously involved for explaining the observations.

Populations in the Brazzaville urban area

Between equatorial Africa and temperate Europe, natural populations exhibit a smooth latitudinal cline for increasing ethanol tolerance and *Adh-F* frequency (David and Bocquet, 1975 ; David *et al*, 1986). Recently, we presented evidence that the overall cline is not linear for *Adh* frequency, and that Mediterranean populations are highly variable, or unstable, over short distances (David *et al*, 1989). In this general pattern, Congolese populations were considered to be uniformly ethanol-sensitive and to harbor a very low frequency of *Adh-F*. However, during the last decade, it has been progressively discovered that this presumed genetic uniformity did not hold in the city of Brazzaville, and that over very short distances, large genetic differences could be found. A careful analysis has progressively unravelled a population structure which is apparently stable over space and time (Vouidibio *et al*, submitted). According to the collecting site, the *Adh-F* frequency varies from 2 up to almost 90 %. Ethanol tolerance also varies from 6 to 15 % and the two traits are highly correlated (Table 2 and Figure 2).

A detailed analysis showed that typical African samples, i.e. with a low ethanol tolerance and a low *Adh-F* frequency, were found in the suburbs and adjacent villages, and also in an urban market where numerous field flies were presumably brought in every day with fruit products. By contrast, populations with high ethanol tolerance and a high frequency of *Adh-F* were found in breweries, where a strong selection due to alcoholic resources is expected. Finally, samples with intermediate values were collected in various parts of the city, more or less distant from breweries.

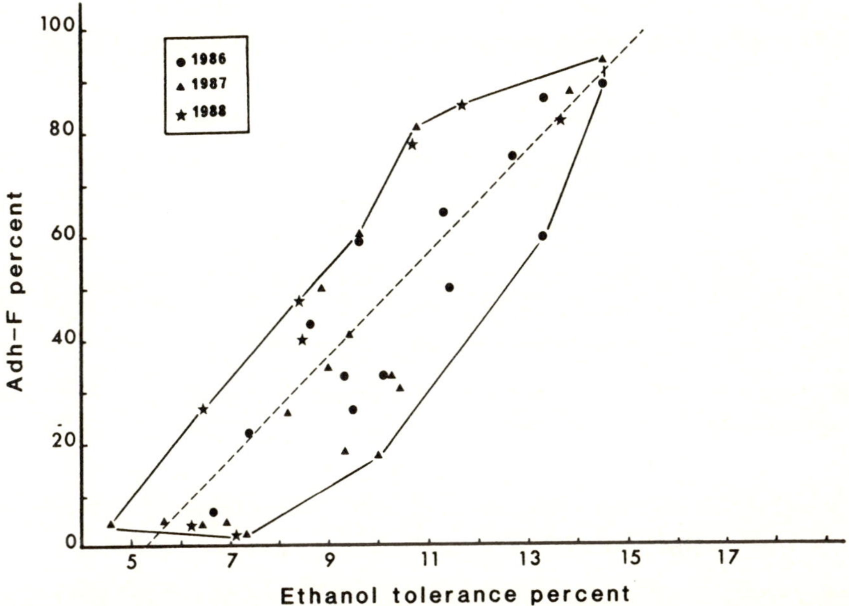

Figure 2 Relationship between ethanol tolerance and Adh-F frequency among samples collected in Brazzaville urban area during years 1986-1988.

In the Brazzaville samples, the genotypic structure was not investigated because, in most cases, allelic frequencies were measured on isofemale lines and not on wild collected flies. Two other enzyme loci were also analysed, Gpdh and G6pd, while Est-6 was not considered. These two diallelic loci also exhibited significant correlated variations with ethanol tolerance and Adh-F frequencies, as indicated in Table 2. More precisely, it has been found that an increase in ethanol tolerance or in Adh-F is correlated with an increase of Gpdh-S or G6pd-F.

The spatial structure found in the Brazzaville area seems difficult to explain by a subdivision of populations into small patches, either completely isolated or with a very restricted gene flow. All the samples were collected in a built-up area with abundant resources, mostly garbage. However, buildings are interspersed with numerous gardens and tree-lined avenues, and these ecological conditions seem very

Table 2. Correlations between ethanol tolerance and allozyme frequencies among samples collected in Brazzaville. r : coefficient of correlation ; n : number of samples ; level of significance : **p<0.01.

traits	r	n
Ethanol toler. - Adh-F	0.88**	38
Ethanol toler. - Gpdh-S	0.50**	35
Ethanol toler. - G6pd-F	0.82**	25
Adh-F - Gpdh-S	0.52**	35
Adh-F - G6pd-F	0.84**	25
Gpdh-S - G6pd-F	0.53**	25

favorable to dispersal and gene flow. The most likely interpretation of the data is that brewery populations maintain their genetic properties because they breed in resources containing a significant amount of alcohol. On the other hand, countryside populations would keep their genetic characteristics because they are, as in other parts of tropical Africa, selected in an opposite direction which implies a low ethanol tolerance and low frequencies of Adh-F, Gpdh-S and G6pd-F. Under this general scheme, samples with intermediate values could result from a diffusion occurring between populations selected in opposite directions.

Inference of ecological patchiness by release-recapture of genetically marked flies

Contrasting with what has been found in southern Spain or in Brazzaville, French *D. melanogaster* populations appear remarkably stable over space and time. Many samples, studied by several investigators and collected in a diversity of habitats including wine cellars, orchards, garbage piles, gardens and houses, all exhibit a high frequency of Adh-F (average 95 %) and very limited variation. For example, a calculation of Fst from 51 different samples gave a value of only 0.02 (David et al, 1989). This observation, which may be extended to other loci, can be interpretated in two ways : either French populations function as a single, panmictic, Hardy-Weinberg population, or the gene frequencies are maintained by some balancing selection.

We tried to falsify the selection hypothesis(Capy et al,

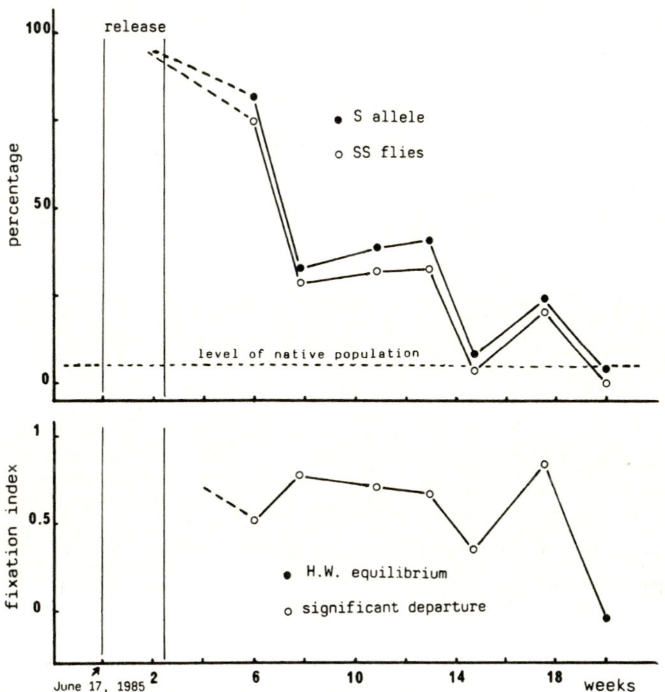

Figure 3. Result of releasing about 50,000 adult flies homozygous for Adh-S in southern France (Bédeille near St-Girons). (from Capy et al, 1987).

1987b) in the following way. Flies homozygous for the rare Adh-S allele were extracted from a natural population of southern France. About 10 independent lines were crossed to produce an outbred population, and about 50,000 such flies, artificially grown on laboratory food, were released in the place from which they originated. The experiment was repeated during 2 successive years with always similar results, as shown in Figure 3 for 1985.

During and just after the release, the native wild population was completely overflowed by the released individuals, and nearly all flies recaptured were homozygous SS. Then the frequency of Adh-S declined, more or less rapidly according to the year, until it came back to the original value of about 5 %. Such experiments showed that the released flies had a clear disadvantage. But this may be due to a diversity of causes, for example inbreeding or a fitness reduction due to artificial breeding in the laboratory. Thus, the results did not demonstrate that Adh polymorphism in nature was maintained by natural selection. However, with respect to this paper, an interesting observation was made. In 1985, the decline of Adh-S was slow and it took 3 months to come back to the initial level (Figure 3). The longevity of Drosophila adults in the wild is not known with precision, but is presumed to be short and around 10 days (Shorrocks and Rosewell, 1986). It is very unlikely to exceed a month in Summer for the longest living flies. So, in the case of the experiment shown in Figure 3, at least two generations elapsed before the Adh-S frequency came back at 5 %. In other words, the released SS flies did reproduce, and after a month, their offspring were recaptured. Surprisingly, the released flies did not hybridize very much with the native population, since all collected samples, during 3 months, were characterized by a strong deficit in heterozygotes and a positive value of Fis (Figure 3). A possible but unlikely explanation would be that the released flies had developed some ethological isolation during their conservation in the laboratory. This possibility was ruled out by complementary experiments (Capy et al, 1987b).

Thus the only interpretation is that the released flies reproduced mostly in habitat-resources not used by native flies and that the adults did not meet and interbreed. This indirect evidence strongly suggests that, even in France, natural populations do not function as a single panmictic unit but may be divided, at least for some time, into small, independent and presumably transient patches.

Concluding remarks

In this paper we present evidence from studies made in France, Spain and Congo, that D. melanogaster populations may exhibit short range genetic variations and are therefore far from panmixia. However, different causes account for these observations.

In French populations there is not evidence, at present, of strong divergent selective pressure in different habitats. So, only the heterozygote deficit found in the release-recapture experiment suggests that larval breeding sites may be subdivided into small and transient entities. Interestingly, a similar observation was made in Italy for the Est-6 locus of wild collected flies (Danieli and Costa, 1977). It was found that, at the beginning of Summer, when population density is low, wild collected samples exhibited a large deficit in heterozygotes, with an Fis index close to 0.60. This value decreased progressively until it almost reached zero in Autumn. The likely interpretation of these data is that, at the beginning of the season, the population density is low and natural resources are scarce and subdivided, consequently collected samples exhibit a strong Wahlund effect. Later, the demographic outburst is accompanied by greater resource reliability and a tendency toward panmixia. Similar observations might be made in France if the genotypic structure of natural populations is monitored for the seasonal variations. However, in all temperate countries with a cold Winter, a main ecological problem remains unsolved : the way natural populations pass through the demographic bottleneck of Winter.

In Spain and Congo, we find that different selective pressures are exerted in different habitats so that the genetic structures which are observed result from the interaction between two opposite adaptive peaks. One of the peaks is most probably due to ethanol selection, although the amount of alcohol in brewery residues is not documented. The other adaptive peak is found in field samples and the possible selective factors are not clear : we do not know for example why Adh-S should be favored in countries with a warm,

Mediterranean or tropical climate, and much remains to be done in this respect.

Interestingly, 3 other loci exhibit correlated variations with Adh, and in all cases the short range pattern matches the long range latitudinal clines (see Lemeunier et al, 1986). The association between Adh-F and Gpdh-S in nature and in the laboratory is well documented (Van Delden, 1984) and can be related to the utilization of NADH in the energy budget (Davis and McIntyre, 1988). The two other associations seem to be new and require some discussion. The relationship between Adh and G6pd may be functional (Vouidibio et al, submitted) : treating the flies with ethanol results in an increase of fatty acid synthesis (Van Herrewege and David, 1987) which needs NADPH as a cofactor. A large proportion of this NADPH is provided by the enzymes of the pentose shunt, among which G6pd. For the association between Adh and Est-6 observed in Spain, we cannot propose any plausible functional interaction, and the problem deserves further study, especially in the Brazzaville area.

Finally, a major problem remains : why populations, in different parts of the world, do not react in the same way to apparently similar selective pressures due to environmental alcohol ? (see the introduction). Answering this question now represents an exciting challenge deserving an international research cooperation.

Acknowledgements : part of these works benefited of grants from PIREN (French CNRS) and IREB.

REFERENCES

Alonso-Moraga A, Munoz-Serrano A (1986) Allozyme polymorphism and linkage disequilibrium of Adh and α-Gpdh loci in wine cellar and field populations of Drosophila melanogaster. Experientia 42:1048-1050

Alonso-Moraga A, Munoz-Serrano A, Rodero A (1985) Variation of alloenzyme frequencies in Spanish field and cellar populations of Drosophila melanogaster. Génét Sél Evol 17:435-442

Alonso-Moraga A, Munoz-Serrano A, Serradilla JM, David JR (1988) Microspatial differentiation of Drosophila melanogaster natural populations in and around a wine cellar in southern Spain. Génét Sél Evol 20:307-314

Anxolabéhère D, Kidwell MG, Périquet G (1988) Molecular characteristics of diverse populations are consistent with the hypothesis of a recent invasion of Drosophila melanogaster by mobile P elements. Mol Biol Evol 5:252-269

Briscoe DE, Robertson A, Malpica (1975) Dominance at Adh locus in response of adult Drosophila melanogaster to environmental alcohol. Nature 255:148-149

Capy P, David JR, Carton Y, Pla E and Stockel J (1987a) Grape breeding Drosophila communities in Southern France : short range variation in ecological and genetical structure of natural populations. Acta Oecol. Oecol gener 8:435-440

Capy P, David JR and Delay B (1987b) Can we modify the genetic structure of a Drosophila natural population by releasing genetically marked flies? Genet. Iberica 39:287-306

Charles-Palabost L, Lehmann M and Merçot H (1985) Allozyme variation in fourteen natural populations of Drosophila melanogaster collected from different regions of France. Génét Sél Evol 17:201-210

Danieli GA and Costa R (1977) Transient equilibrium at the Esterase-6 locus in wild populations of Drosophila melanogaster. Genetica 47:37-41

David JR and Bocquet C (1975) Similarities and differences in latitudinal adaptation of two Drosophila sibling species. Nature, 257:588-590

David JR and Capy P (1988) Genetic variations of Drosophila melanogaster natural populations. Trends in Genetics 4:106-111

David JR, Merçot H, Capy P, McEvey SF and Van Herrewege J (1986) Alcohol tolerance and Adh gene frequencies in European and African populations of Drosophila melanogaster. Génét Sél Evol 18:405-416

David JR, Alonso-Moraga A, Borai F, Capy P, Merçot H, McEvey SF, Munoz-Serrano A and Tsakas S (1989) Latitudinal variation of Adh gene frequencies in Drosophila melanogaster : a Mediterranean instability. Heredity, in press

Davis MB and McIntyre RJ (1988) A genetic analysis of the α.glycerophosphate oxidase locus in Drosophila melanogaster. Genetics 120:755-766

Gibson JB and Wilks AV (1988) The alcohol dehydrogenase polymorphism of Drosophila melanogaster in relation to environmental ethanol, ethanol tolerance and alcohol dehydrogenase activity. Heredity 60:403-414

Endler JA (1986) Natural selection in the Wild. Princeton University Press, Princeton, New Jersey, 336 pp

Hickey DA and McLean MD (1980) Selection for ethanol tolerance and ADH allozymes in natural populations of Drosophila melanogaster. Genet Res 36:11-15

Lemeunier F, David JR, Tsacas L and Ashburner M (1986). The melanogaster species group. In "The genetics and biology of Drosophila", ed by M. Ashburner, H. Carson and J.N. Thompson, Acad Press London, Vol. 3e, pp. 147-256

Marks RW, Brittnacker JG, McDonald JF, Prout T and Ayala FJ (1980) Wineries, Drosophila, alcohol and ADH. Oecologia 47:141-144

McKenzie JA and McKechnie SW (1981) The alcohol dehydrogenase polymorphisme in a vineyard cellar population of Drosophila melanogaster. In "Genetic Studies of Drosophila Populations",ed. by JB Gibson and JG Oakeshott, Australian Nat Univ Press Canberra, pp 201-215

McKenzie JA and Parsons PA (1974) Microdifferentiation in a natural population of Drosophila melanogaster to alcohol in the environment. Genetics 77:385-394

Munoz-Serrano A, Alonso-Moraga A and Rodero A (1985) Annual variation of enzyme polymorphism in four natural populations of Drosophila melanogaster occupying different niches. Genetica 67:121-129

Oakeshott JG, Gibson JB and Wilson SR (1984) Selective effects of the genetic background and ethanol on the alcohol dehydrogenase polymorphism in Drosophila melanogaster. Heredity 53:51-67

Pecsenye K (1989) A comparison of the level of enzyme polymorphism in cosmopolitan Drosophila species between populations collected in distilleries and in their surroundings in Hungary. Génét Sél Evol (in press)

Slatkin M (1987) Gene flow and the geographic structure of natural populations. Science 236:787-792

Shorrocks B and Rosewell J (1986) Guild size in drosophilids : a simulation model. J anim Ecol 55:527-541

Van Delden W (1982) The alcohol dehydrogenase polymorphism in Drosophila melanogaster. Selection at an enzyme locus. Evol Biol 15:187-222

Van Delden W (1984) The alcohol dehydrogenase polymorphism in Drosophila melanogaster, facts and problems. In "Population biology and evolution", ed. by K. Wöhrmann and V. Loeschke, Springer Verlag, Heidelberg, pp. 127-142

Van Herrewege J and David JR (1987) Effects of ethanol on the development of Drosophila melanogaster : comparison of two geographic populations. J Insect Physiol 33:893-898

Vouidibio J, Capy P, Defaye D, Pla E, Sandrin J, Csink A and David JR Short range genetic structure of Drosophila melanogaster populations in an Afrotropical area, and its significance (submitted)

Wright S (1951) The genetical structure of populations. Ann.Eugen 15:323-354

A.2. Experimental

The Variance in Genetic Diversity Among Subpopulations is More Sensitive to Founder Effects and Bottlenecks Than is the Mean: A Case Study

P.M. Brakefield

Section of Evolutionary Biology, Department of Population Biology, University of Leiden, Schelpenkade 14a, 2313 ZT Leiden, The Netherlands

Introduction

There are three widely-recognised effects of genetic drift on polymorphic traits which are associated with 'bottlenecks' or 'founder events' in populations. Rare alleles are expected to be lost, average heterozygosity to decline, and the variance among populations to increase. The theoretical work is summarized in Nei (1987). These effects have been documented many times in laboratory populations (e.g. Buri, 1956; Dobzhansky and Pavlovsky, 1957; Rich, Bell and Wilson, 1979; Wool, 1987). Information from natural populations is less extensive and has often involved allozyme variation, especially on islands or in newly-colonized mainland areas (e.g. Janson, 1987; Black _et al._, 1988; work cited in Wool, 1987). Most studies have, however, tended to concentrate either on the description of the loss of heterozygosity or rare alleles within populations, or on the increase in differentiation among populations. Moreover, many studies involving polymorphic enzyme loci tend to be based on comparatively small samples so that there is uncertainty about whether rare alleles are present or not in populations thought to have experienced genetic drift.

This paper explores the relative contributions of the effects of genetic drift to changes in phenotypic and genetic diversity within and among populations in relation to founder events or bottlenecks of varying magnitude. Data for a homopteran insect which describe the frequencies of melanic phenotypes controlled by a polymorphic gene in large samples from an island archilepago are analysed and used as the basis for a series of simple simulations. The island populations appear to conform to a type of metapopulation structure (Levins, 1970; Harrison, Murphy and Ehrlich, 1988; Wade and McCauley, 1988) with rates of extinction and recolonization, and of inter-population migration probably varying between islands in relation to their size and height. The results reproduce the well-known finding that substantial genetic diversity at polymorphic gene loci only tends to be lost in association with more extreme founder events or population bottlenecks. In contrast, the variance in diversity among populations is much more sensitive to genetic drift.

The island archipelago and collection of samples

The meadow spittlebug _Philaenus spumarius_ is a rather sedentary insect which is abundant in many grassland habitats in Europe. It is common in the Isles of Scilly about 40 km off the southwest coast of England (Fig. 1). In July 1986 vegetation was swept with a stout net on twenty-nine islands

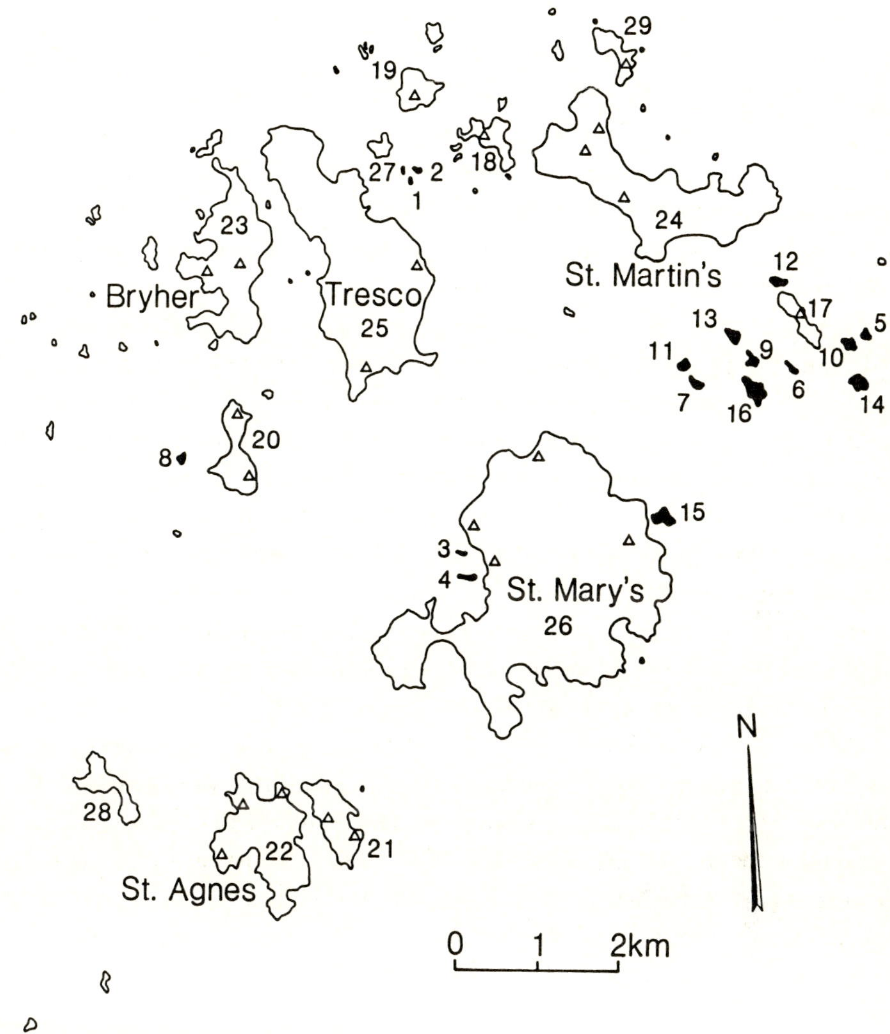

Fig. 1. Map of the Isles of Scilly. All islands which were visited are numbered with those of less than 5 ha in black. Triangles show sampling sites on islands of more than 10 ha

ranging in size from 0.2 to 662 ha. No spittlebugs were collected on two islands (map numbers 27 and 28) and only seven on a third (29). Most of the other islands, which are numbered in size sequence, yielded large samples (mean = 492). In addition, two to four samples were collected at different sites on each of the seven largest islands. The analysis of these samples showed no evidence of any substantial differentiation within islands. Details of this analysis together with further information about the sampling programme and the complete data are given elsewhere (Brakefield, 1989). Evidence is also presented there that island area is as good a predictor of population diversity as combinations of other independent variables describing island height, shape, isolation and ecology. Isolation by long distance is not a feature of the islands. The mean distance to the nearest island for the 21 islands of less than 40 ha and from which large samples were obtained is 0.16 km (range = 0.04 to 0.40 km). Three of the smallest islands are connected to the largest island by sand spits at low tide.

Most of the islands in the archipelago were probably formed some 1500 years ago from a single granitic land mass by a continuing increase in sea level, possibly combined with breaches of an outer ring-shaped area of higher ground by a great storm (Thomas, 1985). Since _P. spumarius_ is univoltine any genetic differences between islands are thus likely to have arisen in a maximum of about 1500 generations.

The study organism and the polymorphism

Philaenus spumarius is polymorphic for eleven widely-distributed non-melanic and melanic phenotypes (see illustrations in Halkka _et al._, 1973). The polymorphism has been investigated extensively in natural populations including those on small islands in the Gulf of Finland. Halkka _et al._, (1970, 1974, 1976) concluded that both selection and genetic drift influence differentiation among these island populations (see also Brakefield, 1989).

The phenotypes can be easily and quickly scored enabling large numbers of individuals to be screened. The three most abundant phenotypes in the Isles of Scilly were the non-melanics: mottled brown *typicus* (overall frequency in females = 73.7%), striped *trilineatus* (19.0%) and pale *populi* (3.1%). Eight melanic phenotypes also occurred. Of these only one, the white-bordered *marginellus* (2.2%), was well-distributed and quite frequent. Melanics accounted for about 4% of all females (209 of 5289) and 0.1% of all males (7 of 7492). This female limitation of melanics is characteristic of the polymorphism throughout much of the species' range including Finland (see Stewart and Lees, 1988). It suggests that the genetics of the polymorphism is also similar to that in Finland (Halkka *et al.*, 1973).

Measures of diversity

Indices of phenotypic and genic diversity were calculated for the individual populations of each sex. Phenotypic diversity was based on a Shannon index where $H_p = -\Sigma p_i . \log_2 p_i$ and p_i is the frequency of the *i*th individual phenotype. The variance of a value of H_p was estimated from the formula given by Hutcheson (1970). Genic diversity (or strictly speaking allelic diversity) was described by a measure of heterozygosity, $H_g = 1 - \Sigma a_i^2$ where a_i is the frequency of the *i*th allele. The following assumptions which are based on the breeding studies of Halkka *et al.*, (1973) and Stewart and Lees (1988) were used to infer allele frequencies from the data for phenotype frequencies: (1) the top dominant T allele in each sex controls the *trilineatus* phenotype; (2) melanic phenotypes are controlled by a single Me allele of intermediate dominance in females and recessive in males (there may also be reduced penetrance) and (3) the *typicus* and *populi* phenotypes are controlled (principally) by the t allele which is recessive in females and of intermediate dominance in males. The second assumption involving a single melanic allele is unlikely to significantly bias estimates of heterozygosity because of the rarity of melanics and the predominance of the marginellus form controlled by the M allele (one of several melanic alleles which have been identified). Estimation of allele frequencies then assumes that populations are in Hardy-Weinberg equilibrium. For the entire sample the estimated allele frequencies are T = 0.1016, Me = 0.0223 and t = 0.8761 in females and T = 0.0883, me = 0.0306 and t = 0.8812 in males.

Average heterozygosity and the distribution of rare alleles

If the islands are grouped into five classes with respect to area there is no indication of any decrease in average heterozygosity with declining size (Table 1).

Melanics were collected on all islands except three of the smallest: islands 2 (0.36 ha), 5 (0.95 ha) and 10 (1.81 ha); the last two being close neighbours (see Fig. 1). The large sample sizes concerned in these cases (230 to 343 females) suggest that melanic alleles were absent in each population. For example, combining across islands 5 and 10 and applying the binomial distribution, the predicted phenotype frequency at which there would have been a greater than 95% chance of collecting at least one melanic female is 0.55% (see also Fig.

Table 1. The mean values of the indices of phenotypic ($\underline{H_p}$) and genic ($\underline{H_g}$) diversity for samples of Philaenus spumarius from islands grouped by area. The number of islands is given in parentheses

Diversity index	Island area, ha				
	0-1 (6)	1-2 (6)	2-5 (4)	10-40 (5)	100-700 (5)
Female $\underline{H_p}$	1.0417	1.0710	1.1984	0.8954	0.9740
Female $\underline{H_g}$	0.1972	0.2405	0.2431	0.1717	0.1987
Male $\underline{H_p}$	1.1746	0.9059	1.1254	0.7074	0.8061
Male $\underline{H_g}$	0.2226	0.1562	0.2067	0.1501	0.1938

2A). The *marginellus* form, and therefore the M allele, was present on all the 23 islands where melanics were found.

Patterns of diversity against island area

Figure 2 plots the indices of phenotypic and genic diversity within female populations against island area. The scatters show similar comet-shaped distributions. Those for males and for the major phenotypes show similar patterns (Brakefield, 1989). The two diversity indices are closely correlated (r = 0.89) and not independent. The scatter plots suggest that the largest islands support populations of similar diversity at this polymorphic locus whereas there is increased variance among the smaller islands. A corresponding analysis of genic diversity (Nei, 1987) shows that differentiation between islands accounts for about 3-4 times more of total diversity for the smaller, than the larger islands (Table 2). However, the fraction of between-population variation is consistently less than 5%. Chi-square analyses of the frequency distributions of phenotypes show that significant heterogen-

Table 2. Analysis of genic diversity in the populations of *Philaenus spumarius* grouped by island area

Island area (ha)	Females			Males		
	H_T	H_S	G_{ST}	H_T	H_S	G_{ST}
0.2-1	0.2070	0.1972	0.0471	0.2287	0.2226	0.0264
1-2	0.2474	0.2405	0.0277	0.1619	0.1562	0.0348
2-5	0.2476	0.2431	0.0184	0.2124	0.2067	0.0270
10-40	0.1731	0.1717	0.0083	0.1513	0.1501	0.0078
100-700	0.2007	0.1987	0.0102	0.1957	0.1938	0.0094

Genic diversity: H_T, total diversity;
H_S, between-island diversity;
G_{ST}, coefficient of between-island differentiation

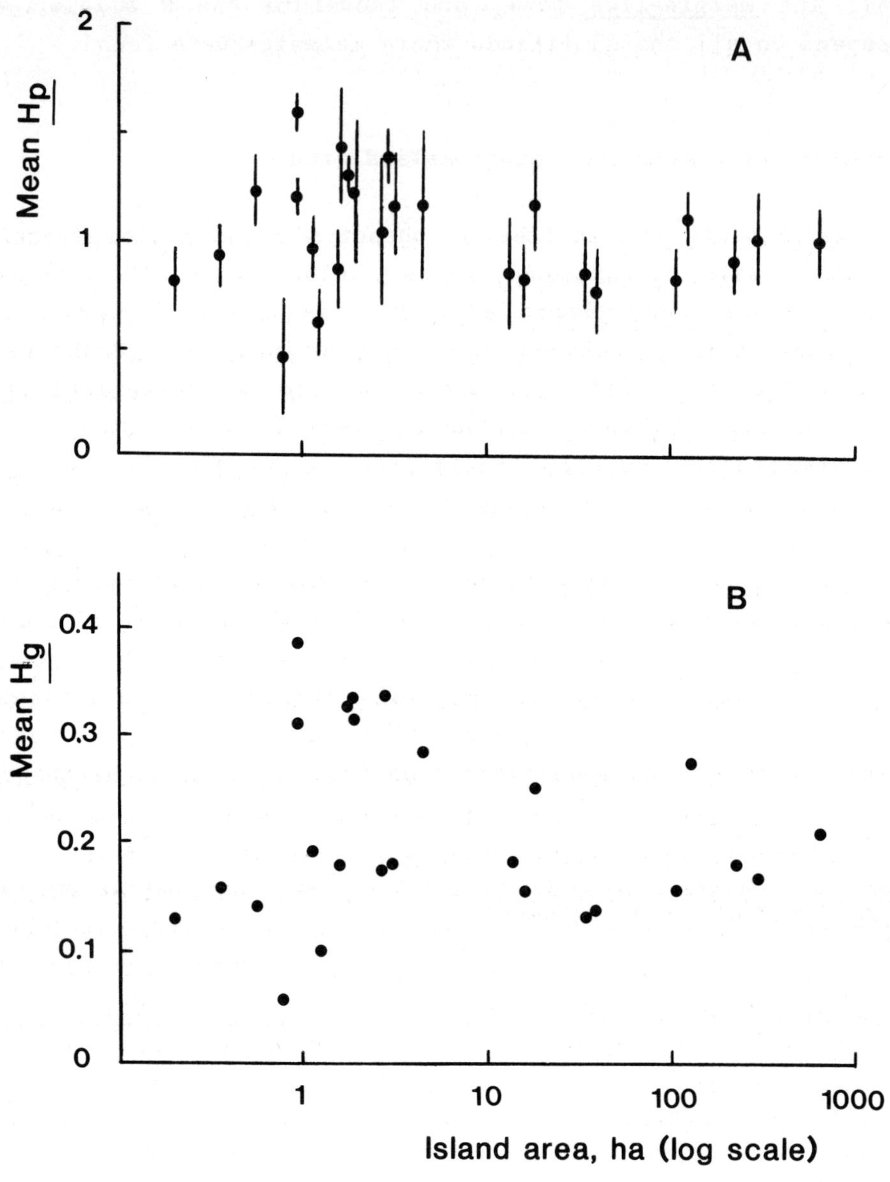

Fig. 2. A. Phenotypic diversity (Hp) and B. Genic diversity (Hg) in samples of female Philaenus spumarius plotted against island area. Bars in A. indicate ± twice the square root of the estimated variance (see text)

eity occurs among populations within each of the five classes of island area (Brakefield, 1989; see also Fig. 2A).

Effects of sample size on diversity indices

The apparent increase in variance among populations on small islands could be an artefact of greater variability in sample size. A bootstrapping procedure was used to examine this possibility by estimating the 95% confidence limits (CLs) for each value of the diversity indices for populations of females. The procedure was: (1) to set up in a microcomputer an Isles of Scilly 'metapopulation' equivalent to the total numbers of females collected of the individual phenotypes; (2) taking the observed sample sizes for females (N) in turn for each island, to 'sample the metapopulation' at random with replacement until N is reached; (3) to calculate diversity indices for that run and store; (4) after 1000 runs to take the 26th and 975th ranked values as the CLs for that island sample.

There is no correlation between the range of the upper and lower confidence limits and the absolute difference between the observed value for diversity and the overall mean (female H_p: r = -0.07; female H_g, r = 0.02). This also holds when the calculated estimates of the variance of the observed values of phenotypic diversity are examined (r = -0.11; see Fig. 2A). Thus, populations which are extreme in diversity are not associated with relatively wide confidence limits and the increased variance among populations on small islands is not an artefact of sample size.

Expected relationships between founder effects and diversity

The bootstrapping procedure was also used to examine the expected relationships between sampling error and the statistics describing diversity at the melanic locus in the Isles of Scilly. Instead of using the observed sample sizes for islands in the simulations, a sequence of sets of 1000 single-generation samples or 'populations' of increasing size (N) was generated. The mean values of diversity and the confidence limits for the diversity indices then represent those expected for populations at the point of a founder event or bottleneck of magnitude given by the different values of N. There are clearly severe limitations in this procedure with respect to simulating the real stochastic processes influencing differentiation in the Isles of Scilly. In at least some cases involving the establishment of a population, colonizers will arrive in several 'waves'. The effects of population history will not be represented at all faithfully by single-generation sampling: they also depend on the time elapsed since the sampling event, on the duration of the bottleneck and the subsequent rate of population increase (Nei et al., 1975, Janson, 1987). Variation in population size between islands and migration between extant populations are also not taken into account. With respect to founder events, the simulations are based on a pure "migrant pool" model where the N colonists are a random sample from the entire (meta)population. A more realistic approach would involve migration assortative by distance. In this case colonists at a particular location are a mix of individuals drawn in some weighted-by-distance way from a few nearby populations. In a "propagule model", the colonists (2N gametes) would come from a single randomly chosen local population. However, there is no information about how the potential source populations should be weighted by distance. Wade and McCauley (1988) examine for a meta-population how the amount of differentiation and the genetic effects of colonization and extinction depend critically upon the applicability of such models and on the relative magnitudes of the number of colonists and the number of migrants moving between extant populations.

Within these types of limitations, the simulations clearly show that in general the confidence limits for measures of diversity among populations are much more sensitive to sampling effects and genetic drift than are the mean values for diversity across populations (Figs 3 and 4). The effects

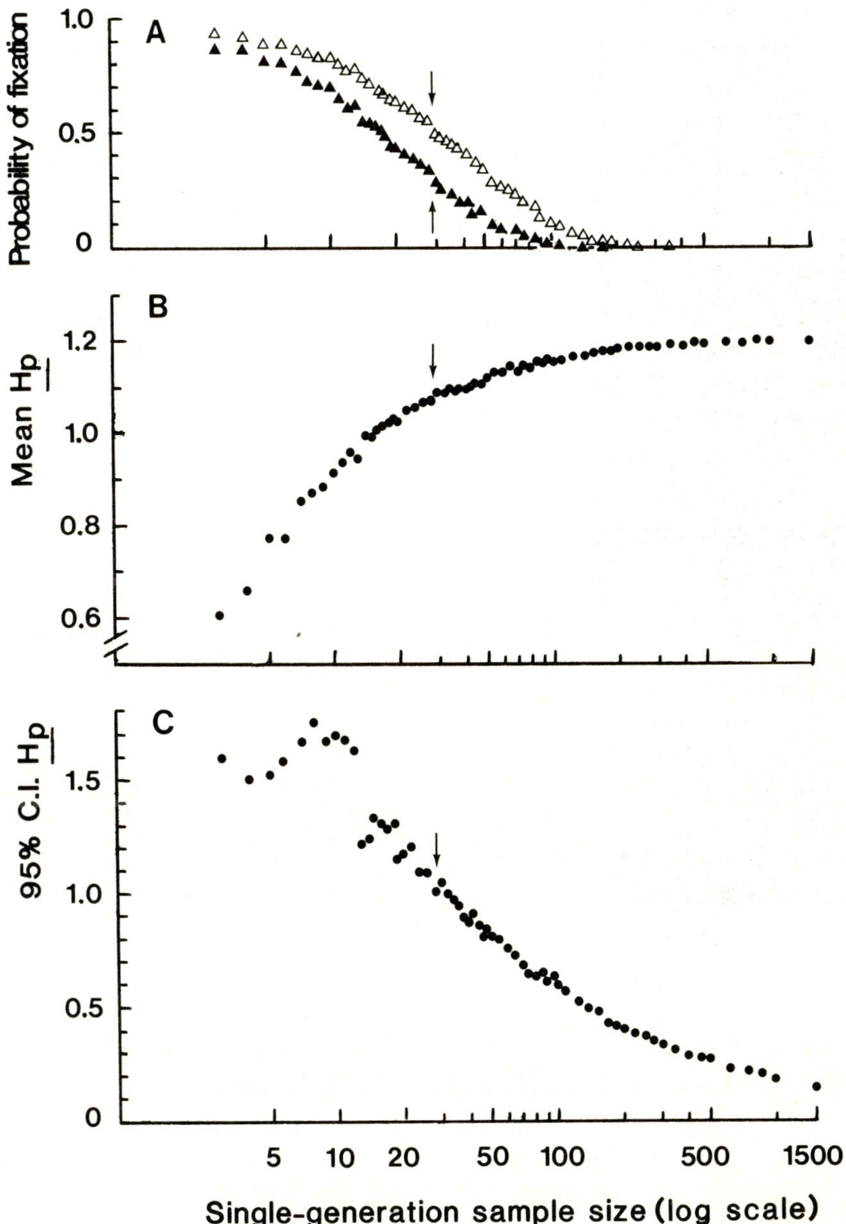

Fig. 3. The results of single-generation sampling effects of varying magnitude on phenotypic diversity as obtained from simulations (see text) using the total Isles of Scilly 'population' of females: A. Proportion of replicates with no melanics (▲) or none of the *marginellus* form (△); B. the mean phenotypic diversity (*Hp*) and C. the 95% confidence interval of *Hp*. Arrows indicate the sampling effect yielding a rate of fixation of non-melanics similar to that observed among the ten smallest islands

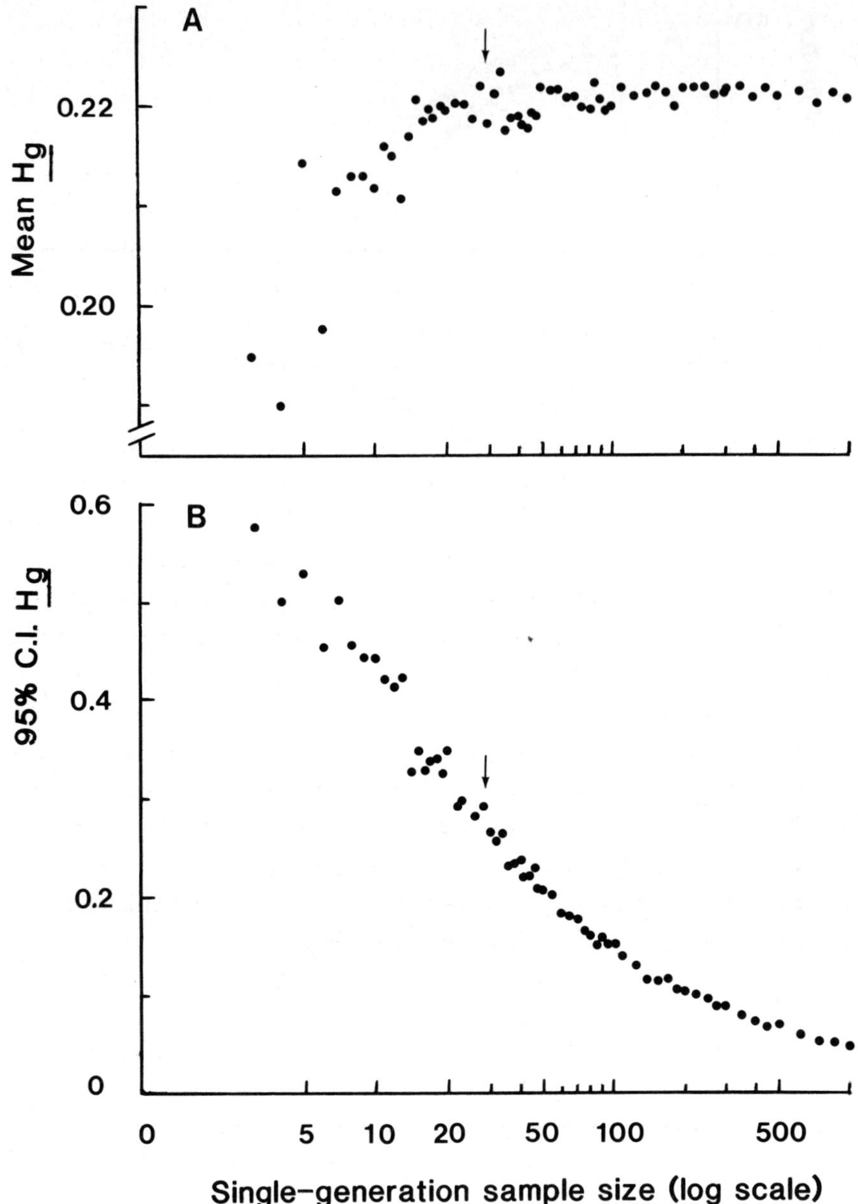

Fig. 4. The results of single-generation sampling effects of varying magnitude on genic diversity as obtained from simulations (see text) using the total Isles of Scilly 'population' of females: A. the mean genic diversity index (Hg) and B. the 95% confidence interval of Hg. Arrows indicate the magnitude of sampling effect yielding a rate of fixation similar to that observed among the smallest islands

on variances, which are not shown, are even more clear-cut (note that an alternative approach would have been to estimate the variance of $\underline{H}_g$ directly by application of equation 8.12 given by Nei, 1987).

Is there any indication of the magnitude of bottleneck or founder events in the Isles of Scilly populations? It has been shown above that melanics were probably absent on three of the ten smallest islands. Fig. 3A shows that a fixation rate of 30% is expected with sample sizes of 30 or a little smaller. Although obviously there will have been variation in population history between islands, this figure can be taken as a crude indication of the average sampling effect equivalent to the events establishing the populations on islands of less than about 2 ha. These events are thus likely to have been equivalent to effective populations sizes of the order of a few tens of individuals, rather than a few individuals or more than fifty or so. The simulations then show that samples of around 30 individuals from the whole 'metapopulation' will have little effect on mean diversity among populations. Application of the equation $\underline{H}_{t+1} = [1-(1/2\underline{N})].\underline{H}_t$ similarly shows that average heterozygosity will only be reduced by 1.67% for a single-generation population size of 30. In contrast, such a sampling effect will greatly increase the variance. It is noteworthy that the predicted 95% confidence intervals for a population size of 30 are comparable to the range of the observed values ($\underline{H}_p$: CI = 1.05, range = 1.12; $\underline{H}_g$: CI = 0.27, range = 0.33).

Discussion - population history and its consequences

The phenotypic and genetic characteristics of populations on the larger islands (>100 ha) in the Isles of Scilly are similar to those on the mainland of southern England and they are probably influenced by a broadly similar regime of visual and climatic selection (Lees, Dent and Gait, 1983). The differentiation found among the populations of small islands after less than 1500 or so generations is consistent with a major influence of genetic drift. Although, there are no clear relationships between diversity and crude indices of spittlebug densities or vegetation type, selection may, of course, also have influenced the differentiation (see Brakefield, 1989).

Several observations suggest that populations on small islands are transient and experience intermittent population crashes (see also Brakefield, 1989):

1) Field observations of spittlebug density, including records of captures per sweep of standard length, suggest that the density of *P. spumarius* was substantially more variable among the smaller islands than in populations on the larger ones which were characterized by a moderate density. The density was low on some small islands while on others, including some with a very simple plant community (e.g. islands 2, 6 and 8), it was extremely high. Colonization of small islands with such an established plant community is likely to lead to rapid increases in population size.

2) *P. spumarius* was absent on the small Crow's Island (27). This was apparently due to an earlier extinction with no successful recolonization since the island was of similar size and habitat to two close neighbours where the spittlebug was abundant. The species was also not found on the intermediate-sized island of Annet (28) although not all the suitable vegetation was searched.

3) There is a trend of increasing island height with island area (r = 0.64). The ten smallest islands range in height from 5 to 26 m with a mean of 14 m (Crow's Island = 5 m). The lowest are more likely to be inundated by seawater during winter storms; a process which may well lead to population crashes or extinctions.

4) The melanic form *lateralis* controlled by the L allele (Halkka *et al.*, 1973) was at a substantial frequency of 14.5% in females on the small Ragged Island (6). It was not found on any of the neighbouring group of 10 islands or in any other samples. It seems likely that the L allele arose as a mutant when this presently very large and high density population on Ragged Island was in an early phase of establishment or

passing through an extreme bottleneck. Alternatively the high frequency may indicate that this allele is favoured by selection: it would be interesting to introduce it elsewhere. The *marginellus* form was present at a very low frequency on Ragged Island (1 of 497 females).

Observations also suggest that the rate of migration between islands and of colonization is probably rather low. The case of Crow's Island which had not been colonized from two close neighbours is a case in point. If the polymorphism is not influenced by strong selection in response to variable environments on the small islands (cf. Ford, 1975), then the differentiation which frequently occurs between closely neighbouring islands (including those joined at low tide) is indicative of low migration rates. Spittlebugs usually move by jumping and flight distances are considered to be short. The saline water surrounding the islands suggests that survival time of spittlebugs drifting on the surface will be low. Passive migration in strong winds or on rafts may be more frequent.

Although all these observations are indirect or circumstantial and studies of population dynamics are required, they suggest that the populations on the smallest islands are transient and that together they form a metapopulation. The population on each island is probably subject to fluctuations in size and extinction-recolonization events at a rate dependent in part on island size, height and successional state of the vegetation. Such events when triggered by catastrophic storms or droughts may tend to occur synchronously in particular parts of the archipelago, or they may occur in a largely independent and more continuous manner when due to factors specific to particular populations (cf. Harrison *et al.*, 1988). Extinctions and extreme bottlenecks have probably not occurred in the history of the very large populations on each of the large islands; their populations representing a stable migrant-pool, or a series of such pools, with respect to the small islands. The subpopulations on each of the large islands are also themselves likely to exhibit the shifting mosaic type of dynamics. The isolated archipelago can, therefore, be considered to be a discrete metapopulation, within which there exist subpopulations varying in their degree of transience.

Unfortunately, there are no data describing how short-lived the different subpopulations are. The comparison of the patterns of diversity with the results of the simulations suggest that the effective number of founders or of individuals at population crashes on the smallest islands is of the order of a few tens of insects and that the genetic drift this would cause could account for the observed variance among the small island populations and the lack of overall loss of heterozygosity. Experimental perturbations of extant populations or experimental colonizations would provide further insights about the roles of population structure and dynamics, and of selection and genetic drift. The results to date emphasize the generally much greater sensitivity of the variance of diversity among subpopulations to the effects of genetic drift than of the overall mean diversity or the average heterozygosity. The work on the Isles of Scilly would be usefully extended by surveys of genotypes and across genetic loci and by studies of quantitative traits. The archipelago provides the potential for valuable studies of the effects of founder events, migration-colonization processes and the transience of subpopulations on genetic diversity within a metapopulation. The generation of variance among more transient subpopulations may have longer term consequences for such a metapopulation in terms of its capacity to respond adaptively to environmental perturbations.

Acknowledgements. I would like to thank Philip Hedrick for his suggestion to follow up further the difference in the sensitivity of the variance and mean statistics to sampling error which was apparent from initial analysis of the field data.

References

Buri P (1956) Gene frequency in small populations of mutant Drosophila. Evolution 10:367-402

Black WC, IV, Ferrari JA, Rai KS, Sprenger D (1988) Breeding structure of a colonising species: Aedes albopictus (Skuse) in the United States. Heredity 60:173-181

Brakefield PM (1989) Genetic drift and patterns of diversity among colour-polymorphic populations of the Homopteran Philaenus spumarius in an island archipelago. Biol J Linn Soc (in review)

Dobzhansky Th, Pavlovsky O (1957) An experimental study of interaction between genetic drift and natural selection. Evolution 11:311-319

Ford EB (1975) Ecological Genetics (4th Edn). Chapman and Hall, London

Halkka O, Halkka L, Raatikainen M, Hovinen R (1973) The genetic basis of balanced polymorphism in Philaenus (Homoptera). Hereditas 74:69-80

Halkka O, Raatikainen M, Halkka L (1974) The founder principle, founder selection, and evolutionary divergence and convergence in natural populations of Philaenus. Hereditas 78:73-84

Halkka O, Raatikainen M, Halkka L (1976) Conditions requisite for stability of polymorphic balance in Philaenus spumarius (L.) (Homoptera). Genetica 46:67-76

Halkka O, Raatikainen M, Halkka L, Lallukka R (1970) The founder principle, genetic drift and selection in isolated populations of Philaenus spumarius (L.) (Homoptera). Ann Zool Fennici 7:221-238

Harrison S, Murphy DD, Ehrlich PR (1988) Distribution of the bay checkerspot butterfly, Euphydryas editha bayensis: Evidence for a metapopulation model. Am Nat 132:360-382

Hutcheson K (1970) A test for comparing diversities based on the Shannon formula. J Theor Biol 29:151-154

Janson K (1987) Genetic drift in small and recently founded populations of the marine snail Littorina saxatilis. Heredity 58:31-37

Lees DR, Dent CS, Gait PL (1983) Geographic variation in the colour/pattern polymorphism of British Philaenus spumarius (L.) (Homoptera: Aphrophoridae) populations. Biol J Linn Soc 19:99-114

Levins R (1970) Extinction. Amer Math Soc 2:75-108

Nei M (1987) Molecular Evolutionary Genetics. Columbia University Press, New York

Nei M, Maruyama T, Chakraborty R (1975) The bottleneck effect and genetic variability in populations. Evolution 29:1-10

Rich SS, Bell AE, Wilson SP (1979) Genetic drift in small populations of Tribolium. Evolution 33:579-584

Stewart AJA, Lees DR (1988) Genetic control of colour/pattern polymorphism in British populations of the spittlebug Philaenus spumarius (L.) (Homoptera: Aphrophoridae). Biol J Linn Soc 34:57-79

Thomas C (1985) Exploration of a drowned Landscape. Batsford, London

Wade MJ, McCauley DE (1988) Extinction and recolonization: Their effects on the genetic differentiation of local populations. Evolution 42:995-1005

Wool D (1987) Differentiation of island populations: A laboratory model. Am Nat 129:188-202

Part B Evolutionary Mechanisms

Mobile Genetic Elements and Quantitative Characters in *Drosophila*: Fast Heritable Changes Under Temperature Treatment[1]

V.A. Ratner and L.A. Vasilyeva

Institute of Cytology and Genetics, Novosibirsk 630090, USSR

INTRODUCTION

In 1928, when analysing the phenotypical expression of the mutation *venae transversae incompletae* (vti), which causes interruption or disappearance of a radial wing vein in *Drosophila funebris*, N.V. Timofeeff-Ressovsky showed that a change in the temperature regime at different stages of pupal ontogenesis can make this mutation change rapidly in penetrance and expressivity. The author stated that there exist at least two temperature-sensitive periods, during early pupal stages (L1 or L2), at which a rapid change in temperature may affect in the strongest manner the phenotypical expression of a character in the individuals subjected to this treatment. In his studies, Timofeeff-Ressovsky did not investigate how a change of the character would behave in further generations and he just stopped at the phenogenetic description of the phenomenon.

Svetlov and Korsakova (Svetlov, 1962; Svetlov and Korsakova, 1962, 1965, 1966a,b, 1972) took the next step by studying the characters of three recessive mutations of *Drosophila melanogaster*, namely, *vestigial* (vg), II chromosome, 67 cM; *forked* (f), I chromosome, 56.7 cM; *eyeless* (ey), IV chromosome, 2.0 cM. All three characters determined by these genes, rudimentary wings, anomalous bristles and reduced eyes, respectively, not only changed their phenotypical manifestation immediately after temperature treatment but also inherited the acquired state for as long as dozens of generations under cultivation at optimum temperature(25^0C) without additional treatment (the *forked* gene for 20, and with the *eyeless* gene, for more than 67 generations). The basic method in these studies was that virgin females and, in some tests, larvae of different ages (Svetlov and Korsakova, 1966a,b) were subjected to short-term temperature

[1]Dedicated to the 90th anniversary of the birth of N.V. Timofeeff-Ressovsky, great Russian geneticist who has left an indelible trace in the memory of everyone who ever knew him.

stress treatments (either heat shock to 38^{0}C, or cooling to +7^{0}C) and subsequent rearing of the larvae at 30^{0}C.

Temperature-sensitive periods were found for various mutations. For the mutation *vestigial*, heating the females (39^{0}C) for one hour decreased the size of the wing, i.e., it increased the expression of the vg mutant phenotype. Rearing larvae or especially virgin females (3-4 days post-eclosion) at the temperature of 30^{0}C sharply increased the size of the wing, i.e., it decreased the expression of the vg mutant phenotype (Svetlov and Korsakova, 1962). Two temperature-sensitive periods have been stated for the *forked* mutation: the first, at the ontogenetic stage of 6 to 7 days before egg laying, the second, at the fifth day of pupal life (Svetlov and Korsakova 1965, 1966a,b). As for the *eyeless* mutation, heating at the 1st to 3rd day for 60 to 90 min and cooling of larvae at prepupal age (t = +5^{0}C for 150 min) caused a heritable increase in expressivity from 39% to 70-80% of flies with the *eyeless* phenotype. This effect remained inheritable for over 67 generations (Svetlov and Korsakova, 1972). Since it was impossible to give any plausible genetic explanation to the mechanism of temperature action, the authors ventured to call the actual phenomenon as the "protracted modification". No genetic analyses were carried out.

In the 1970's temperature effects were also studied by Bucheton and co-authors (Bucheton et al., 1976; Bucheton, 1978, 1979; Bucheton and Picard, 1978) in the *SF* line of *D.melanogaster*, whose females have decreased fertility. Heat treatment at the end of ontogenesis (probably, at the *vitello* stage) increased the probability of larvae to hatch, and decreased fertility at earlier stages of development. These temperature and age-dependent modifications of fertility expression were shown to be partially inherited. Other authors (Khristolyubova and Auslender, 1967; Neel, 1940; Tarasoff and Suzuki 1970) have reported similar heritable effects of temperature on development. However, they remain genetically unstudied.

In 1966, one of us (L.A. Vasilyeva) began to study the system of genetic determination of the quantitative character of *D. melanogaster radius incompletus (ri)*. The character was chosen based upon the advice of Timofeeff-Ressovsky, who had noticed a strong dependence of an interrupted vein on the conditions of rearing. The first stage of the work concerned the study of selection dynamics, the initiation of "selection"

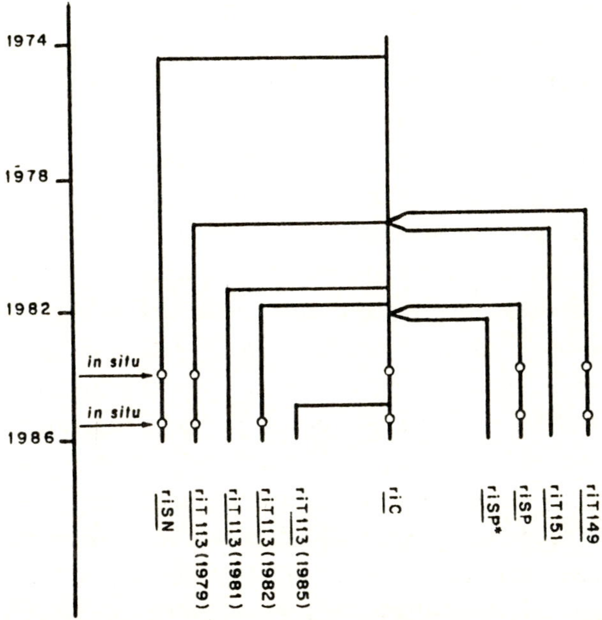

Figure 1: A genealogical tree of the lines used. The control line, *riC*, is placed at the center; to the left: the similar (by phenotype) lines *riSN* and *riT113*; to the right: the similar (by phenotype) lines *riSP* and *ri149* (151). The arrows indicate the years when the in situ hybridization was done.

lines, and the genetic analysis of the polygenic system of character determination. In 1979 Vasilyeva began temperature treatment of *ri* mutants from the control line *riC*. The result was the discovery of two temperature-sensitive periods, where temperature treatment caused heritable changes that have been maintained in the "temperature" lines for over 200 generations. Those lines were also subjected to a detailed genetic analysis, which revealed the multiple nature of the genetic changes in the flies after temperature treatment.

Fig. 1 presents the actual "genealogy" of five *Drosophila melanogaster* ri-lines studied and Fig. 2 the phenotypes of the quantitative character in separate lines. It should be noted that lines *riSN* and *riT113* and lines *riSP* and *riT149* show respective pair-wise phenotypical similarities.

A new turning point took place in the early 1980's when Gvozdev, Kaydanov and their colleagues (Georgiev and Gvozdev, 1980; Gvozdev et al., 1981; Pasyukova et al., 1986; Gvozdev and Kaidanov, 1986) demonstrated that the expression of some quantitative characters, such as viability and sexual activity of males, from different lines that had been subjected to selection, is connected with the pattern of localization of mobile genetic elements (MGEs) in *Drosophila* polytenic chromosomes. Later it was shown by Mackay and colleagues (Mackay 1985, 1988; Shrimpton et al., 1987) that, in the P-M system, insertional variability in the number of abdominal and sternopleural bristles may be induced in *Drosophila* by way of disgenic crossing. Following selection during 16 generations for an increase or decrease of these quantitative characters, it resulted in the change of the P-element localization pattern as well as in the appearance of novel localizations. Biemont and Terzian (1988) carried out selection for high or low viability in fly lines from a natural population, and after 18 generations found novel specific localizations of *mdg-1*.

In 1983 we started analogous studies on our "selection" and "temperature" fly lines *ri*. From the first it became clear that the expressivity of this quantitative character was tightly correlated with the pattern of MGE localization. What was new, however, was that the correlation was associated not only with the "selection" but also with the "temperature" lines (Vasilyeva et al., 1987a,b; 1988). These results prompted us to reevaluate the traditional notions of the genetic basis of determination of the quantitative characters in *Drosophila* and supplement it with the notion that MGEs may modify the expression of polygenes in affecting the expression of quantitative characters (Vasilyeva et al., 1985; Ratner and Vasilyeva, 1987).

THE POLYGENIC SYSTEM OF THE *ri* CHARACTER UNDER GENETIC ANALYSIS: RESULTS

The *ri* oligogene has been positioned at 47.0 cM of *Drosophila* chromosome 3 (Lindsley and Grell 1967). The normal allele ri^+ determines the formation of a complete radial vein, while the *ri* mutation cuts the vein into two (proximal and distal) fragments (see Fig. 2a). The length of the fragments measured at the microscope are good quantitative characters.

To analyze the polygenic system genetically, we used the initial control line *riC*, and two "selection" lines obtained from that one by way of protracted selection: 1) *riSN*, a result of minus-selection, which had almost lost both of the fragments, and 2) *riSP*, a result of plus-selection, which had completely restored the vein (see Fig. 2b,c).

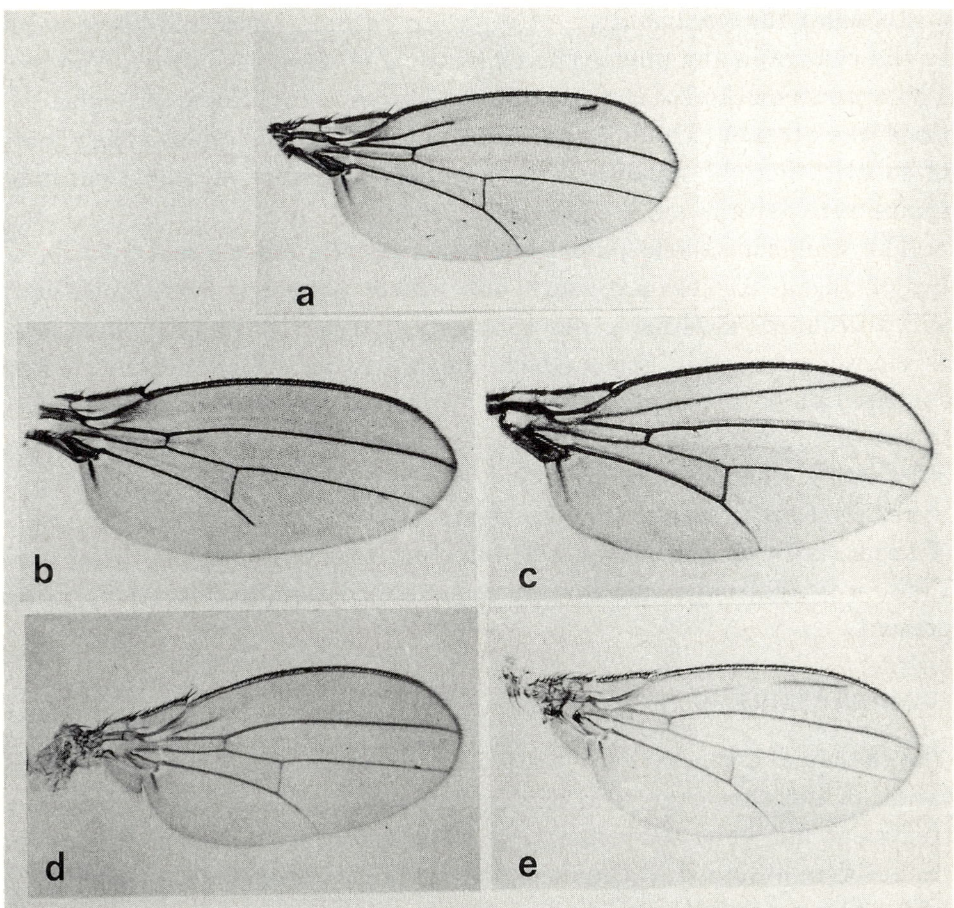

Figure 2: The female wings from different lines of *Drosophila melanogaster* : a) starting and control form (*riC*); b) the result of selection for decreasing the radial vein (*riSN*), distal fragment is totally eliminated; c) the result of selection for increasing the radial vein (*riSP*), the radial vein is totally restored; d) the result of single stepwise temperature treatment (29°C to 18°C) of the line *riC* at the stage of 113 ± 5 h after egg laying (at 29°C - early pupa), the subsequent generations being cultivated at 25°C (*riT113*); e) the result of the same treatment at the stage of 149 ± 5 h after egg laying (at 29°C - late pupa), the subsequent generations being cultivated at 25°C (*riT149*).

To asses the contribution of separate chromosomes and their parts to the change in the phenotype expression of these characters, Thoday-Thompson's method of chromosome (or chromosomal fragment) substitutions was utilized (Thoday and Thompson 1974). The phenotype effects of substitutions of separate chromosomal fragments or whole chromosomes were determined.

The wing radial vein phenotype is genetically determined by a number of oligogenes (among which only the *ri* gene has been localizaed) and a group of modifier genes (polygenes) (see Fig. 7). All three long chromosomes contain genes which contribute to the phenotype. Their contribution to the expression of the proximal fragment is additive, but to the distal one, non-additive. In chromosomes 1 and 2, not less than 9 "effective" regions were identified, where polygenes could be localized. Seven of them influence the expression of the proximal fragment and 2 of them (both on chromosome 2) influence expression of the distal one. Thus, a typical polygenic system of genetic determination seems to be present.

TEMPERATURE EFFECTS IN F_0

The change of a temperature regime significantly influences the manifestation of the character *ri* in that generation in which the temperature treatment has been subjected (F_0). The most instructive and obvious results were obtained at the step-wise rearing temperature from 29^{0}C to 18^{0}C. The procedure consisted of flies from the control line, *riC*, being allowed to lay eggs during an hour and then the vials placed into an incubator at 29^{0}C. After the 107th hour of cultivation, three replicate vials were hourly shifted to another incubator set at 18^{0}C. The length of both vein fragments were measured on adult flies. The results of the experiment are presented in Fig. 3.

First of all, we found that the pupal stage up to the imago is sensitive to temperature changes with regard to the manifestation of the character in F_0, but most of these changes were not heritable. However, there are two narrow periods within this stage where sharp changes of temperature may cause some heritable phenotypic changes (see Fig. 3). Qualitative consequences of the temperature effects may be summarised as follows: 1) At a constant lowered temperature in the F_0 (18^{0}C), the proximal

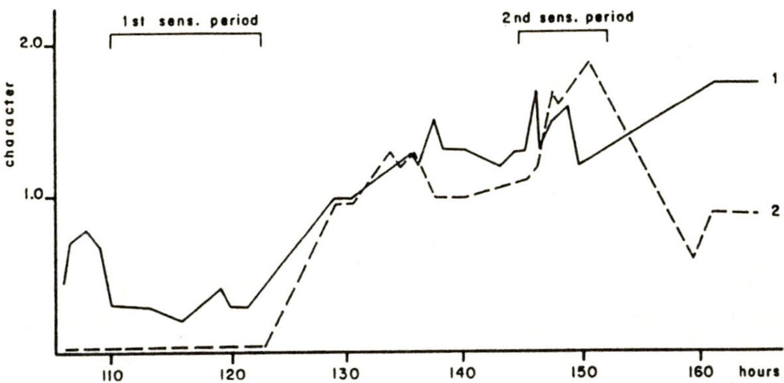

Figure 3: The influence of temperature treatment (29°C to 18°C) upon the length of the fragments of the radial vein in the F_0 (Vasilyeva et al. 1987a, 1988): – – –– is the proximal fragment, —- is the distal one. In terms of axis X: the stage of ontogenesis at 29°C when treatment was in action. Two temperature-sensitive intervals are indicated: The 1st (113 ± 5 h) and the 2nd (149 ± 5 h), where temperature treatment leads to the arising of heritable characters in the F_1, etc.

fragment lengthens, and the distal one shortens as compared to those at normal temperature (25°C). Neither change was heritable.

2) At a constant elevated temperature in the F_0 (29 °C), the proximal fragment shortens, and the distal one lengthens, as compared to normal temperature (25 ° C). Neither change was heritable.

3) Stepwise change in temperature (29 °C to 18 °C) before the 1st sensitive period (113 ± 5 h) results in an F_0 phenotype which is the same as that observed with a constant temperature treatment of 18°C. Neither change was heritable.

4) Stepwise change in temperature (29°C to 18°C) within the interval between the 1st and 2nd sensitive periods, i.e., 113 ± 5 h and 149 ± 5 h, resulted in the F_0 proximal fragment having the same length, as it has at 29°C, and the distal one as at 18°C. Neither change was heritable.

5) Stepwise change in temperature (29°C to 18°C) after the 2nd sensitive period (149 ± 5 h) causes the same phenotype in F_0 as it did at the constant temperature of 29°C. Again neither change was heritable.

6) Stepwise change in temperature (29°C to 18°C) during the 1st sensitive period (113 ± 5 h) resulted in an abrupt lessening of the proximal fragment and absolute disappearance of the distal one in the F_0. This

effect was found to be repeated and stabilized in following generations at the temperature of 25^{0}C.
7) Stepwise change in temperature (29^0 to 18^{0}C) during the 2nd sensitive period (149 $\pm$ 5 h) resulted in the proximal fragment to decrease in length in the F_0. In the following generations it increased and stabilized at a higher level than in the control.

The peculiarities of temperature effects in the F_0 most probably indicate that at the pupal stage, the formation of the vein runs on its own from both ends - distal and proximal - and depends to a great extent on growth temperature . When the growth temperature undergoes a general constant change, the speed of formation of vein and wing fragments changes disproportionally to each other, and, as a consequence, the phenotype acquires some changes that are not heritable along the germ line of cells. Aside from this, the fact that the phenotype can change separately on different fragments in response to stepwise temperature changes between the 1st and 2nd sensitive intervals indicates that modifier genes of the proximal fragment in somatic tissues are active during the 1st sensitive period (113 $\pm$ 5 h) at 29^{0}C and those of the distal one near the 2nd sensitive period (149 $\pm$ 5 h) at 29^{0}C. The effect of these groups of modifiers are apparently pleiotropic.

HERITABLE TEMPERATURE EFFECTS

As was mentioned above, we have identified two narrow temperature sensitive pupal periods in which sharp changes of temperature lead to significant changes of the F_0 phenotype.

Perhaps the most surprising result is that the inherited effects appeared after a single temperature stress-like treatment and they occur massively, since they are inherited in most flies of the F_1 and of the next generations. As a result, we have obtained two"temperature" lines, *viz, riTII3* and *riTI49*, that have inherited the changed phenotypes for over 200 generations (see Figs. 1, 2 d,e).

"Temperature" lines were genetically analyzed together with the *riC, riSN* and *riSP* lines by the method of chromosomal substitution. On the whole, the results we obtained for the "temperature" lines are about the same as for the "selection" lines: i.e., all three large chromosomes contribute to the changes in character expression, though the relative contributions of the chromosomes are different from those in the "selection" lines. Therefore, it may be argued that in the "temperature" lines *multiple genetic changes, dispersed throughout the genome*, arose as a result of a one-time temperature action.

In addition, it should be emphasized that when treatments are done with the same temperature change, but at different sensitive periods, the *riC* line yields many lines with quite different phenotypes and chromosome contributions altering the phenotype. Thus, it seems impossible that the actual genetic changes are due to ordinary mutations; rather, they seem to be connected with some special temperature-sensitive states of the chromosomes of germ cells at specific pupal stages.

The temperature effects are reproducible. Thus, we have succeeded in producing some new "temperature" lines of *riT113* again in 1979, 1982, 1985 and 1986 (see Fig. 1) that showed very similar phenotypical characteristics. At the same time, it should be noted that not every treated fly culture from a control line acquired the described properties of the "temperature" lines. In other words, the penetrance of inherited "temperature" effects is incomplete. We believe that this fact is due to some experimental variations. The initial aim of the "temperature" experiments was to isolate the most phenotypically distinct "temperature" fly lines suitable for a genetic analysis. That is why from the treated cultures those that changed most and those that remained most stable in their offspring were chosen. They were the source of the "temperature" lines *riT113*, *riT149*, etc. Moreover, fly cultures are never synchronized for stages of development. Therefore, a variability in values, which appears as incomplete penetrance, is inevitable.

In our studies all F_1 replicate cultures changed significantly in the same direction: those treated during the 1st sensitive period moved towards a decrease in the mean length of the vein fragments, and those treated during the 2nd sensitive period, moved towards an increase. We designated them as the I and II groups of replicates, respectively. Up to 50% of the replicates from the I group lost their distal fragment in the F_1 (i.e., penetrance being zero), whereas the rest of them contained 1-2 flies with the distal fragment (i.e. penetrance being 1-2%). This latter group was discarded. In the following generations of some cultures, individual flies with a distal fragment appeared. Such cultures were likewise discarded. In summary, only 2-4% of replicated cultures have maintained the changed phenotype over several dozens of generations. They constitute the *riT113* lines, maintained by random crossing for over 200 generations and keeping the changed phenotype, the penetrance of a distal fragment being zero.

In the F_1, all the replicates of the II group had a displacement towards an increase of the mean phenotype, but a part of them contained some flies with no distal fragment (high penetrance). Such replicates were discarded. In the next generations, many replicates also yielded individual flies with no distal fragment. All those cultures were rejected as well. In the long run only 1-2% of replicates have managed to maintain the mean changed phenotypes over dozens of generations. Further, they formed the *riT149(151)* lines, maintained for over 200 generations, without losing the changed phenotype, the penetrance of the distal fragment being 100%.

Thus, temperature treatment was accompanied by some very rigid selection for the most contrasting replicates. However, since changes in mean phenotypes from each group were present in all the culture replicates, i.e., a mass response to the temperature treatment, then the role of that selection was in fact to increase the contrasting range of the mean phenotypes (i.e., to increase their expressivity) and to fix either zero (in the *riT113* lines) or 100% (in the *T149* lines) penetrance of the distal fragment. Genetic sources of incomplete penetrance of the distal fragment in a control and in other cases may be acted by the genetic interaction between polygenic alleles (dominance, recessiveness), by recombinational shuffling of polygenic systems when crossing, or by asynchrony in development of individuals in cultures.

In the long run, we can state that the polygenic system of determination of the quantitative character considered has non-Mendelian genetic properties: a one-time temperature treatment results in mass (in a population) and multiple (in the chromosomes of individuals) heritable changes.

CYTOGENETIC HERITABLE EFFECTS

The genetic and temperature effects described above make one think that their underlying molecular mechanism should cause multiple, dispersed, genetic changes in all *Drosophila* chromosomes. Among the molecular-genetic mechanisms studied, various properties of the Mobile Genetic Elements (MGE) meet this requirement to the highest extent. MGEs occupy a considerable part of the *Drosophila* genome (Khesin, 1984; Rubin, 1983; Ananyev, 1984) and are capable of sudden transposition inside the genome (Gerasimova et al., 1984). They also have an important influence upon the functions of adjacent genes (Khesin, 1984; Ananyev, 1984). Gvozdev, Kaidanov and their collaborators (Gvozdev et al., 1981; Pasyukova et al., 1986; Gvozdev and Kaidanov, 1986) showed that the pattern of MGE localization in *Drosophila* chromosomes influences the expression of fitness components. They put forward the hypothesis that the "hot spots" of localization of *copia*-like MGE are connected with the polygenes having control over fitness (Gvozdev and Kaidanov, 1986). In our case the lines had no essential differences in fitness. So, only the degree of expressivity and penetrance could be implicated. All five lines, namely, *riC* (control), two "selection" lines *riSN* and *riSP*, and two "temperature" lines, *riT113* and *riT149*, were used for a comparison analysis of localization of *copia*-like MGE, *viz, mdg-1 (Dm-225), mdg-2 (Dm-412), mdg-3 (Dm-58), mdg-4 (gypsy)* and *copia*. The DNA of the vectors that contained these MGE fragments were *in situ* hybridized with polytene chromosomes of larvae from those lines. In the long run, the pattern of MGE localization was revealed for all the lines.

The basic results were obtained while working with *mdg-2 (Dm-412)* (Vasilyeva et al., 1987b, 1988). In each line, a sample of 7-9 individuals were investigated and the linear pattern of MGE localization was revealed for each larva by using Bridges map (see Lindsley and Grell, 1968). Furthermore, by the methods of comparison of linear sequences, measures of distance were estimated between them. From the matrix distance, with the use of the methods of matrix cluster analysis (Sneath and Sokal, 1973), trees were built to express the similarity between the lines through their patterns of MGE localization. Fig. 4 presents such a tree for *mdg-2*.

The most interesting result is that lines with a similar pattern in MGE localization were clustered due to the manifestation of their quantitative traits: i.e. *riSN* and *riT113, riSP* and *riT149*, respectively. It is interesting that the most similar lines were obtained in quite different ways: "selection" lines, by means of a prolonged selection for a quantitative

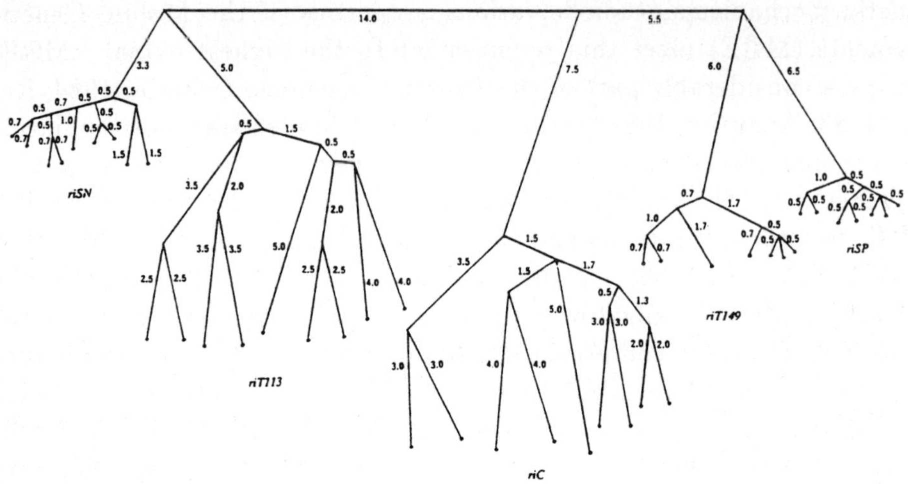

Figure 4: The tree of the similarity of patterns of *mdg-2* localization for the control (*riC*), two "selection" (*riSN* and *riSP*) and two "temperature" (*riT113* and *riT149*) fly lines. Constructed by S.A. Zabanov (Vasilyeva et al., 1987 b, 1988).

character, whereas "temperature" lines were got by temperature treatment during the sensitive ontogenetic periods. The result of selection is decided by artificial selection, i.e., it is under the investigator's control. The outcome of the temperature treatment is predetermined by some inner mechanisms of the sensitivity of germ-line cells, and is increased by a hard phenotypic selection for expressivity and penetrance. Therefore, pair-wise coherence of two "selection" and two "temperature" phenotypes is, on the whole, random. However, it enabled the dependence between the phenotype and the pattern of MGE localization to be revealed.

It may be argued (Vasilyeva et al., 1987b, 1988; Ratner and Vasilyeva, 1987) that, from the standpoint of population mechanisms (genetic drift, accumulation of MGE transpositions, etc.), this connection has its logic. It was reproducible at replicate breeding of "temperature" lines in 1979, 1982, 1985 and 1986. We have at least three arguments in favour of the non-random nature of the differences.

First, when rearing all the lines together, a complex of anti-drift measures was created to impede random fixation or loss of separate variants of MGE localization. When the flies in a separate replicate total N > 100 and the initial frequency of the variant is p = 0.5, the mean time of the variant to be fixed is:

$$\bar{t} = \frac{4N_e(1-p)}{p} ln(1-p) \cong 2.8N_e > 280$$

generations (Crow and Kimura, 1970; Ratner et al., 1985). However, once in 30-50 generations, the shuffling of flies from 20-40 replicates takes place, i.e., random fluctuations of frequencies are levelled off, and the probability of random variants to become fixed is negligible (Vasilyeva et al., 1987b, 1988). In this case, the role of the genetic drift in the beginning of the differences between the lines becomes small.

Second, the experience of phylogenetic analysis of polynucleotide and protein sequences gives proof that trees of similarity, constructed from them, coincide in most cases, or, at least, are rather close to the trees of natural phylogenetic processes. The basis of this coincidence is essentially the evenness of random accumulation of fixed substitutions, neutral or adaptive (Ratner et al., 1985). The fixations of neutral substitutions behave as if they formed the "background" of the evolution of sequence families. The share of fixed adaptive substitutions may increase or decrease, depending on the selection conditions, whereas the neutral ones keep on accumulating themselves, contributing to molecule divergence. In our case we have two trees, independently constructed: a genealogical tree of lines (see Fig. 1) reflecting the actual process of their breeding, and a similarity tree of lines (see Fig. 4) reflecting the degree of their similarity by the pattern of *mdg-2* localization. A comparison between these trees proves that they are quite different. That means that while breeding and microevolving, these lines had no uniform accumulation of differences in MGE localization, i.e. localization of changes was non-random and, probably, very uneven.

Finally, the similarity between lines in the pattern of MGE localization may be either "residual", since all the lines have the same origin (see Fig. 1), or newly acquired, if the spectra of change are similar and non-random in these lines. To solve this alternative, one should create the spectra of differences between all the daughter lines and a control, *viz, riC*, and then compare them. Comparisons made for these spectra show (Vasilyeva et al., 1987b, 1988) that the most phenotypically similar lines also have the most similar spectra of change of MGE localization. Thus, the *riSN* and *riT113* lines have 27 common changes from 35-33 as compared to *riC*; the

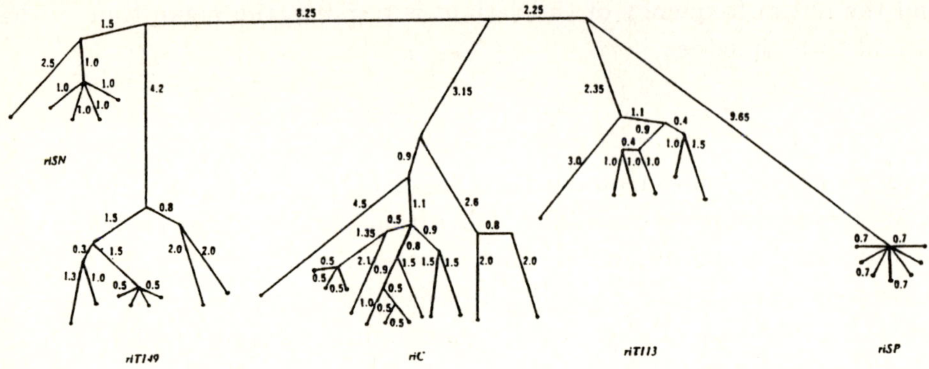

Figure 5: The tree of similarity of patterns of *mdg-1* localization for the same flies as of Fig. 4. Constructed by S.A. Zabanov (Zabanov et al., 1989).

riSP and *riT149* lines have 17 common changes from 30-23 substitutions. Therefore, the vast majority of spectra of changes is to a great extent non-random as to its position: it is canalized.

Other *copia*-like MGE have rather similar properties (Zabanov et al., 1989). The patterns of *mdg-1, mdg-3* and *copia* chromosomal localization are almost different for each line. However, the trees of pattern comparisons are similar to each other (see Fig. 5-6). At any rate, among the lines studied, the most similar lines are always the same pairs: *riSN* and *riT113, riSP* and *riT149*. The spectra of transpositions of MGE of daughter lines are to a great extent non-random as compared to the control, *riC*, and embrace rather similar sets of positions, when the phenotypes are similar; *mdg-4 (gypsy)* showed none of the mentioned features. The number of positions where it is localized in the *Drosophila* genome is not large and for that reason no similarity tree to correspond was constructed. Experimental material for *mdg-3* is not available so far; that is why we do not report the corresponding tree, though, by previous data, the main results confirm the general view.

Thus, it is becoming clear, that the pattern of MGE localization in chromosomes is tightly related to expressivity and penetrance of the quantitative character *radius incompletus*, and, as it may be judged by the published studies, also to the properties of many other quantitative

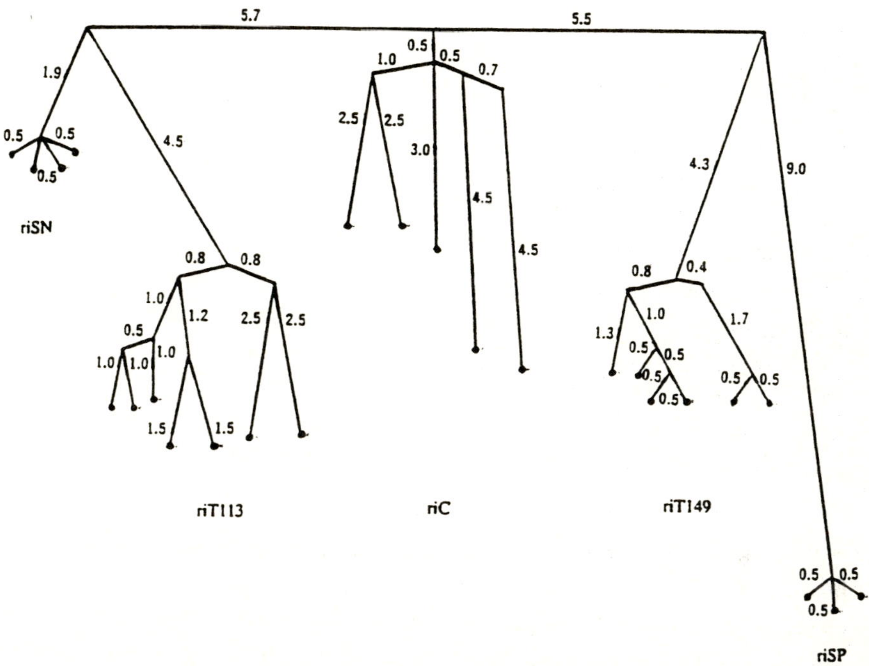

Figure 6: The tree of pattern similarities of the mobile element *copia* localization for the same fly lines as in Figs. 4 and 5. Constructed by S.A. Zabanov (Zabanov et al., 1989).

characters. All these characters are qualified by a common property: their response to selection, i.e., they are limiting sections in ontogenetic form-creation as well as in selection at a populational level. Therefore, one might believe that here we are dealing with the genomic system of the MGE influence upon expression and penetrance of any genetic trait, concerning either fitness or not.

A WORKING MODEL OF A POLYGENIC SYSTEM OF THE DETERMINATION OF A QUANTITATIVE CHARACTER AND TEMPERATURE EFFECTS

Consider a number of known properties of *copia*-like MGE. First of all, each MGE family has many dozens of localization sites in the chromosomes (Khesin, 1984; Rubin, 1983), these positions being rather stable in each *Drosophila* line. They can differ rather notably between diverged lines (Ananyev 1984). The probability of spontaneous transposition is not high, about 10^{-4} per position, per genome, per generation (Ananyev, 1984). It was shown that the so-called "molecular memory" exists: after MGE excision, one long terminal repeat (LTR) stays in its place to secure a replicate insertion of another MGE copy from the same family at a higher probability (Mizrokhi et al., 1985). "Transposition bursts" occur when as a result of a genetic action, or spontaneously, in certain germ-line cells, mass MGE transpositions take place (Gerasimova et al., 1984). It is shown that in many properties (transposition bursts, response to heat shock, etc) *mdg-1, mdg-2, mdg-3* and *copia* behave in the same way, whereas *mdg-4* shows independent properties (Junakovic and Angelucci, 1986). In sequenced MGE, various signs of punctuation and management were found: promoters (Khesin, 1984), enhancer-like sites (Shakhmuradov et al., 1986), regions similar to heat shock regulatory sites (Kapitonov et al., 1987; McDonald et al., 1987), etc. Copia-like MGE were shown to be retroposons, i.e., they are replicated through a free RNA copy with a reversed transcription (Ilyin et al., 1985; Mossie et al., 1985). Junakovic and his colleagues (Junakovic et al., 1986; Junakovic and Angelucci, 1986) used Southern blotting for restriction DNA fragments of some *Drosophila* lines and hybridized with probes for the presence of *copia*-like MGE. They showed that after heat shock (37^0C for 90 min) the spectrum of selected fragments changes considerably in the next generation. This is an independent confirmation of "burst-like" and mass character of MGE transposition under temperature treatment. It was also shown that *copia* and *mdg-1* are more mobile than *mdg-4*. On the whole, *copia*-like MGE mobilize themselves synchronously, which gives an idea about the correlated induction of their transpositions. In addition, Strand and McDonald (1985) showed that *copia*-like elements are responsive to heat shock by transcription activation. However, the restriction fragments containing MGE have not been compared with the regions of the chromosome map yet. It should be noted that heat-shock regulatory sites are inducible enhancers and their mechanism of action is considered positive by definition, yet it may be variable and quite complicated (Maniatis et al., 1987).

The list of quantitative characters in which a dependence between their expression and the MGE pattern of localization is documented has become vast. Mackay et al., (Mackay, 1988; Shrimpton et al., 1987) showed that these properties are inherent in the number of abdominal and sternopleural bristles. Selection for changes in character after dysgenic crossing in the P -M system entailed alterations of the pattern of P-factor localization. In this case, as in ours, the character was not directly linked with fitness.

Fig. 7 reports basic data on the localization of *D. melanogaster* oligogenes, "effective regions", genetic markers and *copia*-like MGE on the cytological map. This report demonstrates, in general, how complicated the genetic system of determination of the quantitative character *radius incompletus* is.

A statistical analysis of the patterns of the *mdg-1, mdg-2* and *copia* localization showed the following: The total distribution of the MGE sites over the chromosome arms (*X, 2L, 2R, 3L, 3R*) is uniform in most cases, though the *3R* arm statistically has a surplus of sites, especially of those concentrated near the centromere (see Fig. 7); the distributions of separate MGE over the chromosome arms do not differ statistically from each other, and they are independent as to the patterns of localization; the total distribution of the number of sites for all three MGE over enlarged segments of Bridges' map statistically differs from a Poisson distribution ($\mu = 1.6$, $\sigma = 0.8$, . i.e., $\mu \neq \sigma$) i.e., it is not random. Overall, these data do not contradict the expected properties, yet they add nothing essentially new to earlier conclusions.

We can now generalize the concept of a quantitative character in *Drosophila* based on the data obtained. We postulate that the regularities unveiled have a common significance for different quantitative characters of *Drosophila* . The quantitative character we have chosen (the interrupted vein of the wing) is not peculiar as to its dependence on the patterns of *copia*-like MGE localization. These properties must be attributes of many other quantitative characters of *Drosophila* controlled by oligogenes and polygenic modifiers. Thus, the polygenic system and the MGE start to have an influence upon character expression provided that one of the oligogenes becomes a limiting factor of expression (in our case, it is the *ri* oligogenic mutation). If fitness is involved, then the system is sensitive to natural selection. On the whole, one can assert that

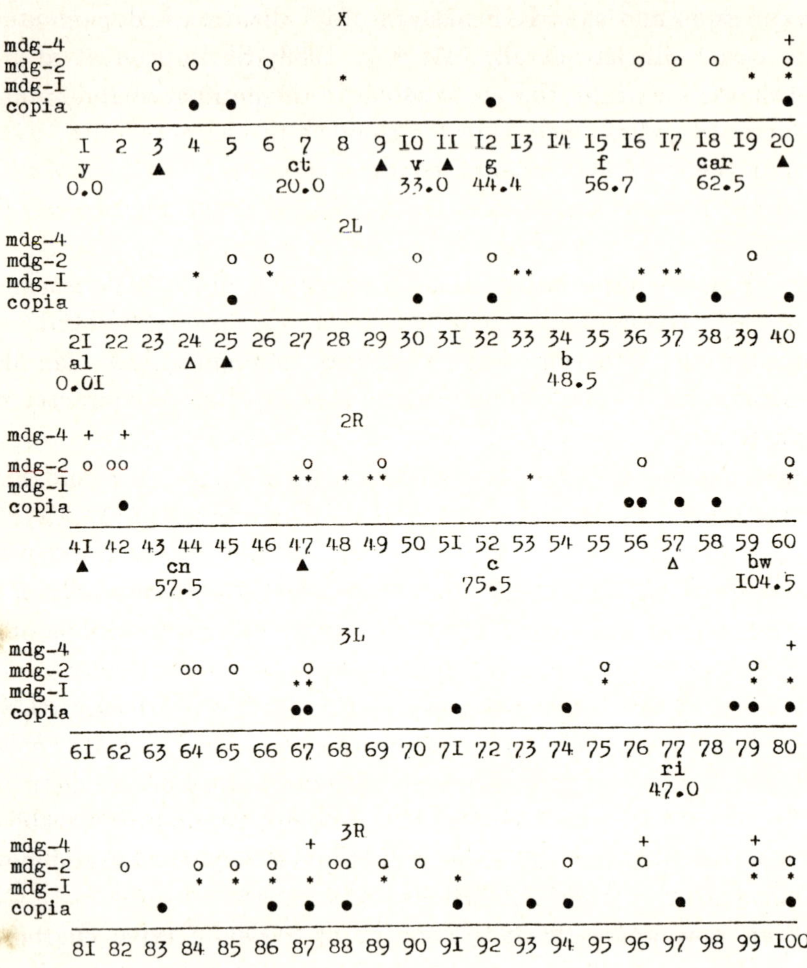

Figure 7: General picture for localization of hybridization sites: • - *copia*, ∗ - *mdg-1*, ○ - *mdg-2*, + - *mdg-4*. △ "effective regions" for the distal fragment of the radial vein by the data of genetic analysis, △ - "effective regions" for the proximal fragment. Double signs imply that there are two different MGE localization inside the segment. Segments of Bridges' cytological map for *D. melanogaster* polytenic chromosomes are indicated; the *ri* locus and the markers highlighting the regions of the map are given below. The *ri* locus and its vicinity in chromosome 3 contain no localized *copia*-like MGE.

limiting sections are evolving selectively, and non-limiting ones neutrally.

Thus, we assume a typical genetic system of the determination of quantitative characters in *Drosophila* to consist of three groups of genetic components:

1) **Oligogenes** (major-genes or genes of main effect), that are necessary for character formation. If an oligogene becomes limiting in a morphogenetic process of character formation (being either a normal or mutant allele), its expression would become dependent on polygene influence.

2) **Polygenic modifiers**, each of them not being necessary for character formation, but in total they are able to change expressivity essentially. It is important that polygenic modification shows up only with limiting oligogenes. In other words, the phenomenon of penetrance and expressivity, introduced to genetics by N.V. Timofeeff-Ressovsky (1928), takes place only in this case. Polygenes appear to enjoy their allelic variety and to be localized in a dispersed manner inside the genome. At least, at present, we have no evidence indicating their clustering. Most probably the role of polygenes is played by some genes that have their own direct effect but whose pleiotropy influences the expression of other systems of limiting oligogenes through the metabolic, morphogenetic and/or other systems.

3) **Mobile genetic elements**, having in each case a rather stable pattern of localization in the genome, is fixed for a long time by "molecular memory". It is not obligatory that the MGE pattern coincide with the topography of oligogenes and polygenes. Most probably the MGE patterns are more or less randomly superimposed on the topography of the localization of polygenes; despite this, the relationship is quite stable. MGE contain various functional sites, including enhancers and heat-shock regulatory sites, by means of which they influence the phenotypes of genes far from the place of their localization and also undergo external inducing actions, for instance, temperature changes. MGE act as standard migrating modifiers of the polygene contribution to MGE quantitative characters.

In this case the MGE pattern superimposed on the topography of the localized polygenes determines the group of polygenes that make a defined contribution to the change of character. Each polygene, however, does not contribute much to the oligogene expressivity; the allelic variety of the contribution appears to be small, because, had it been otherwise, then the polygenes would have been determined just as well as the oligogenes. So, if the MGE modifier effect upon the polygene contribution to character acts in a suppressing way, then its frame will be much narrower than in the absence of MGE. In other words, such contributions are still less notable, and the role of MGE is poor. On the other hand, if MGE make the polygenic contribution to be dozens of times *higher*, then their role will be more essential. The enhancers are known to bring the transcription of the neighbouring genes from dozens, to hundreds of times higher (Khoury and Gruss, 1983; Maniatis et al., 1987) and the heat-shock regulatory sites are necesary, in a stress-like situation, for considerable activation of the transcription of the heat-shock genes subjected to them (Strand and McDonald, 1985).

Thus, it can be readily assumed that MGE essentially influence the contribution to those characters attached to just those polygenes that are notably activated by them. But since the role of these polygenic alleles proves to be much less intense than the role of MGE, then the main contribution to the expressivity and penetrance of the character turns out to be made mainly by MGE. In other words, MGE act as if they were specifying the basic functions of the polygenes, their allelic variety easily being neglected.

Thus, we instantly run across the "effective" idea that MGE and polygenes are simply the same thing. Indeed, these are different genetic entities, yet only polygenes activated by the neighbouring MGE make notable contributions. Therefore, as a first approximation, it is sufficient to consider the polygenes as mobile elements and to estimate their contribution only. As for other pleiotropic manifestations of the polygenes, they, if not limiting, will stay slight.

The temperature effects may be assumed to be connected with the impact produced by a stress temperature through the heat shock system on the MGE ability of transcription and transposition. Such phenomena have been described for *copia*-like MGE (Junakovic and Angelucci, 1986; Junakovic et al., 1986) and other cases. This makes mass transpositions (penetrance) and transposition multiplicity (expressivity) possible events in most individuals, and in separate individuals, respectively.

Non-randomness of temperature transpositions (i.e., of different spectra of the MGE patterns of localization before and after temperature treatment) may be connected with the existence of some strong constraints on transposition localization during the sensitive periods of ontogenesis. These constraints (transposition canalization) must exist in the germ-line cells and be different in different sensitive periods. These restrictions, most probably, should run at the level of interactions or condensation of chromosomes. In this case, the allowed transpositions should correspond to the superimposing of the MGE pattern of localization on the pattern of allowed transitions. Transpositions will only take place provided that the two patterns, if any, partly coincide. Therefore, there will be cases when temperature treatment causes no transpositions. We have revealed such effects. Temperature treament of some isogenic lines of *Drosophila* that have a particular and specific form of MGE localization patterns, failed to induce transposition.

Comparing the "selection" and "temperature" lines, we can see that their phenotypical similarity is always accompanied by the similarity of the MGE localization patterns, despite different ways of producing these lines. It becomes possible if spontaneous transpositions (i.e., not induced by temperature treatment) in the "selection" lines obey the same constrictions as those in the "temperature" lines, but arise and accumulate gradually. A difference in results after selection and temperature treatment (the phenotypes being similar) is that selection changes both the MGE pattern and the allelic variety of polygenes (the latter appears not to be essential), whereas temperature treatment during one generation renders active only the MGE transpositions and never involves the polygenic alleles. By combining these ways of affecting phenotype and pattern, one can try and separate the effects of just MGE from those of the allelic variety of polygenes.

Consider the working model in question from long population-evolutionary views. It is known that in many complicated *Drosophila* oligogenes (*white, scute, bithorax, etc*), a considerable (perhaps the biggest) share of variability inheres exactly in MGE insertions (Rubin, 1983), securing the limitation of these genes in the systems of expression of the corresponding characters. On the other hand, limiting mutations immediately become a subject of natural selection (if they do influence fitness) or artificial selection and laboratory genetic study, as strictly observed alternative changes. On this background, the modifying effect of different MGE through the polygenes influencing the limiting olygogenes that have arisen shows brightly.

The *Drosophila* genome, containing up to 10% of MGE from various families, now should be considered a system of various patterns of MGE localization. It is capable of jointly or separately mass-inducing transpositions, the system being able to rearrange itself in response to stress-like external actions (temperature and others) during the sensitive periods of ontogenesis. It entails multiple genetic after-effects for *Drosophila* populations: a sharp change in the species norm for limiting characters, the range of the next mutations, recombinations, etc. Such events are supposed to be essential when the population finds itself in a long-term stress which yields mass transpositions. In different cases these transpositions can be both random and stiffly canalized as to their localization. However, they bear a novel variant of genetic variability for the polygenic systems, when expression may essentially change as a result of mass transpositions.

Due to a lack of synchronism in the development of individuals in the population, this variant seems to involve just a little part of the population, and even in this case the probability of such events is by a factor of some dozens greater than the probability of the usual mutations. Under the conditions that a population is divided into autonomous subpopulations, demes, and that the occupation of new ecological niches with sharply changed living standards takes place, some novel forms induced by an external stress may become founders of new populations with the phenotype sharply changed by the limiting quantitative characters. It is clear that in this case both adaptive and random variants of fast rearrangements are possible. In the long run, it is these events which become one of the main components of the variability and evolution of the *Drosophila* genome. Perhaps, changing the pattern of MGE localization is one of the mechanisms of speciation.

Thus, the problem of penetrance and expressivity of genetic systems, introduced to genetics 60 years ago by N.V. Timofeeff-Ressovsky (1928), has shown its striking profundity at the molecular level and its close relationship with the problem of individual and population dynamics of the system of MGE families.

REFERENCES

Ananyev EV (1984) Molecular Cytogenetics of Mobile Genetic Elements of *Drosophila melanogaster*. Itogi nauki i tehniki, ser. Molekularnaya biologia v.20 VINITI-Press Moscow 65-100 (Russian)

Biemont C, Terzian C (1988) *Mdg-1* mobile element polymorphism in selected *Drosophila melanogaster* populations. Genetica 76: 7-14

Bucheton A (1978) Non-Mendelian female sterility in *Drosophila melanogaster*: influence of ageing and thermal treatment. I. Evidence for a partly inheritable effects of these two factors. Heredity 43: 357-369

Bucheton A (1979) Non-Mendelian female sterility in *Drosophila melanogaster*: influence of ageing and thermal treatment. II. Action of thermal treatment on the sterility of SF females and on the reactivity of reactive females. Biologie Cellulaire 34: 43-50

Bucheton A, Picard G (1978) Non-Mendelian female sterility in *Drosophila melanogaster*: hereditary transmission of reactive levels. Heredity 40: 207-223

Bucheton A, Lavige JM, Picard G, L'Heritier Ph (1976) Non-Mendelian female sterility in *Drosophila melanogaster*: quantitative variations in the efficience of inducer and reactive strains. Heredity 36: 305-314

Crow JF, Kimura M (1970) An introduction to population genetics theory. Harper & Row, Pub. New York

Georgiev GP, Gvozdev VA (1980) Mobile dispersed genes of eucariotes. Vestn AN SSSR 8: 19-27 (Russian)

Gerasimova TI, Mizrokhi LY, Georgiev GP (1984) Transposition bursts in genetically unstable *Drosophila melanogaster*. Nature 309: 714-716

Gvozdev VA, Belyaeva ESp, Ilyin IV, Amosova IS, Kaidanov LZ (1981) Selection and transposition of mobile dispersed genes in *Drosophila melanogaster*. Cold Spring Harbor Symp Quant Biol 45: 673-686

Gvozdev VA, Kaidanov LZ (1986) Genome variability induced by mobile element transposition, and fitness of *Drosophila melanogaster* individuals. Jurn Obschey Biologii 47: 51-63 (Russian)

Ilyin YV, Arkhipova IR, Gorelova GV, Shuppe NG (1985) Discovery of intermediates of reverse transcription of RNAs of mobile dispersed genes *mdg-1* and *mdg-3* in *Drosophila* cells. Molekularnaya biologia 19: 162-172 (Russian)

Junakovic N, Angelucci V (1986) Polymorphism in the genomic distribution of *copia*-like elements in related laboratory stocks of *Drosophila melanogaster*. J Mol Evol 24 I:83-88

Junakovic N, Di Franco C, Barsanti P, Palumbo G (1986) Transposition of *copia*-like nomadic elements can be induced by heat shock. J Mol Evol 24 I:89-93

Kapitonov VV, Kolchanov NA, Shakhmuradov IA, Solovyev VV (1987) Presence of regions homologous to regulatory sites of heat shock in mobile elements. Genetika 23: 2112-2119 (Russian)

Khesin RB (1984) Inconstancy of genome. Nauka Publs Moscow 578 (Russian)

Khouri G, Gruss P (1983) Enhancer elements. Cell 33: 313-314

Khristolyubova NB, Auslender JE (1967) Properties of inheritance of functional changes of *Drosophila* salivary gland chromosomes. Genetika 2: 76-79 (Russian)

Lindsley DL, Grell EH. (1967) Genetic variations of *Drosophila melanogaster*. Carnegie Institution of Washington Publication No. 627

Mackay TFC (1985) Transposable element-induced response to artificial selection in *Drosophila melanogaster*. Genetics 111: 351-374

Mackay TFC (1988) Transposable element-induced quantitative genetic variation in *Drosophila*. In: Weir ES, Eisen EJ, Goodman MM, Namkoong GE (eds) The 2nd International Conference on Quantitative. Sunderland Sinauer, pp 219-235

Maniatis T, Goodbourn S, Fisher JA (1987) Regulation of inducible and tissue- specific gene expression. Science 236: 1237-1245

McDonald JF, Strand DJ, Lambert ME, Weinstein IB (1987) The responsible genome: evidence and evolutionary implications. In: Raff RA, Raff EC, Liss AR (eds) Development as an evolutionary process. New York, pp 239-263

Mizrokhi LY, Pryjmyagi AF, Ilyin YV, Gerasimova TI, Georgiev GP (1985) Molecular mechanism of transpositional memory in system of *mdg-4* cut locus of *Drosophila melanogaster*. Dokl AN SSSR 285: 1458-1460 (Russian)

Mossie KG, Young MW, Varmus HE (1985) Extrachromosomal DNA forms of *copia*-like transposable elements, F-elements and middle repetitive DNA sequences in *Drosophila melanogaster*. J Mol Biol 182: 31-43

Neel GV (1940) The interpretation of temperature, body size and character expression on *Drosophila melanogaster*. Genetics 25: 225-250

Pasyukova EG, Belyaeva ES, Kogan GL, Pavlova MV, Kaidanov LZ, Gvozdev VA (1986) Concerted transpositions of mobile genetic elements coupled with fitness changes in *Drosophila melanogaster*. Mol Biol Evol 3: 299-312

Ratner VA, Vasilyeva LA (1987) Quantitative character in *Drosophila*: genetic, ontogenetic, cytogenetic and population aspects. Genetika 23: 1070-1081 (Russian)

Ratner VA, Zharkikh AA, Kolchanov MA, Rodin FN, Solovyov BB, Shamim BB (1985) Problems of the theory of molecular evolution. Nauka Publ. Novosibirsk (Russian)

Rubin GM (1983) Dispersed repetitive DNAs in *Drosophila*. In: Schapiro JNI (ed) Mobile genetic elements. Acad Press, pp 329-361

Shakhmuradov IA, Kolchanov NA, Solovyev VV, Ratner VA (1986) Enhancer-like structures in moderately repeating sequences of eucaryotic genomes. Genetika 22: 347-367 (Russian)

Shrimpton AE, Mackay TFC, Brown AJL (1987) A molecular genetic analysis of the response to selection for bristle number in *Drosophila*. In "Abstr of 2nd International Conference on Quantitative Genetics" Raleigh 102

Sneath PHA, Sokal RR (1973) Numerical Taxonomy. The Principles and Practice of Numerical Classification. Freeman & Co San Francisco

Strand DJ, McDonald JF (1985) *Copia* is transcriptionally responsive to environmental stress. Nucl Acids Res 13: 4401-4410

Svetlov PG (1962) The problem of pathogenesis of hereditary and non-hereditary defects of normal development in the ligth of general laws of ontogenesis. Vestnik AN SSSR 11: 13-18 (Russian)

Svetlov PG, Korsakova GF (1962) Dependence of the wing size of vestigial mutants of *Drosophila melanogaster* on temperature conditions of development on the larval and proembryonal stages of ontogenesis. Doklady AN SSSR 145: 922-925 (Russian)

Svetlov PG, Korsakova GF (1965) About dependence of characters of forked mutation at descendants of females of *Drosophila melanogaster* on temperature treatments. Doklady AN SSSR 165: 214-216 (Russian)

Svetlov PG, Korsakova GF (1966 a) Influence of short heating of females of *Drosophila melanogaster* with forked mutation on expressivity of character of this mutation at the range of consequent generations. Doklady AN SSSR 168: 191-194 (Russian)

Svetlov PG, Korsakova GF (1966 b) Prolonged modifications in the experiments with temperature treatments on larvae of forked mutants of *Drosophila melanogaster*. Doklady AN SSSR 170: 439-442 (Russian)

Svetlov PG, Korsakova GF (1972) Inheritance of expressivity changes of eyeless mutation in *Drosophila melanogaster* induced by temperature treatment at special periods of development. Ontogenez 2: 347-355 (Russian)

Tarasoff M, Suzuki DT (1970) Temperature-sensitive mutations in *Drosophila melanogaster*. VI. Temperature effects on development of sex-linked recessive lethals. Devel Biol 23(3): 492-509

Thoday JM, Thompson JN Jr (1976) The number of segregation genes implied by continuous variation. Genetics 45(3): 335-344

Timofeef-Ressovsky NV (1928) The influence of temperature on formation of venae transversae of the wing in one genovariation of *Drosophila funebris*. Jurn Experim Biologii ser A 4: 199-214 (Russian)

Vasilyeva LA, Zabanov SA, Ratner VA (1985) On the possible role of Mobile Genetic Elements (MGE) in determination and evolution of quantitative character. In Internat Sympos "Biological Evolution" Bary (Italy) 84-86

Vasilyeva LA, Ratner VA, Zabanov SA (1987 a) Expression of quantitative character *radius incompletus*, temperature effects and localization of mobile elements in *Drosophila*. I. Properties of investigated populations. Genetika 23: 71-80 (Russian)

Vasilyeva LA, Zabanov SA, Ratner VA, Zhimulev IF, Protopopov MO, Belyaeva ES (1987 b) Expression of quantitative character *radius incompletus*, temperature effects and localization of mobile elements in *Drosophila*. II. Mobile genetic elements Dm-412. Genetika 23: 81-92 (Russian)

Vasilyeva LA, Zabanov SA, Ratner VA, Zhimulev IF, Protopopov MO, Belyaeva ES (1988) Expression of a quantitative character *radius incompletus*, temperature effects and localization of the mobile genetic elements Dm-412 in *Drosophila melanogaster*. Gen Sel Evol 20(2): 65-85

Zabanov SA, Vasilyeva LA, Ratner VA (1989) Expression of quantitative character *radius incompletus*, temperature effects and localization of mobile elements in *Drosophila*. III. Mobile genetic elements *mdg-1* and *copia*. Genetika 25 (in press) (Russian).

The Potential Evolutionary Significance of Retroviral-Like Transposable Elements in Peripheral Populations

J.F. McDonald

Department of Genetics, University of Georgia, Athens, GA 30602, USA

1 Introduction

Evolutionists have long speculated that peripheral populations (i.e., genetically isolated populations, founded by a few individuals and located on the margin of a species' range) may be especially conducive to the emergence of new species and other evolutionary novelties (e.g., Mayr 1954; Carson 1959). This speculation is based in part on observation and in part on theory. Peripheral populations are frequently observed to be associated with morphological and behavioral phenotypes which significantly differ from those characteristic of populations making up the continuous body of a species' range (e.g., Mayr 1942; Soule 1966). The theoretical explanations which have been offered to account for the putatively unique properties of peripheral populations differ in detail but generally consist of ecological and genetic components (Mayr 1960; Carson and Templeton 1984, Carson 1985). Populations located at the periphery of a species range are often ecologically marginal. This means that phenotypes which are adaptive throughout the main or central range of the species may not be optimally or even adequately suited to ecological conditions on the periphery. Thus, there may be significant selective pressure exerted on peripheral populations to evolve new adaptive phenotypes.

As important as these ecological differences may be, they are not generally considered sufficient or, in some cases, even necessary for the emergence of evolutionary novelties. The key factor, according to many theoreticians, that inclines peripheral populations to rapid and dramatic changes in genotype is that they are generally founded by a single or a few individuals and, once established, are genetically isolated from the continuous range of the species (Mayr 1954; Carson 1959, 1971, 1982; Templeton 1981). It should be noted in passing, however, that not all models of speciation and the origin of evolutionary novelties presume the backdrop of a peripheral population. Indeed, there is good experimental evidence, especially in plants (e.g., Gotlieb 1984; Grant 1981; Levin 1983; Lewis 1973) and some insects (e.g., White 1978; Bush 1974, 1975), indicating that the emergence of new species and other evolutionary novelties can arise within other genetic and ecological contexts.

We are currently at a transitional stage in the development of evolutionary theory. Recent years have witnessed the interweaving of newly acquired knowledge on the workings of the eukaryotic genome into the existing fabric of evolutionary thought. In many instances, the process has been relatively smooth and the new molecular findings have served to confirm and complement pre-existing notions (e.g., Ayala 1976; Nevo 1986). In other instances, however, attempts at integration have been more problematic and have led to challenges of long-standing evolutionary paradigms (e.g., Steele 1979; McDonald 1983; Syvanen 1984; Bush and Howard 1986; Dover 1986; Pollard 1987; Cairns et al. 1988; Hall 1988).

Recent studies on the molecular biology of retroviral-like transposable elements (RLEs) indicate that this class of transposons may be of particular evolutionary significance (McDonald et al. 1987; McDonald 1989). In this paper, I attempt to summarize certain newly discovered properties which suggest that RLEs may play a role in catalyzing the formation of new species and/or the occurrence of other evolutionary novelties. The evolutionary impact of RLEs may be especially pronounced within the unique ecological and genetic context of peripheral populations.

2 Some evolutionary significant properties of retroviral-like elements

2.1 Retroviral-like elements are the most abundant and widely distributed class of eukaryotic transposable elements

The largest and most widely distributed group of eukaryotic mobile elements consists of those whose structure and presumed mode of replication parallels that of mammalian retroviruses. RLEs have been found to be represented within the genomes of vertebrates (Kuff et al. 1983; Benveniste 1985; Callahan et al 1985), invertebrates (Finnegan 1986; Finnegan and Fawcett 1986), plants (Johns et al. 1985) and yeast (Roeder and Fink 1983).

The numbers of RLEs present within genomes can vary significantly. For example, inbred and wild-type subspecies of Mus musculus have been estimated to contain a haploid number of approximately 1000 RLEs per genome (Cole et al. 1981; Kuff et al. 1983). In contrast, laboratory strains of the yeast

Saccharomyces cerevisiae typically contain fewer than 100 RLEs (Roeder and Fink 1982). Generally, the RLEs present within genomes can be grouped into structurally homologous families. In Drosophila melanogaster, for example, there are at least 50 families of RLEs and each family is represented by between 10 and 100 randomly distributed structurally homologous members per haploid genome (Rubin 1983).

2.2 Retroviral-like elements are a significant source of regulatory and developmental mutations

RLEs are a well-recognized source of mutation (Lambert et al. 1988). Indeed, in Drosophila melanogaster, most of the morphologically detectable spontaneous mutations which have been examined on the molecular level are the result of the insertion of a RLE (Green 1988; Sankaranarayanan 1988; Finnegan 1985; see Table 1). Consistent with this are the results of preliminary surveys which indicate that RLE insertion variants account for a significant amount of the allelic variation present in natural populations of Drosophila melanogaster (Leigh Brown 1983; Aquadro et al. 1986).

TABLE 1. Proportion of spontaneous mutations in Drosophila melanogaster which are due to the insertion of a retroviral-like element (RLE) [after Sankaranrayanan, 1988]

Locus	Proportion of mutants due to RLE insertion	RLE (no. mutants)
v (vermillion)	4/5	412 (3), B104 (1)
ct (cut)	28/28	gypsy (27), copia (1)
ry (rosy)	3/5	calypso (1), B104 (2)
f (forked)	3/4	gypsy (3)
su(s) (suppressor-of-sable)	5/7	gypsy (5)
Bx (Beadex)	4/4	gypsy (2), B104 (2)
bx (bithorax)	8/9	gypsy (7), B104 (1)
sc (scute)	2/2	gypsy (2)
Antp (Antennapedia)	2/5	B104 (2)

Retroviral-like element insertion mutants are frequently characterized by the acquisition of novel regulatory phenotypes and thus may constitute a genetic mechanism by which new patterns of developmental and regulatory networks are acquired over evolutionary time (McDonald 1989). In higher eukaryotes like Drosophila and mice, mutant RLE insertion alleles are often associated with new developmental and/or tissue-specific patterns of expression (Strand and McDonald 1989; Wilson et al. 1988; Parkhurst and Corces 1986). For example, our laboratory has recently characterized a naturally occurring Drosophila melanogaster alcohol dehydrogenase (adh) allele which contains a copia retroviral-like transposable element inserted 240 bp upstream from the distal (adult) adh transcriptional start site (Strand and McDonald 1989). Our results demonstrate that patterns of adh expression are quantitatively disrupted at life-stages and in tissues where copia is actively expressed. In addition, we found that levels of adh expression in flies homozygous for the variant allele can be modulated by chromosomal genes identified as putative trans-regulators of copia expression (Mount et al. 1988; Zachar et al. 1987; Rutledge et al. 1988). These types of RLE mediated regulatory changes may have dramatic phenotypic effects, particularly if the mutant gene encodes a product which is involved in early stages of morphological development. For example, in Drosophila melanogaster, 8 out of 9 bithorax (homeotic/developmental) gene mutations which have been characterized on the molecular level have been shown to be associated with a RLE insertion (Peifer and Bender 1986; Sankaranarayanan 1988, see Table 1).

2.3 Retroviral-like transposable elements are transcriptionally and transpositionally induced by environmental and genomic stress

Latent vertebrate retroviruses are known to be subject to activation by viral infection (e.g., Reed and Rapp 1976; Gendelman et al. 1986; Mosca et al. 1987; Greene et al. 1988), radiation (e.g., Weiss et al. 1971; Rowe et al. 1971; Tennant and Rascati 1980) and other forms of environmental and genomic stress (e.g., Lowy et al. 1971; Robinson et al. 1976; Rascati and Tennant 1978). Recently, data have been accumulating which suggest that at least some families of RLEs may also be stress inducible. For example, our laboratory has shown that heat-shock and heat-shock mimetic treatments significantly increase transcript levels of the Drosophila copia RLE in adult flies and in transfected mammalian cell lines (Strand and McDonald

1985; McDonald et al. 1987; see also Braude-Zolotarjova and Schuppe 1987). In addition, we have recently found that exposure of *Drosophila* to gamma radiation significantly increases levels of *copia* transcripts (McDonald et al. 1988; Strand and McDonald, unpublished results). This later observation is consistent with earlier findings that the numbers of retroviral-like particles in cell nuclei is dramatically increased in *Drosophila* which have been exposed to gamma radiation (Philpott et al. 1969). These nuclear retroviral-like particles have subsequently been shown to contain *copia* RNA and to be associated with reverse transcriptase activity (Shiba and Saigo 1983; Becker et al. 1987). The yeast RLE, *Ty*, has likewise been shown to be transcriptionally induced by ultraviolet light and DNA damaging chemicals (Rolfe et al. 1986; McEntee and Bradshaw 1988).

Since the transposition of RLEs is believed to involve reverse transcription of an RNA intermediate (Boeke et al. 1985), the fact that at least some families of RLEs are transcriptionally responsive to environmental stress suggests that rates of retrotransposition may be stress-responsive as well (McDonald et al. 1988). Consistent with this prediction, it has recently been demonstrated that chemically stressed yeast cells experience a significant elevation in levels of *Ty* transcripts and a correlated increase in the rate of *Ty* element-mediated insertion mutations (McEntee and Bradshaw 1988). It should be noted, however, that the full mechanistic relationship between RLE transcript levels and rates of retrotransposition is far from being understood. For example, recent studies indicate that formation of retroviral-like particles may be a rate-limiting step in the retrotransposition process (Hull and Covey 1986). These findings coupled with recent data indicating that formation of retroviral-like particles may, at least in part, be regulated at a post-transcriptional level (Strand et al. 1989; McDonald et al. 1988; Curcio et al. 1988; Youngren et al. 1988), suggest that increased levels of RLE transcripts *per se* may not be sufficient to increase rates of retrotransposition.

It remains to be demonstrated whether or not all stress-induced increases in RLE transcript levels result in a correlated increase in rates of retrotransposition. Likewise, it is presently unknown whether the types of environmental stresses periodically experienced by peripheral populations are sufficient to induce increased rates of retrotransposition. Nevertheless, the existing evidence is suggestive and leaves open the possibility that rates of retroviral-element mediated mutations may be

TABLE 2. Suppressor genes and suppressible retroviral-like element insertion mutants in *Drosophila*, yeast and mice [compiled from: Strand and McDonald 1989; Rutledge et al. 1988; Winston 1988; Sweet 1983]

Species	Suppressor gene	Suppressible mutant	Retroviral element	Enhanceable mutant
Drosophila	su(f)	lz^1, f^1, f^5, bx^{34e} adh^{RI42}	gypsy copia	$w(a), lz^{37}$
	su(Hw)	lz^1, l, f^1, f^5, Bx^2 $bxd, y^2, Hw^1, ct^6,$ ct^k, sc^1	gypsy	lz^{37}
	su(pr)	$pr^{bw}, pr^1, lz^1, f^1, f^5$ lz^{34}, lz^{2k}	412	$Hw^1, ct^k, bx^3,$ bx^{34e}
	su(s)	s^1, sp^1, pr^1, v^1	412	lz^1, ct^k, f^1 $bx^3, bx^{34e},$ $lz^{34}, lz^k,$ lz^{37}
	su(wa)	$w(a), adh^{RI42}$	copia	$lz^k, lz^{34},$ $lz^8, lz^{27},$ $ct^6, f^1,$ bx^{34e}, bx^3
	e(e)	$lz^1, f^1, f^5, bx^{34e},$ lz^h	gypsy	w(e)
yeast	spt1, spt2 spt5	his4-912, lys-2173R2	Ty	not determined
	spt3, spt6 spt7, spt8	his4-912, lys-2173R2, his4-917	Ty	"
mice	dsu	d (dilute)	MuLV	"

particularly labile in peripheral and other environmentally unstable populations.

2.4 The phenotypic effects associated with retroviral-like element insertion mutations may be modified by the action of host-encoded suppressor genes

One of the most interesting characteristics of RLE insertion mutations from the evolutionary perspective is the fact that their phenotypes are not rigidly determined but may be substantially modified by the action of a number of host-encoded suppressor and enhancer genes (e.g., Winston 1988; Kubli 1986, Rutledge et al. 1988; see Table 2). Laboratory studies of RLE insertion mutants in yeast (Winston et al. 1987), mice (Sweet 1983) and Drosophila (Parkhurst and Corces 1986) have demonstrated that the phenotypes associated with this class of genetic variants may, depending upon the genetic background in which the mutant allele is expressed, vary dramatically or only slightly from that of the progenitor "wild-type" allele. It follows that if natural populations maintain suppressor alleles in non-trivial frequencies, the possibility exists that regulatory variants representing novel and perhaps dramatic phenotypic potentials may be harbored in natural populations and at least partially shielded from the action of natural selection. Moreover, the kinds of bottlenecks postulated to be associated with the founding of peripheral populations could result in substantial changes in suppressor allele frequencies resulting in the sudden appearance of RLE insertion variants whose mutant phenotypes were effectively suppressed within the parental population. The likelihood of such scenarios will be better judged as more information becomes available on the frequency of suppressor alleles in natural populations.

2.5 Retroviral-like element expression is variable among natural populations

Our laboratory has recently completed a world-wide survey of copia element transcriptional variation among natural populations of Drosophila melanogaster (Csink and McDonald 1989). The results indicate that levels of copia transcripts vary by 100-fold among flies representing geographically diverse populations of D. melanogaster and that this variation is not correlated with variability in copia copy number. Moreover, analysis of transcript levels among inter-population hybrids demonstrates that at least some of this variation is attributable to the action of both dominantly and recessively inherited trans-acting controls (Csink and McDonald 1989).

There are at least two reasons why naturally occurring variability in RLE expression may be of potential evolutionary significance. First, recent studies of RLE insertion mutants in yeast and Drosophila indicate that transcriptional interference between the inserted element and the mutant gene may contribute to the novel regulatory patterns associated with RLE insertion variants (Strand and McDonald 1989; Winston 1988; Parkhurst and Corces 1986). Thus, the observation that there is significant variability in levels of copia expression among natural populations suggests that there may also be significant variation among populations in the ability to modify the phenotype of copia insertion variants (see Section 2.3). This hypothesis is currently under investigation in our laboratory.

The second reason why naturally occurring variation in RLE expression may be of evolutionary significance relates to its possible influence on mutation rates. If, as recent evidence suggests, RLEs transpose via reverse transcription of RNA intermediates (Boeke et al 1985), then variability in RLE transcript levels among individuals and populations may contribute to variability in rates of retrotransposition and associated mutagenic potential as well. Consistent with this hypothesis is the recent demonstration that increased transcription of a single Ty element also increases its frequency of transposition (Boeke et al 1985). Also consistent are those previously cited studies which have established a correlation between stress induced increases in Ty transcript levels and rates of Ty element transposition (McEntee and Bradshaw 1988). However, if rates of retrotransposition are subject to regulation at the post-transcriptional level (see Section 2.3), the relationship between levels of RLE transcripts and rates of retrotransposition may not always be easily predicted (McDonald et al. 1988; Curcio et al. 1988).

3 Possible evolutionary role of retroviral-like elements in peripheral populations

As alluded to earlier, peripheral populations are believed to be especially condusive to rapid and dramatic evolutionary change. The fact that they are likely to be founded by one or a few individuals and to be subsequently isolated from the parental gene pool, implies that peripheral populations are, from the outset, genetically unique. This situation, coupled with the distinct and possibly substantial selective pressures often experienced by peripheral populations is believed by some to be the primary driving force behind the emergence of new species and other evolutionary novelties.

3.1 The genetic architecture of evolutionary novelties is consistent with established characteristics of retroviral element insertion mutants

Although the precise nature of the genetic changes which underlie the emergence of evolutionary novelties is currently the subject of some debate (e.g., Carson and Templeton 1984; Barton and Charlesworth 1984), genetic studies of the process, especially among Hawaiian Drosophila has shed empirical light on the issue (Carson et al. 1982). For example, Templeton (1977) and Val (1977) have shown that the dramatic morphological differences which exist between the rapidly diverged Hawaiian picture-wing species D. silvestris and D. heteroneura, are primarily determined by a major segregating unit interacting strongly with several modifier genes. It is interesting to note that this type of genetic architecture (Type II in Templeton's terminology) is wholly consistent with what has been observed for RLE insertion variants, i.e., a novel regulatory or developmental phenotype associated with a single mutant allele which is subject to phenotypic suppression (or enhancement) by several unlinked but epistatically interacting modifier genes (Parkhurst and Corces 1985; Strand and McDonald 1989; Rutledge et al 1988; Parkhurst and Corces 1986; Chang et al 1986; Zachar et al 1985).

The hypothesis that retroviral-element insertion variants may play a role in the rapid evolutionary changes associated with peripheral populations is also consistent with the results of laboratory studies designed to mimic the effects of founding events. For example, Templeton exploited parthenogenetic strains of D. mercatorum to generate the most extreme founder event possible from a natural, outcrossing sexual population, i.e., diploidization based on a single haploid genome (Templeton 1979). The results of these experimentally induced founder events were the rapid establishment of strong pre- and postmating barriers. The genetic basis of one of these barriers (aa) was extensively investigated by Templeton (1979) and shown to be attributable to a major X-linked segregating unit that engages in strong epistatic (suppressor and enhancer) interactions with X-linked, Y-linked and autosomal modifier genes. Additional studies have shown that the major segregating unit is associated with preferential amplification of 28S ribosomal genes containing insertions of 5-10 kb in length. These insertions have been found to disrupt the ribosomal genes' normal patterns of transcription (DeSalle and Templeton 1983).

Regardless of whether retroviral-element insertion variants and their associated modifier loci are ultimately found to be the basis of the specific examples cited above, it remains significant that the molecular phenotype which is typically associated with RLE insertion mutants is fully consistent with the type of change in genetic architecture postulated to be the basis of the rapid emergence of at least some evolutionary novelties.

3.2 Some environmental and genomic stresses experienced by peripheral populations are known to increase mutation rates and to be activators of retroviral-like transposable elements

As outlined above, there is a growing body of evidence indicating that RLEs may be responsive to various environmental and genomic stresses. Since populations located on the periphery of a species range are typically ecologically marginal as well, it is reasonable to hypothesize that rates of retrotransposition in peripheral populations may be at least periodically elevated in response to environmental stress contributing to the emergence of new species and/or other evolutionary novelties.

Although data directly bearing on this issue are as yet unavailable, the hypothesis is nevertheless consistent with longstanding observations that spontaneous mutation rates are elevated when organisms are exposed to the kinds of environmental stresses frequently experienced by peripheral populations. For example, it has been shown in Drosophila that as the nutritional content of the food supply decreases, the spontaneous mutation rate significantly increases (e.g., Newberne and Zeiger 1978; Herskowitz 1963). Other environmental stresses which might be periodically experienced by ecologically marginal populations such as temperature stress (e.g., Plough and Ives 1935; Grossman et al. 1933; Goldschmidt 1929) and viral infections (Gershenson 1986; Golubovsky and Plus 1982) have also been associated with significantly elevated rates of spontaneous mutation. Many of these environmental stresses are known to be activators of retroviral-like transposable elements (Section 2.3).

Since RLE expression can be regulated by host-encoded and often recessively-inherited trans-acting alleles (Section 2.4), the bottlenecks associated with founder events could result in the initial establishment of populations having a high frequency of genomes which are prone to increased rates of retrotransposition (e.g., populations having a high frequency of mutant repressor alleles). There is certainly ample evidence that inbreeding can result in elevated mutation rates and associated increases in levels of phenotypic variability (e.g., Bryant et al. 1986a,b; Lints and Bourgois 1982; see also paper by Bryant et al. in this volume). Interestingly, some instances of selective inbreeding have actually been associated with the movement of RLEs (Pasyukova et al. 1981; Gvozdev 1980; see also paper by Ratner et al. in this volume).

4. Summary and conclusions

RLEs are the largest and most widely distributed class of eukaryotic transposable elements. The significance of RLEs as mutagenic agents is reflected in the fact that the majority of all morphologically detectable spontaneous mutants in Drosophila have been associated with the insertion of RLEs.

RLEs display properties which may contribute to rapid evolutionary change especially within the context of peripheral populations. The fact that RLEs may be activated by environmental stresses and inbreeding effects which are reasonably expected to be experienced by populations located on the periphery of a species range, suggests that these elements may contribute to the major genetic restructurings postulated to occur within isolated populations (Carson 1982; Templeton 1980; Mayr 1954). Analysis of laboratory induced and naturally occurring RLE insertion mutants indicates that RLE insertions into non-coding regions of genes frequently impart novel regulatory and developmental patterns of expression which may be subject to suppression and/or enhancement by several unlinked chromosomal loci. The fact that the architecture of the genetic changes associated with the rapid emergence of new species and other evolutionary novelties is consistent with the established genetic consequences of RLE insertion events, supports the hypothesis that RLEs may play a significant role in effecting evolutionary change within peripheral populations.

Acknowledgements: I am grateful to my colleagues S. Voss, H. Naveira, J. Avise and M. Tracey for reading and commenting on earlier versions of this manuscript. Many thanks also to Professor Gateff and Deutscher Akademischer Austauschdienst for supporting my sabbatical leave in Germany during 1988.

References

Ayala FJ (ed) (1976) Molecular evolution. Sinauer Press, Sunderland, MA.

Aquadro CF, Deese SF, Bland MM, Langely CH, Laurie-Ahlberg CC (1986) Molecular population genetics of the alcohol dehydrogenase gene region on D. melanogaster. Genetics 114:1165-1190

Barton NH and Charlesworth B (1984) Genetic revolutions, founder effects and speciation. Ann. Rev. Ecol. Syst. 15:133-164

Becker J-L, Barre-Sinoussi F, Dormont D, Best-Belpomme M and Chermann J-C (1987) Characterization of the purified RNA dependent DNA polymerase isolated from Drosophila. Cell. Mol. Bio. 33:225-235

Benveniste RE (1985) The contributions of retroviruses to the study of mammalian evolution. In: MacIntyre (ed) Molecular evolutionary genetics. Plenum Press, New York, pp 359-418

Boeke J, Garfinkel D, Styles C, and Fink G (1985) Ty elements transpose through an RNA intermediate. Cell 40:491-500

Braude-Zolotarjova T and Schuppe NG (1987) Transient expression of hsp-CAT1 and copia-CAT1 hybrid genes in D. melanogaster and D. virilis cultured cells. Dros. Infor. Ser. 66:33

Bryant E, McCommas S, and Combs L (1986) The effect of an experimental bottleneck upon quantitative genetic variation in the housefly. Genetics 114:1191-1211

Bryant E, Combs L, and McCommas, S (1986) Morphometric differentiation among experimental lines of the housefly in relation to a bottleneck. Genetics 114:1213-1223

Bush GL (1974) The mechanisms of sympatric race formation in the true fruit flies. In: White (ed) Genetic mechanisms of speciation in insects. Australian and New Zealand Book Co., Sydney, pp 3-23

Bush GL (1975) Modes of animal speciation. Ann. Rev. Ecol. Syst. 6:334-364

Bush GL and Howard DJ (1986) Allopatric and non-allopatric speciation: Assumptions and evidence. In: Karline and Nevo (eds) Evolutionary processes and theory. Academic Press, New York, pp 411-438

Cairns J, Overbaugh J and Miller S (1988) The origin of mutants. Nature 335:142-145

Callahan R, Chiu IM, Wong JF, Tronick SH, Rue B, Aaronson SA, and Schlon J (1985) A new class of endogenous human retroviral genomes. Science 228:1208-1211

Carson HL (1959) Genetic conditions which promote or retard the formation of species. Cold Spring Harbor Symp. Biol. 24:87-105

Carson HL (1971) Speciation and the founder principle. Univ. Mo. Stadler Genet. Symp. 3:51-70

Carson HL (1982) Speciation as a major reorganization of polygenic balances. In: Barigozzi (ed) Mechanisms of speciation. Allan Liss Press, New York, pp 411-433

Carson HL (1985) Unification of speciation theory in plants and animals. Syst. Biol. 39:678-686

Carson HL and Templeton AR (1984) Genetic revolutions in relation to speciation phenomena: The founding of new populations. Ann. Rev. Ecol. Syst. 15:97-131

Chang D, Wisely B, Huang S and Voelker R (1986) Molecular cloning of suppressor-of-sable, a Drosophila melanogaster transposon-mediated suppressor. Mol. Cel. Biol. 6:1520-1528

Cole M, Ono M, and Huang RC (1981) Terminally redundant sequences in cellular intracisternal A-particle genes. J. Virol. 38:680-687

Csink AK and McDonald JF (1989) Copia expression is variable among natural populations of Drosophila (in preparation)

Curcio MJ, Sanders NJ, and Garfinkel DJ (1988) Transpositional competence and transcription of endogenous Ty elements in Saccharomyces cerevisiae: Implications for regulation of transposition. Mol. Cel. Biol. 8:3571-3581

DeSalle R and Templeton A (1983) Molecular basis of the abnormal abdomen phenotype in Drosophila mercatorum. Genetics 104:s21

Dover GB (1986) The spread and success of non-Darwinian novelties. In: Karlin and Nevo (eds) op cit., pp 199-238

Finnegan DJ (1985) Transposable elements in eukaryotes. Int. Rev. Cytol. 93:281-326

Finnegan D and Fawcett D (1986) Transposable element in Drosophila melanogaster. Oxf. Surv. Eukaryotic Genes 3:1-

Gendelman HE, Phelps W, Feigenbaum L, Ostrove J, Adachi A, Hawley PM, Khoury G, Ginsberg, HS and Martin, MA (1986) Transactivation of the human immunodeficiency virus long terminal repeat sequence by DNA viruses. Proc. Natl. Acad. Sci., USA 83:9759-9763

Goldschmidt R. (1929) Experimentlle Mutationen und das problem der sorgennannten Parallelinduktion: Versuche an Drosophila. Biol. Zbl. 49:437-448

Golubovsky M and Plus N (1982) Mutability studies in two Drosophila melanogaster isogenic stocks, endemic for C Picornavirus and virus-free. Mut. Res. 103:29-32

Gershenson SM (1986) Viruses as environmental mutagenic factors. Mut. Res. 167:203-213

Gotlieb LD (1984) Genetics and morphological evolution in plants. Amer. Natur. 123:681-709

Grant V (1981) Plant speciation. Columbia Univ. Press, New York

Green MM (1988) Mobile DNA elements and spontaneous gene mutation. In: Lambert, McDonald, Weinstein (eds.) Eukaryotic transposable elements, Cold Spring Harbor Press, Cold Spring Harbor, N.Y., pp 41-50

Greene, WC, Dukovich M, and Wano Y, and Siekevitz M (1988) HTLVOI, HIV and human T cell growth. In: Lambert, McDonald and Weinstein (eds.) op cit., pp 309-318

Grossman E and Smith T (1933) Genetic modifications in Drosophila induced by heat irradiation. Am. Nat. 67:429-436

Gvozdev VA, Belyaeva E, Ilyin Y, Amosova I, Kaidanova L (1981) Selection and transposition of mobile dispersed genes in Drosophila melanogaster. Cold Spring Harb. Symp. Quant. Biol. 45:673-685

Hall B (1988) Adaptive evolution that requires multiple spontaneous mutations. I. mutations involving an insertion sequence. Genetics 120:887-897

Herskowitz I (1963) An influence of maternal nutrition upon the gross chromosomal mutation frequency recovered from the X-rayed sperm of Drosophila melanogaster. Genetics 48:703-710

Hull R and Covey S (1986) Genome organization and expression of reverse transcribing elements: variations and a theme. J. Gen. Virol. 67:175-179

Johns M, Mottinger J and Freeling M (1985) A low copy number copia-like transposon in maize. EMBO J. 4:1093-1102

Kubli E (1986) Molecular mechanisms of suppression in Drosophila. Trends Genet. 2:204-208

Kuff EL, Feenstra A, Lueders K, Smith L, Hawley R, Hozumi N, and Shulman M (1983) Intracisternal A-particle genes as movable elements in the mouse genome. Proc. Natl. Acad. Sci., USA 80:1992-1996

Lambert ME, McDonald JF, and Weinstein IB (1988) Eukaryotic transposable elements as mutagenic agents. Cold Spring Harbor Press, Cold Spring Harbor, NY

Leigh Brown AJ (1983) Variation at the 87A heat-shock locus in Drosophila melanogaster. Proc. Natl. Acad. Sci., USA 80:5350-5354

Lints F and Bougois M (1982) A test of the genetic revolution hypothesis of speciation. In: Lakovaara (ed) Advances in genetic, development and evolution of Drosophila. Plenum Press, New York, pp. 423-436

Levin DA (1983) Polyploidy and novelty in flowering plants. Amer. Natur. 122:1-25

Lewis H (1973) The origin of diploid neospecies in Clarkia. Amer Natur. 107:161-170

Lowy DR, Rowe WP, Teich N and Hartley JW (1971) Murine leukemia virus: High frequency activation in vitro by 5-iododeoxyuridine and 5-bromodeoxyuridine. Science 174:155-156

Mayr E (1942) Systematics and the origin of species. Columbia Un. Press, New York

Mayr E (1954) Change of genetic environment and evolution. In: Huxley, Hardy, Ford (eds) Evolution as a process. Macmillian, New York, pp 157-180

Mayr E (1960) The emergence of evolutionary novelties. In: Tax (ed) The evolution of life. The Un. Chicago Press, Chicago, pp 349-380

McDonald JF (1989) Regulatory evolution: Ten years later. Bioscience (in press)

McDonald JF, Strand DJ, Brown MR, Paskewitz SM, Csink AR and Voss SH (1988) Evidence of host-mediated regulation of retroviral element expression at the post-transcriptional level. In: Lambert, McDonald, Weinstein, op. cit.

McDonald JF (1983) Molecular basis of adaptation: A critical review of relevant ideas and observations. Ann. Rev. Ecol. Syst. 14:77-102

McDonald JF, Strand DJ, Lambert ME, and Weinstein IB (1987) The responsive genome: Evidence and evolutionary implications. Allan Liss Press, New York

McEntee K and Bradshaw V (1988) Effects of DNA damage on transcription and transposition of Ty retrotransposon of yeast. In: Lambert, McDonald and Weinstein (eds) op cit., pp 245-254

Mosca JD, Bednarik DP, Ray NB, Rosen CA, Sodroski JF, Haseltine WA and Pitha PM (1987) Herpes simplex virus type-1 can reactivate transcription of latent human immunodeficiency virus. Nature 325:67-70

Mount S, Green M, and Rubin G (1988) Partial revertants of the transposable element-associated suppressible allele white-apricot in Drosophila melanogaster: Structures and responsiveness to genetic modifiers. Genetics 18:221-234

Nevo E (1986) Mechanisms of adaptive speciation at the molecular and organismal levels. In: Karlin and Nevo (eds) op cit., pp. 439-474

Newberne P and Zeiger E (1978) Nutrition, carcinogenesis and mutagenesis. Genetics 27:519-536

Pasyukova E, Cogan G, Ivoleva O, Kaidanova L, and Gvozdev V (1985) Coordinated changes in location of mobile elements in the genome of Drosophila melanogaster reflecting the results of directed selection for quantitative characters (in Russian) Doklad. An SSSR 283:1476-1480

Parkhurst S and Corces VG (1985) Forked, gypsys and suppressors in Drosophila. Cell 41:429-437

Parkhurst S and Corces VG (1986) Interactions among the gypsy transposable element and the yellow and the suppressor-of-hairy-wing loci in D. melanogaster. Mol. Cel. Biol. 6:47-52

Peifer M and Bender W (1986) The anterobithorax mutations of the bithorax complex. EMBO J 5:2293-2303

Philpott D, Weibel J, Atlan H, and Miguel J (1969) Virus-like particles in fat body, oenocytes and central nervous tissue of D. melanogaster imagoes. J. Invertbr. Pathol. 14:31-34

Plough H and Ives P (1935) Induction of mutations by high temperature in Drosophila. Genetics 20:42-69

Pollard J (1987) The movable genome: Weismann's doctrine and new models for speciation. Biol. Forum 80:11-54

Rascati RJ and Tennant RW (1978) Induction of endogenous murine retroviruses by hydroxyurea and related compounds. Virology 87:208-211

Reed CL and Rapp F (1976) Induction of murine p30 by superinfecting herpesviruses. J. Virol. 19:1028-1033

Roeder G and Fink G (1983) Transposable elements in yeast. In: Shapiro (ed) Mobile genetic elements. Academic Press, New York, pp 300-311

Robinson HL, Swanson CA, Hruska JF, and Crittenden LB (1976) Production of unique C-type viruses by chicken cells grown in bromodeoxyuridine. Virology 69:63-74

Rowe WP, Hartley JW, Lander MR, Pugh WE, and Teich N (1971) Noninfectious AKR mouse embryo cell lines in which each cell has the capacity to be activated to produce infectious murine leukemia virus. Virology 46:866-876

Rubin GR (1983) Dispersed repetitive DNAs in Drosophila. In: Shaprio (ed) op cit., pp. 329-332

Rolfe M, Spanos A, and Banks G (1986) Induction of yeast Ty element transcription by ultraviolet light. Nature 319:339-340

Rutledge B, Mortin M, Schwarz E, Thierry-Mieg D, and Meselson M (1988) Genetic interactions of modifier genes and modifiable alleles in Drosophila melanogaster. Genetics 119:391-397

Sankaranarayanan K. (1988) Mobile genetic elements, spontaneous mutations and the assessment of genetic radiation hazards in man. In: Lambert, McDonald and Weinstein (eds) op cit., pp. 319-336

Shiba T. and Saigo K (1983) Retrovirus-like particles containing RNA homologous to the transposable element copia in D. melanogaster. Nature 302:119-124

Soulé M. (1966) Trends in the insular radiation of a lizard. Amer. Natur. 100:47-64

Steele EJ (1979) Somatic selection and adaptive evolution. Un. Chicago Press, Chicago

Strand DJ and McDonald JF (1985) Copia is transcriptionally responsive to environmental stress. Nuc. Ac. Res. 14:4401-4410

Strand DJ and McDonald JF (1989) Insertion of a copia element 5' to the D. melanogaster gene is associated with altered patterns of developmental and tissue-specific expression. Genetics (in press)

Sweet HO (1983) Dilute-suppressor, a new suppressor gene in the house mouse. J. Hered. 74:305-307.

Syvanen M (1984) The evolutionary implications of mobile genetic elements. Ann. Rev. Genet. 18:271-293

Templeton A (1977) Analysis of head shape differences between two interfertile species of Hawaiian Drosophila. Evolution 31:630-641

Templeton A (1979) The unit of selection in Drosophila mercatorum. II. Genetic revolutions and the origin of co-adapted genomes in parthenogenetic strains. Genetics 92:1265-1288

Templeton A (1981) Mechanisms of speciation -- a population genetic approach. Ann. Rev. Ecol. Syst. 12:23-48.

Tennant RW and Rascati RJ (1980) Mechanisms of carcinogenesis involving endogenous retroviruses. In: Slaga (ed) Modifiers of chemical carcinogenesis, Vol 5. Raven Press, New York, pp 185-205

Val FC (1977) Genetic analysis of the morphological differences between two interfertile species of Hawaiian Drosophila. Evolution 31:611-629

Weiss RA, Friis RR, Katz E, and Vogt PK (1971) Induction of avian tumor viruses in normal cells by physical and chemical carcinogenesis. Virology 46:920-938

White MJD (1978) Modes of speciation. Freeman Press, San Francisco

Wilson MC, Policastro PF, and Fredholm M (1988) Regulation of expression and transposition of murine endogenous retroviral elements. In: Lambert, McDonald and Weinstein (eds) op cit., pp. 131-144

Winston F (1988) Genes that affect Ty-mediated gene expression in yeast. In: Lambert, McDonald, Weinstein (eds) op. cit., pp 145-153

Winston F, Dollard C, Malone B, Clare J, Kapakos J, Farabaugh P and Minehart P (1987) Three genes are required for trans-activation of Ty transcription in yeast. Genetics 115:649-656

Youngren SD, Boeke JD, Sanders NJ, and Garfinkel DJ (1988) Functional organization of the retrotransposon Ty from Saccharomyces cerevisiae: Ty protease is required for transposition. Mol. Cel. Biol. 8:1421-1431

Zachar Z, Davison D, Garza D, and Bingham P (1985) A detailed developmental and structural study of the transcriptional effects of insertion of the copia transposon into the white locus of D. melanogaster. Genetics 111:495-515

Zachar Z, Chou T-B, and Bingham P (1987) Evidence that a regulatory gene autoregulates splicing of its transcript. EMBO J. 6:4105-4111

Paradoxes of Molecular Coevolution in the rDNA Multigene Family

J.M. Hancock and G.A. Dover

Department of Genetics, University of Cambridge, Downing Street,
Cambridge CB2 3EH, United Kingdom

The Problem

It is generally assumed that selection underlies the evolution of coadapted gene complexes required for the establishment of coordinated phenotypes. The molecular coevolution between functional regions of the rDNA multigene family and other genetic components, as well as between different regions within the rDNA genes, requires a more elaborate explanation because the evolutionary dynamics of multigene families are subject to a variety of molecular mechanisms of DNA turnover, as well as natural selection. We describe paradoxes of molecular coevolution in the rDNA for which there are no clear-cut solutions as to why and how they became established.

Concerted Evolution and Molecular Drive: Pattern and Process

As more information becomes available on the molecular organization of the genes making up multigene families, more apparent paradoxes arise, in the sense that the accumulation of non-random sequence patterns are not easily explicable either in the context of a neo-Darwinian framework of mutation and natural selection, or in terms of neutral drift. The original paradox presented by ribosomal DNA (rDNA) and other multigene families was: how could individual members of a gene family evolve in concert, that is retain high levels of sequence identity, if they were independently subjected to the processes of mutation and natural selection? The answer to this was to suggest that "cross-talk" occurs between individual members of a gene family by the mechanisms of genomic turnover, in particular gene conversion and unequal crossingover. The realization that these processes act two or more orders of magnitude faster than the rate of point mutation, and that, in sexual species, the homogenization consequences of

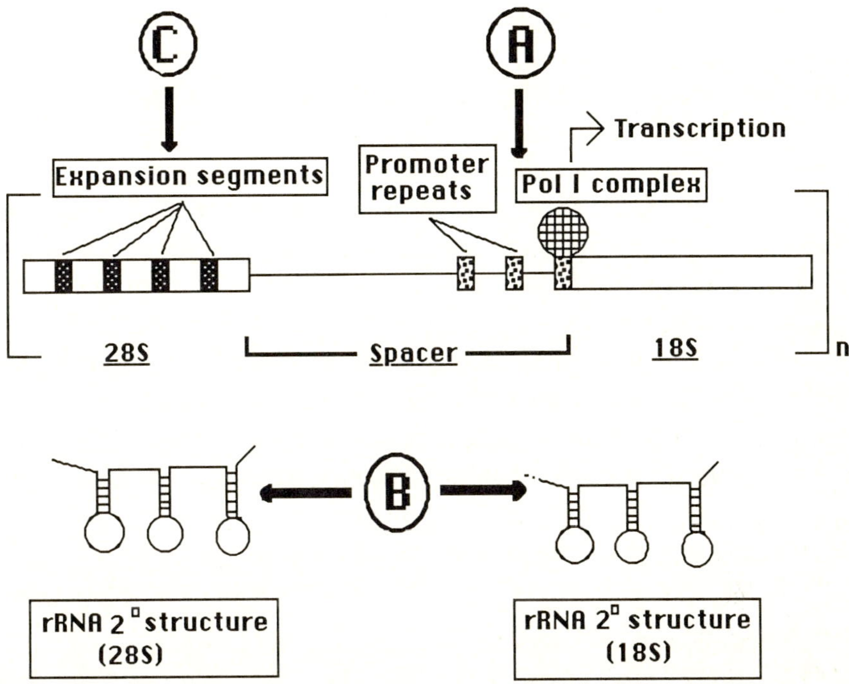

Figure 1. Schematic representation of a typical eukaryotic repetitive rDNA unit consisting of the genes for the 28S rRNA (of the large subunit, LSU) and the 18S rRNA (of the small subunit, SSU), separated by the intergenic spacer, IGS. Most species have several hundred copies of the unit in tandem arrays, often on non-homologous chromosomes. Three instances of molecular coevolution (A,B,C) are depicted. A: The coevolution of promoters and RNA polymerase I cofactors to ensure intermolecular compatibility. Note: promoters may exist as multiple copies within each rDNA unit. B: The coevolution of mutations involved in the maintenance of RNA stem-loop structures. Note: there are approximately 140 stem-loop structures making up the 28S and 18S rRNAs. C: Coevolution of expansion segments in the 28S rRNA gene.

such cross-talk mechanisms within individuals would inevitably and concomitantly spread mutations through a population of individuals, led to the proposal of a "third force" in evolution (molecular drive), operationally distinct from selection and drift, for altering the average genetic composition of a population (Dover, 1982).

The concerted evolution observed within multigene families is particularly well characterized in the rDNA family in a number of species. In Drosophila the rDNA consists of approximately 500 copies of a repeat unit which share very similar gene sequences both within the two rDNA arrays (located on the X and Y chromosomes) and between individuals in a breeding population (Coen *et al* , 1982a,b). The most distinctive differences that can be detected between individual rDNA repeats is in the numbers of copies in an array of subrepeats located within the intergenic spacer (IGS) of each rDNA unit. Stochastic fluctuations in repeat copy-number are the direct consequence of unequal crossingover taking place between individual repeat units, leading to the detection of a number of length variants of the rDNA unit. In contrast to the near identity in sequence within and between rDNA arrays within a given Drosophila species, comparison of the transcription promoter regions and of the promoter duplications contained within each subrepeat of the IGS between Drosophila species show clear differences (Tautz *et al,* 1987). These differences are reflected in the inability of the RNA polymerase I complex of *D.melanogaster* to make use of the promoter of *D.virilis* in a cell free system (Kohorn & Rae, 1982). It has been proposed that this observation reflects a process of "molecular coevolution" between two components of the transcription machinery: the promoter region and a protein factor which binds to it (Dover & Flavell, 1984), (see Fig. 1,a). Species-specific factors associated with RNA polymerase I have been identified in a wide variety of species (see Sollner-Webb & Tower, 1986).

Molecular Coevolution: The Case of Promoters and Polymerases

Dover & Flavell (1984) suggested that the process by which molecular co-evolution occurred in this case was one of selection of alleles of a putative single gene encoding a transcription factor whose products were best able to interact functionally with a gradually changing pool of promoter variants. Because of the stochastic nature of the unequal crossingover mechanism, the relative proportions of promoter variants would change with time, as some variants were lost and others appeared and started to spread through the family and through the population. At any one time, there would be a number of promoter variants existing within a population, and, under idealized conditions of freely cross-talking genes and freely mating individuals, to similar extents within each individual. To preserve the transcriptional function of the gene family under conditions of a gradually varying composition of promoter variants, appropriate alleles of the transcription factor gene, (corresponding to different DNA-binding sites which could recognize different promoter variants) would be subjected to selection. Whichever promoter variant came to predominate in a particular population, a compatible transcription factor, if available, would become established. If no compatible polymerase alleles were available, then no molecular coevolution could take place, and selection would presumably operate against individuals with too many of the new promoters. Thus the process of coevolution is one which preserves the function of a molecular system involving two (or more) components, because the sequences of the individual interacting components can evolve in unison.

In the above example, it is possible to provide a reasonable explanation of the observed interspecific incompatibility between promoters and RNA polymerase I factors based on an interaction between natural selection and molecular drive. However, the rDNA presents other examples of molecular co-evolution which are more difficult to explain, (Fig. 1,b and c). These involve the phenomena of compensatory mutation within the rRNA genes, required to maintain correct secondary structural folding of the rRNAs, and the coevolution of the expansion segments within the rRNA genes.

Molecular Coevolution: The Case of Compensatory Mutation

As more rRNA sequences have become available, it has become plain that a considerable number of point mutations have been accommodated within their genes during evolution, but that nevertheless the corresponding rRNAs in different species have a common secondary structural core. Detailed analysis of sequences and secondary structures show that mutations which occur on one strand of a secondary structural stem are, in general, compensated by a second mutation occurring in the complementary strand, so that the secondary structure of the region is conserved by "canonical" base pairing. This process, which is known as "compensatory mutation", has become a well-established tool for the construction of consensus secondary structure models: pairs of nucleotides which consistently show a pattern of compensatory change are considered to be involved in hydrogen bonding in the ribosome (Gerbi, 1985) (see Fig. 1,b). Given the availability of large numbers of rRNA gene sequences, particularly of the small subunit rRNAs (SS-rRNAs) (see Dams *et al*, 1988; Gutell & Fox, 1988), detailed analyses of compensatory mutations have led to models of rRNA secondary structure which are in very good agreement with experimental data on the secondary structure of rRNAs in the native ribosome (Moazed *et al*, 1986). Indeed the approach has even been applied in an attempt to identify possible tertiary interactions (Gutell *et al*, 1986).

From an evolutionary and populational point of view compensatory mutation poses a significant conceptual problem. If one simply traces the sequences of individual rRNAs through time, the origins of the process appear to be self explanatory - a first mutation occurs which has to be compensated by a second to retain function. In the real world of populations and multigene families this explanation is inadequate. First, if a mutation occurs within one unit of the multigene family, and is then compensated by a second mutation, why does this new combination of mutations spread through the rDNA to become the predominant form in the species? Furthermore, when does this compensatory mutation arise? Does the appearance of a first mutation predispose the gene to the appearance of a second mutation, or is the time which elapses before the occurrence of the second mutation predictable purely on the basis of rates of random point mutation? Secondly, at what level might selection act to fix a unit with a pair of such mutations? Does a single point mutation within a single rRNA molecule result in a lesion in a stem large enough to produce a significantly less fit phenotype? Why would a single change in one of several hundred genes, all presumably contributing to the final pool of nascent rRNAs, be subject to selection? In the buffered systems of multigene families, the homogenization and fixation consequences of turnover need to be considered.

The determination of the sequences of the genes encoding the *D.melanogaster* rRNAs gave us the opportunity to consider these questions in more detail (Hancock *et al*, 1988, Tautz *et al*, 1988). The *D.melanogaster* rRNA genes are far more AT-rich than any others thus far sequenced, raising the question how can such an extreme change in base composition be accommodated within a conserved secondary structure.

It is clear from secondary structure modelling that the *D.melanogaster* rRNAs conform to general models of the secondary structure of eukaryotic rRNAs (Hancock *et al*, 1988). It has also been possible, from available sequences of the 5' end of 18S rRNA genes of three other *Drosophila* species and of the Tsetse fly *Glossina morsitans morsitans* (Tautz *et al*, 1987; Cross & Dover, 1987), to follow the evolution of a single case of compensatory mutation in stem E9-1 of the 18S rRNA. In this case there is a G-C base pair in *G.m.morsitans* and *D.virilis*, a G-U pair in *D.melanogaster*, and an A-U pair in *D.orena*. As *D.melanogaster* and *D.virilis* shared a common ancestor 35-65 million years ago (MYA) (Beverley & Wilson, 1984), while *D.melanogaster* and *D.orena* shared a common ancestor c. 5-15 MYA (Ashburner *et al*, 1984), it appears that a first mutation occurred in the lineage leading to *D.melanogaster* and *D.orena* between 65 and 15 MYA, whilst a second, compensating mutation has occurred in the *D.orena* lineage within the last 15 million years or so. This suggests that pairs of compensatory mutations may not arise at the same time as each other, and that there may be a significant lag phase before the appearance of a second mutation. This proposal is supported by the frequent observation of G-U base pairs in conserved secondary structural stems, and even of completely non-complementary pairs of bases in otherwise apparently conserved stems which are well supported by experimental evidence (see for example Mougel *et al*, 1988).

The frequent occurrence of non-compensatory mutations in rRNA sequences gives us some insight into the mechanisms involved in the spread of compensatory mutations. It must be possible for a rDNA unit with a single, uncompensated mutation to spread through the entire rDNA gene family as a consequence of repeated unequal crossovers. An interesting corollary of the spread of such mutations, however, is that the total number of units bearing the uncompensated mutation present in the population as a whole increases greatly as the mutant rDNA unit spreads. The variant unit therefore becomes a larger "target" for further mutations, increasing by orders of magnitude the probability of the compensatory mutation occurring in one of the variant units. For example, if a single unit is homogenized so that it makes up 100 units in every fly, and is present in every individual of a population of 10,000, the probability of it acquiring a compensatory mutation (and, of course, any other mutation) is increased by six orders of magnitude. Thus the probability of a compensating mutation occurring will increase as the first mutation spreads through the gene family.

However, the above argument begs an important question: at what stage does selection start to act? There is certainly strong evidence that point mutations at certain highly conserved sites in the rRNAs can have a major effect (see for example Rottmann *et al,* 1988). The sites involved in compensatory mutations are essential only to the extent that they maintain secondary structure. Furthermore, there is considerable evidence that secondary structural stems can remain stable even in the presence of mismatches. The significance of this is that a species can tolerate a number of mismatches in its rRNA at any one time. This provides the necessary flexibility in the system which allows the species to wait for compensatory mutations. Presumably, as a given rRNA gene accumulates mutations and as that gene spreads in the family, there will be a point at which the rRNA for which it codes contains an unstable stem or other defect which will make it inactive, and subject the individual to negative selection. The logical consequence of this argument is that we might expect the majority of rRNA genes in any individual to encode rRNAs which are close to the limit of viability, that is harbouring mutations that are below the number required to affect rRNA functions. Compensatory mutation could then be seen as an almost necessary process of repair which ensures that individuals contain rRNA genes which produce active rRNAs.

Does this mean that the generation of compensatory mutations is an active, directed process rather than a passive one? If so it would have to involve an entire new class of mechanisms. One possibility would involve the screening of rRNA transcripts for mismatches combined with reverse transcription. Other mechanisms which could be envisaged might involve proof-reading of rRNA genes against active rRNAs, or repair of DNA secondary structure. In the first case, genes encoding defective rRNAs would need to be identified, their sequences compared with those of non-defective rRNAs (or the genes encoding them) and repaired in a biased manner so that their function was restored. If potentially active and inactive genes could be identified (as seems to be the case with 28S rRNA genes containing the type I insertion in Drosophila, which are not transcribed *in vivo* (Long & Dawid, 1979)), direct conversion of inactive by active genes could be envisaged. In the second case, under conditions of torsional stress that might extrude secondary structure in DNA, the rRNA genes would fold into rRNA-like structures on both strands, in which mismatches within stems could be repaired. Indeed, in the light of recent evidence that RNA can be "edited" in the absence of a DNA template (reviewed by Eisen, 1988), such sequence correction might even take place at the RNA level and feed back to the rRNA genes.

The above picture becomes further complicated when considering the establishment of compensatory mutation in the expansion segments of the genes, which are themselves coevolving.

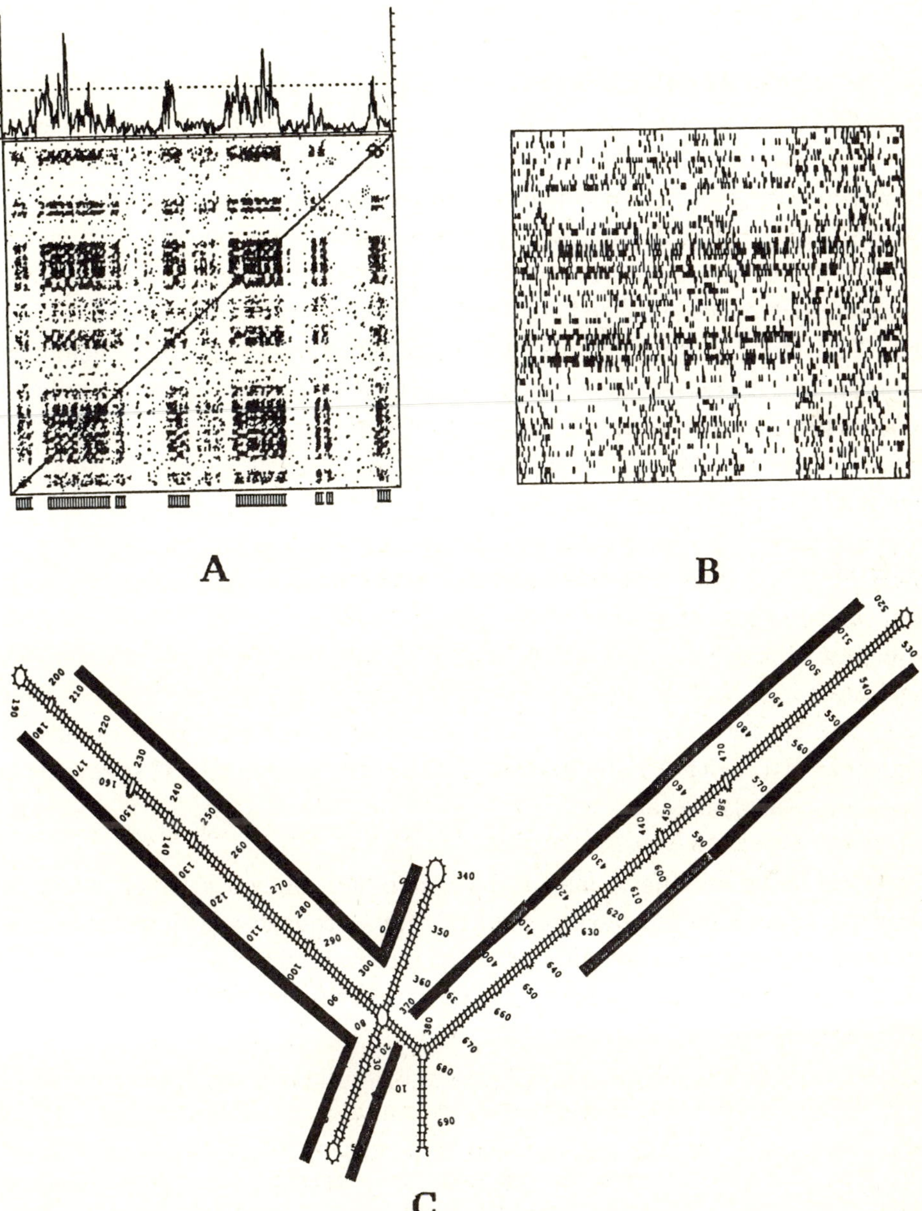

Figure 2. A: DIAGON plot and simplicity profile of a human 28S rRNA gene. Expansion segments are indicated by hatched bars (bottom). The stringency of match is 19 of 35 bp. A simplicity profiles (reflecting regions of DNA with high and low density of short repetitive motifs) is aligned with the sequence in the horizontal direction (top). The horizontal dotted line superimposed upon the simplicity profile represents the highest simplicity score of the randomization runs (see Tautz *et al*, 1988; Hancock & Dover, 1988 for details).(continued)

Molecular Coevolution: The Case of Expansion Segments.

A third level at which molecular coevolution appears to be taking place in the rRNA genes involves regions of the rRNA which have no known counterpart in bacterial rRNAs: the variable regions or expansion segments (see Fig. 1,c). When eukaryotic rRNAs are aligned against their prokaryotic equivalents (for example *D.melanogaster* 18S rRNA against *E.coli* 16S rRNA), it immediately becomes obvious that there is a considerable length discrepancy. For example human 28S rRNA is 5025 nucleotides long, while *E.coli* 23S rRNA is only 2904 nucleotides long. The differences in lengths can be accounted for almost entirely by additional stretches of sequence at specific sites in the eukaryotic rRNA genes: seven in the SS-rRNAs and 13 or 14 in the LS-rRNAs. These regions, which are variously known as "expansion segments" (Gerbi, 1985; Hancock & Dover, 1988), or variable regions (Gorski *et al*, 1987), vary greatly in length between species, although they are virtually constant between individual genes in a single species (Gonzalez *et al*, 1985; 1988; Maden *et al*, 1987). In the following discussion we will refer to these structures as "expansion segments" in the sense that they represent expansions of the rRNA molecules with respect to the 16S and 23S rRNAs of *E.coli*. This does not imply that they arose as insertions, as will become clear later. A detailed investigation of the properties of expansion segments in LS-rRNAs of eukaryotes (Hancock & Dover, 1988) reveals a number of remarkable characteristics.

The first characteristic is that in the vertebrates, and to a lesser extent in Drosophila, dot matrix comparisons of LS-rRNA sequences against themselves show a considerable amount of internal repetition within the sequences (Fig. 2a). This sequence repetition manifests itself on two levels, both of which involve

---

Figure 2 (cont'd). Expansion segments are internally repetitive and similar in sequence to each other, in contrast to the core segments.

B: The distribution of all 64 triplets within the human 28S rRNA gene, starting from the top line TTT and ending with the bottom line AAA. Each vertical bar is a triplet. Expansion segments share an overabundance of GC-rich motifs (CCC, CCG, GCC, CGC, CGG, GCG, GGC, GGG), whereas the core segments generally contain a more random distribution of most triplets.

C: Secondary structure of expansion segment D8 of human 28S rRNA, depicting mutually exclusive blocks of slippage-generated motifs. CCC-rich regions (black bars) base pair with regions rich in GGG and related trinucleotide motifs (grey bars). Compensatory slippage (see text) ensures, like compensatory point mutations, the conservation of secondary structure despite sequence divergence.

only the expansion segments. First, the expansion segments are internally repetitious structures at the sequence level, in contrast to the conserved, core regions between them. Secondly, and more remarkably, all of the expansion segments of the LS-rRNA of a particular vertebrate species share large numbers of sequence motifs which result in similarity between all points in one expansion segment and all points in the other. The set of expansion segments within the 28S rRNA genes of a species are coevolving.

The second characteristic is that expansion segments deviate very considerably in base composition from the rest of the rRNA sequence. For example, the rRNAs of mammals are fairly GC-rich (66-69% G+C, see Hancock & Dover, 1988), but their expansion segments are consistently more GC-rich than the rest of the molecule (72-79% G+C), while the reverse is true of *D.melanogaster* rRNAs, which are AT-rich and have expansion segments which are even more AT-rich.

The third characteristic of expansion segments becomes clear when they are analysed using a computer program developed in this laboratory which searches DNA sequences for a property known as cryptic simplicity, that is scrambled permutations of short repetitive sequence motifs (Tautz *et al*, 1986). This analysis involves passing a 64 bp window along the DNA sequence under analysis and scoring for repetitions of tri- and tetranucleotide repeats within the window to give a value for the "simplicity" of the sequence contained in the window for each position along the sequence. A score is generated which represents the overall simplicity of the whole sequence along which the window has passed, which is then compared to a score generated using ten or a hundred randomized sequences, each 10 kilobases long, of the same base composition as the natural sequences. Sequences which have a simplicity value that is statistically significantly higher than those of the random sequences are considered to be simple sequences. This simplicity will be high if the sequence contains either tandem repeats of a single motif (pure simplicity), or different short direct repeats scrambled one with another (cryptic simplicity). When this program is applied to rDNA sequences which contain expansion segments, the sequences register as simple sequences (reflected by the simplicity "profile" depicted in Fig. 2a, - see legend for details), suggesting a greater representation of certain motifs than would be expected from the sequence base composition. Only the expansion segment sequences are simple however; the sequences representing the conserved cores of the rRNAs are not. Furthermore, the longer the LS-rRNA sequence, or the longer the expansion segment sequence, the higher the simplicity of the sequence.

Taken together, these results suggest that the changes in length that are observed in expansion segments reflect the accumulation of sequences within them which themselves result in elevated levels of simplicity. All such changes are closely coordinated between expansion segments. It is likely that sequences which show high levels of simplicity are generated by slippage replication or similar mechanisms (so-called "slippage-like mechanisms", Tautz *et al* , 1986). This suggests that the changes in length of the expansion segments also result from the actions of such mechanisms in the rRNA genes as the copy-numbers of particular motifs increase or decrease.

Compensatory Slippage

How do the slippage-generated simple sequences of expansion segments relate to the previous discussion of compensatory mutation, given the evidence that there is a similar secondary structure for any given expansion segment within kingdoms of organisms (Michot & Bachellerie, 1987; Hancock *et al,* 1988)? Recent evidence suggests that the accumulation of slippage-generated products within expansion segments does not occur at random, but that there is considerable constraint upon both the locations and types of motifs that can be accommodated. Analysis of the distribution of triplet motifs within vertebrate expansion segments shows a non-random distribution of sequence motifs, with blocks of complementary motifs alternating along the primary sequence (Fig. 2b) (Hancock & Dover, manuscript in preparation). These blocks of motifs correspond to complementary strands of secondary structural stems, and to regions of growth of these stems (as noted by Michot & Bachellerie, 1987), suggesting that slippage-generated motifs can only accumulate in such a way as to preserve a compact secondary structure (Fig. 2c). This in turn can only occur if accumulation of slippage generated products is gradual, and occurs at sites in the DNA corresponding to both strands of the rRNA secondary structure at more or less the same time. There is now evidence that the rRNA genes of primates show a degree of intraspecific variability which is concentrated in their expansion segments (Gonzalez *et al,* 1985; 1988; Maden *et al*, 1987). This variability manifests itself as variability in copies of short repetitive motifs. We can thus envisage a model of the evolution of new elements of secondary structure by the gradual accumulation of additional copies of already repeated motifs by slippage replication, followed by their gradual spread through the population, to produce the observed intraspecific variability at these sites. Compensation for such motifs would also be produced by slippage at complementary repetitive motifs: a process of "compensatory slippage" analogous to the process of compensatory mutation.

A complete and satisfactory explanation of the evolution of expansion segments is still not forthcoming. Why are the motifs generated by slippage-like mechanisms similar in all expansion segments of a particular species (and indeed between the expansion segments of species of vertebrates and those of rice 25S rRNA; Hancock & Dover, 1988)? Does this shared motif composition reflect some shared functional requirement? Although expansion segments were originally considered to be functionless structures which were tolerated only because they did not disrupt ribosome function (Gerbi, 1985), there is now evidence that their absence from LS-rRNAs can affect the function of the ribosome (R.J. Planta, personal communication). However, the functions of individual expansion segments do not explain their sequence-relatedness unless they either interact with each other or alternatively unless all expansion segments interact with some other ligand. Although expansion segments tend to be located on the surface of the ribosome, there is no evidence for any such interactions.

A second plausible explanation for the evolution of expansion segments is purely mechanistic. It has been pointed out that slippage is a type of feed-back mechanism (Dover & Tautz, 1986), which will tend to amplify particular pre-existing repetitive sequence motifs once they start to accumulate. This is to be expected because if slippage is an homology-dependent process, then motif variants will tend to be excluded. Perhaps the sequence relatedness of expansion segments reflects the action of slippage-like mechanisms on similar original motifs within each segment. Biases in the occurrence of the acceptance of mutations are probably responsible for changes in the overall base composition of core regions (Hancock *et al*, 1988), and could in turn generate particular sequence motifs upon which slippage would focus in the expansion segments. An alternative possibility is that slippage within and between rDNA repeat units occurs over longer distances than has been previously considered. If one expansion segment could slip with another which had a similar sequence composition to start with, all the expansion segments in a unit would evolve in a concerted manner, with the processes of molecular drive ensuring that the new variants were homogenized throughout the rDNA array and spread through the population. Finally, mechanisms other than slippage might also be involved in the process. For example, microconversion processes acting between and within expansion segments could generate sequence similarity across the set of segments.

Although it is evident from the above discussion that expansion segments in vertebrates have undergone a process of expansion by the accumulation of slippage-generated motifs, the changes in size of expansion segments are restricted both in the taxa involved (non-chordate animal taxa show no patterns of expansion segment cross-similarity and little sequence simplicity of their LS-rRNAs) and in the particular expansion segments involved (many expansion segments remain small even in human 28S rRNA, which is composed of 52% expansion segment, while the SS-rRNA expansion segments vary much less in size than the LS-rRNA expansion segments) (see Hancock & Dover, 1988). Whilst the latter probably reflects different levels of constraint in different part of the ribosome, the former may reflect a general reduction in constraint in chordate, and particularly vertebrate, taxa. Thus it cannot be assumed that expansion segments are derived solely by slippage, and that the absence of expansion segments represents the ancestral state. Indeed, there are compelling arguments to suggest that the presence of expansion segments is a remnant of the early assembly of long rRNA genes from once separated sub-components or minigenes (Boer & Gray, 1988; Dover, 1988). From the theoretical viewpoint, early organisms would have been incapable of the faithful replication and transcription necessary to maintain active rRNA genes 2 kb or more in length, and would more likely have assembled their rRNAs from smaller units until their replication and transcription machinery became more efficient (Clark, 1987). Recent observations of bizarre organizations of rRNA genes in organelle genomes of several species show them to be highly fragmented in many cases (for example see Boer & Gray, 1988), which is strongly indicative of such a fragmentary origin for rRNA genes (for a short review see also Dover, 1988).

Summary

There are a number of levels at which molecular coevolution takes place within the rDNA multigene family. The extent to which these systems have been characterized theoretically and experimentally vary considerably, from promoter-cofactor coevolution (for which cell-free systems now exist in a number of laboratories), to the expansion segments, about the function of which little is still currently known. In addition, the evolution of these systems may involve some mechanisms, such as RNA/DNA feedback and RNA proofreading (see above), which are as yet still experimentally uncharacterized.

REFERENCES

Ashburner M, Bodmer M & Lemeunier F (1984) On the evolutionary relationship of *D.melanogaster*. Dev Genet 4: 295-312

Beverley SM & Wilson AC (1984) Molecular evolution in Drosophila and the higher Diptera.II. A time scale for fly evolution. J Mol Evol 21: 1-13

Boer PH & Gray MW (1988) Scrambled ribosomal RNA gene pieces in *Chlamydomonas rheinhardtii* mitochondrial DNA. Cell 55: 399-411

Clark CG (1987) On the evolution of ribosomal RNA. J Mol. Evol. 25: 343-350.

Coen ES, Thoday JM & Dover GA (1982) Rate of turnover of structural variants in the rDNA gene family of *Drosophila melanogaster*. Nature 295: 564-568

Coen ES, Strachan T & Dover GA (1982) Dynamics of concerted evolution of ribosomal DNA and histone gene families in the melanogaster species group of Drosophila. J Mol. Biol. 158: 17-35

Cross NCP & Dover GA (1987) Tsetse fly rDNA: an analysis of structure and sequence. Nuc. Acids Res. 15: 15-30

Dams E, Hendriks L, Van de Peer Y, Neefs J-M, Smits G, Vandenbempt I & De Wachter R (1988) Compilation of small ribosomal subunit RNA sequences. Nuc Acids Res 16: r87-r173

Dover GA (1982) Molecular drive: a cohesive mode of species evolution. Nature 299: 111-117

Dover GA & Flavell RB (1984) Molecular coevolution: DNA divergence and the maintenance of function. Cell 38: 622-623

Dover GA & Tautz D (1986) Conservation and divergence in multigene families: alternatives to selection and drift. Phil Trans R Soc Lond B 312: 275-289

Dover GA (1988) rDNA world falling to pieces. Nature 336: 623-624

Eisen H (1988) RNA editing: who's on first? Cell 53: 331-332

Gerbi SA (1985) Evolution of ribosomal DNA. in "Molecular Evolutionary Genetics" (MacIntyre RJ, ed) (Plenum, New York) pp 419-517.

Gonzalez IL, Gorski JL, Campen TJ, Dorney DJ, Erickson JM, Sylvester JE & Schmickel RD (1985) Variation among human 28S ribosomal RNA genes. Proc Natl Acad Sci USA 82: 7666-7670

Gonzalez IL, Sylvester JE & Schmickel RD (1988) Human 28S ribosomal RNA sequence heterogeneity. Nuc Acids Res 16: 10213-10224

Gorski JL, Gonzalez IL & Schmickel RD (1987) The secondary structure of human 28S rRNA: the structure and evolution of a mosaic rRNA gene. J Mol Evol 24: 236-251

Gutell RR & Fox GE (1988) A compilation of large subunit RNA sequences presented in a structural format. Nuc Acids Res 16: r175-r269

Gutell RR, Noller HF & Woese CR (1986) Higher order structure in ribosomal RNA. EMBO J 5: 1111-1113

Hancock JM & Dover GA (1988) Molecular coevolution among cryptically simple expansion segments of eukaryotic 26S/28S rRNAs. Mol Biol Evol 5: 377-391

Hancock JM, Tautz D & Dover GA (1988) Evolution of the secondary structures and compensatory mutations of the ribosomal RNAs of *Drosophila melanogaster*. Mol Biol Evol 5: 393-414

Kohorn BD & Rae PMM (1982) Accurate transcription of truncated ribosomal DNA templates in a Drosophila cell-free system. Proc Natl Acad Sci USA 79: 1501-1505

Long EO & Dawid IB (1979) Expression of ribosomal DNA insertions in *Drosophila melanogaster*. Cell 18: 1185-1196

Maden BEH, Dent CL, Farrell TE, Garde J, McCallum FS & Wakeman JA

(1987) Clones of human ribosomal DNA containing the complete 18S-rRNA and 28S-rRNA genes. Biochem J 246: 519-527

Michot B & Bachellerie J-P (1987) Comparisons of large subunit rRNAs reveal some eukaryote-specific elements of secondary structure. Biochimie 69: 11-23

Moazed D, Stern S & Noller HF (1986) Rapid chemical probing of conformation in 16S ribosomal RNA and 30S ribosomal subunits using primer extension. J Mol Biol 187: 399-416

Mougel M, Philippe C, Ebel J-P, Ehresmann B & Ehresmann C (1988) The *E.coli* 16S rRNA binding site of ribosomal protein S15: higher-order structure in the absence and in the presence of the protein. Nuc Acids Res 16: 2825-2839

Rottmann N, Kleuvers B, Atmadja J & Wagner R (1988) Mutants with base changes at the 3' end of the 16S RNA from Escherichia coli. Construction, expression and functional analysis. Eur J Biochem 177: 81-90

Sollner-Webb B & Tower J (1986) Transcription of cloned eukaryotic ribosomal RNA genes. Ann Rev Biochem 55: 801-830

Tautz D, Trick M & Dover GA (1986) Cryptic simplicity in DNA is a major source of genetic variation. Nature 322: 652-656

Tautz D, Tautz C, Webb D & Dover GA (1987) Evolutionary divergence of promoters and spacers in the rDNA family of four Drosophila species. J Mol Biol 195: 525-542

Tautz D, Hancock JM, Webb DA, Tautz C & Dover GA (1988) Complete sequences of the rRNA genes of *Drosophila melanogaster*. Mol Biol Evol 5: 366-376

Two Ways of Speciation

N.N. Vorontsov and E.A. Lyapunova

N.K. Koltzov Institute of Developmental Biology, USSR Academy of Sciences, Moscow 117334, USSR

INTRODUCTION

Speciation is a multifactorial process. Causal factors, forming species in space and time, are different. Natural selection and neutralism, genetic drift and isolation, geographical and chromosomal speciation and allopatric and sympatric ways are not exclusive or alternative explanations of evolutionary processes. All these factors, as well as forms and ways of speciation have a distribution in various groups and in various stages of evolution.

It was thought until recently that chromosomal mutations play only a minor role in the process of speciation (Mayr, 1963). Evolution is described as a process of change in allelic frequencies in many textbooks. It was presumed that speciation takes place as a consequence of geographical isolation, and the divergence of isolated populations is explained on the basis of minor mutations or on the basis of the change of allelic frequencies. A system of isolating mechanisms was thought to develop as a barrier only during the final stage of speciation.

In contrast to Mayr's conception of geographical speciation by changes of allelic frequencies in separate populations, the first author of this publication proposed the conception of two pathways of speciation (Vorontsov, 1960):

1) The "traditional" or "ordinary" pathway, which is associated with the gradual accumulation of interpopulational differences and culminates in the development of reproductive isolation;

2) The "genetic" pathway, which starts with the development of reproductive isolation as a consequence of chromosomal rearrangements and culminates in the divergence of allelic frequencies and ecological differences, as well. Chromosomal speciation is a speciation without changes in genetic information -without changes in DNA- but with principal changes in probability of gene flow between forms with similar gene pools but with different chromosomal complements.

Various groups of animals and plants are characterized by the predominance of different speciation mechanisms. Some groups such as frogs among Amphibia, jerboas (Dipodidae), among rodents and most Pinnipedia, for example, are characterized by an amazing uniformity of their chromosome complements. It is obvious that divergence and maintenance of reproductive barriers in these groups is ensured by different, non-chromosomal mechanisms. Frogs, for example, have great differences in sexual songs, jerboas in structure of genitalia, etc. Behavioral, ecological and seasonal isolation plays a special role in the realization of reproductive isolation in many amphibian, bird, pinniped and other groups.

Not all types of chromosomal rearrangements interrupt gene flow between populations with different karyotypes. Differences in the heterochromatic regions of chromosomal sets, such as deletions or duplications of heterochromatic arms of chromosomes, have small isolating effects. Other groups of chromosomal mutations, such as inversions and translocations of euchromatic parts of chromosomes, have great reproductive influence. As a result of such reproductive separation, reproductively isolated populations with different karyotypes and primary identical gene pools initiate diversification not only in chromosomal characters, but also in allelic frequencies.

It is clear now, that chromosomal speciation is very important, but not the only way of evolution. There has been an intensive discussion in the last two decades between gradualists and punctualists. This discussion was stimulated by a publication of Eldredge and Gould (1972). The problem was the subject of a special symposium in Dijon (Chaline, 1983). To resolve the problem by either punctualism or gradualism is an oversimplification. In our earlier (Timoffeeff-Ressovsky, Vorontsov and Yablokov, 1969) and recent publications (Vorontsov, 1980, 1987, 1988) it was emphasized that "gradual" ("common") and "explosive" ("punctuated", "sudden") speciation are not mutually exclusive, both forms being observed in evolution. Here, we will present processes of subspeciation by the action of natural selection under isolation as evidence of gradual speciation, and data on sudden speciation based on the chromosomal mechanisms of speciation.

GRADUAL SPECIATION

We cannot present the gradual way of speciation without the demonstration of clinal patterns in intraspecific variation and without the study of step cline zones.

Colour Polymorphism of Lady Beetle *Harmonia (=Leis) axyridis* **Pall.** The remarkable intraspecific polymorphism of elytral colour, peculiar to many species of *Coccinellidae*, long attracted the attention of population geneticists and zoologists. Population genetics of Asian species of *Harmonia* (*Leis* in recent taxonomy of *Coccinellidae*) *axyridis* lady beetles was a subject of the classical works of Dobzhansky (1924, 1937) and other authors (Tan,1946; Komai, 1956; Vorontsov, 1983, 1987; Vorontsov and Blekhman, 1986). Polymorphism of the elytral colour of *H. axyridis* is the result of a combination of genotypic and phenotypic variation and the process of mosaic dominance. Elytral colour is controlled by alleles of one gene. As a result of the process of mosaic dominance, colour alleles look like a multiple allelomorphic series. Figure 1 depicts the results of phenogeographical and genogeographical analyses. (Vorontsov, 1983, 1987; Vorontsov and Blekhman, 1986). In most parts of the species range allelic frequencies change clinally. But between the Korean peninsula and Japan in the Tsusima Strait, we can see a step cline zone. Interruption of the gradual pattern of clinal variability in this area is the result of the separation of the Japanese islands from the Asian continent nearly 18,000 years ago. Because in this part of its range *H. axyridis* can yield two generations per year, we can evaluate the biological time of origin of the large allelic differences (0.25 versus 0.85) in 36,000 generations. The same separation time between various islands of Japan cannot disprove the selective character of the gradual change of allelic frequencies from south to north. This may be explained by an intensive gene flow across such water barriers as the Songar Strait (between Hondo and Hokkaido: ca. 18 km) in contrast with Tsusima Strait (200 km).

Genetic and Morphological Variability of Differently Aged Isolates of Two Species of Ground Squirrels: *Citellus undulatus* **and** *C.parryi.* These two very interesting related species were studied in a series of mutual Soviet-American investigations (Hoffmann et al., 1974; Nadler et al., 1974; Lya-

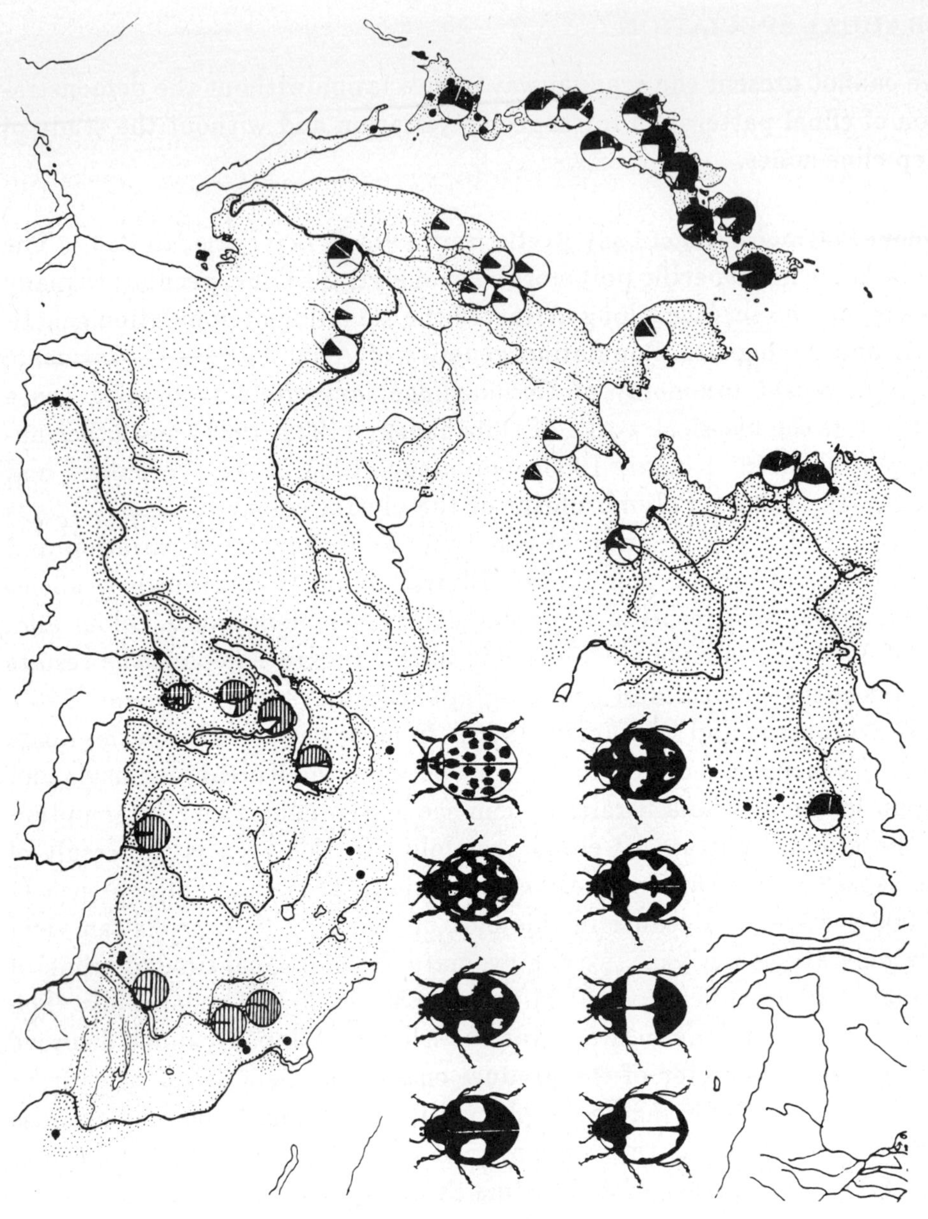

punova and Vorontsov, 1970; Vorontsov and Lyapunova, 1970; Vorontsov et al., 1978, 1980, 1987, etc.). *C. undulatus* occupies central Siberia ranging from Dzhungarian Ala-Tau and Altay in the west to Jakutien and the Amur Basin in the east. *C. parryi* is distributed in the Amphiberingian area; this Arctic ground squirrel inhabits northeast Siberia, Alaska and northern Canada. Vorontsov and Lyapunova (1969) suggested that *C. undulatus* had occupied Siberia from the Eopleistocene, while *C. parryi* was the last Late Pleistocene (Wurm, Wisconsin) migrant from America to Asia. This view is confirmed by the higher level of the intraspecific variation of *C.undulatus* when compared to Siberian populations of *C.parryi.*

Isolation of *C. undulatus* and *C. parryi* from their common ancestor may have taken place nearly 10^6 generations (=years) ago. During this period, these species have not retained a single common allele of the transferrin locus. The age of the Siberian ground squirrel *C. undulatus* may be estimated as 0.6-1.0 x 10^6 generations. *C. parryi* cannot have migrated from America to Asia before the beginning of the Wisconsin Glaciation, i.e., 0.4-0.5 x 10^5 generations ago. The comparison of morphological variability between *C. undulatus* and Siberian populations of *C. parryi* shows a fundamental difference in the levels of the subspecific diversity of these two species: some populations and subspecies of *C. undulatus* (*C.u.jacutensis, C.u. menzbieri*) have a high level of morphological differentiation as a result of their long evolution and subspeciation in Asia. In contrast, most parts of Siberian subspecies of *C. parryi* do not show real morphological differences (Vorontsov et al., 1980). This does not apply to the isolated subspecies *C.p. janensis* from the Yana river and the semiisolated subspecies *C.p. steinegeri* from the Kamtchatka

---

Figure 1: Phenogeography of colour polymorphism in *Harmonia axyridis* Pall lady-beetles. (Coccinellidae) (Vorontzov and Blekhman, unpublished data). Black sector: *conspicua*; crossed lines: *spectabilis*; horizontal lines: *axyridis*; white: *succinea*. Insert: Variants of elytral colour in lady-beetles *Harmonia axyridis* Pall. First row: Common morphotypes (from left to right): *conspicua, spectabilis, axyridis, succinea* (series of multiple alleles). Numbers of black points in *succinea* (from 0 to 19) are determined by nongenetical factors. Second row: Rare alleles. The frequency of rare alleles indicates the level of air pollution. Near a large industrial cellulose factory at the Lake Baikal, the frequency of two rare alleles increases up to 70 times.

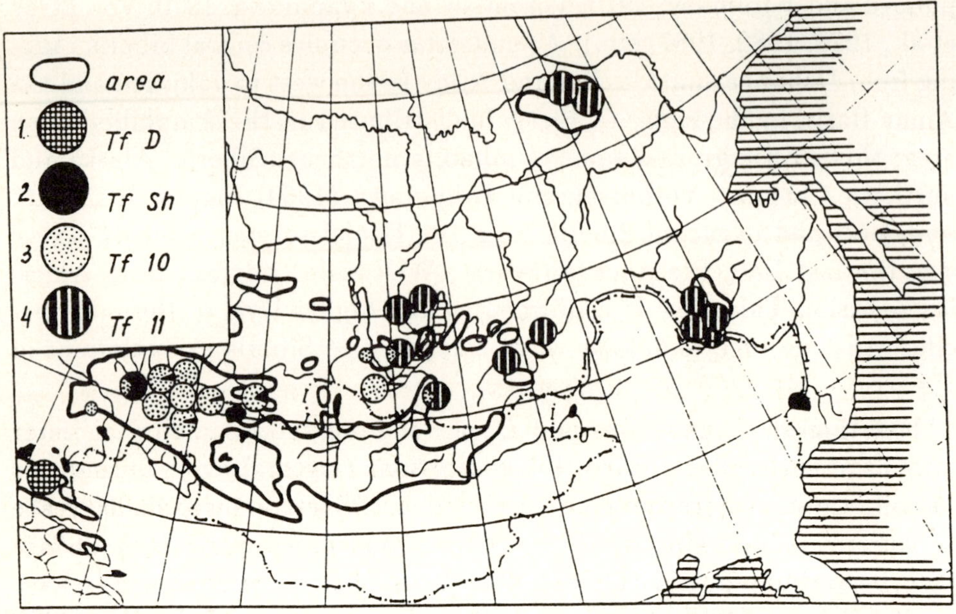

Figure 2: Map of distribution of the Siberian long-tailed ground squirrel *Citellus (=Spermophilus) undulatus* Pall. (Sciuridae) and transferrin allelic frequencies (Vorontsov et al., 1980). This species is known in Siberia from Eopleistocene. All peripheral isolates have transferrin monomorphism in contrast to populations from the central part of the area (see Fig. 3).

peninsula.

Studies of transferrin (Tf) genogeography in *C. undulatus* demonstrated that peripheral isolates are homogeneous in tranferrin types (Fig. 2). Transferrin polymorphism is typical only of the central part of the area of *C. undulatus*. (Fig. 3). Thus, all the peripheral isolates of *C. undulatus* are characterized by Tf monomorphism and a high degree of morphological divergence.

The genogeography of Tf in North American populations of *C. parryi* (Fig. 4) is essentially different from that of northeast Siberia (for details see Nadler et al., 1974; Hoffmann et al., 1974; Vorontsov et al., 1987). The agreement of these results concerning low variability of *C. parryi* in N.E. Asia with the pattern of clinal changes in Tf frequencies is noteworthy. The subspecific separation of American populations of *C. parryi* agrees well with the pattern of sharp interruption of clines according to Tf frequencies in a number of regions of N. America. Genogeographic data indicate a more complex history of American *C. parryi* as compared to Asian populations, which matches well with the complex Late Pleistocene history of the landscapes of N. America.

C. parryi populations of the Beringian region (including St. Lawrence Island), despite isolation by the Bering Strait for ca. 12,000 generations, have maintained their genetic commonality. It is interesting to note that in the period after the melting of the Canadian ice sheet, during which no fewer than 8,000 generations of *C. parryi* had elapsed, the genetic separation of southern and northern populations of *C.p. plesius* living on different sides of the ice sheet did not disappear. The territorial separation of southern and northern populations existed for a maximum of 33,000 generations (the start of the Wisconsin Glaciation was 45,000 years ago, the end of the melting of the ice cover was 12,000 years ago). Differences originating during this period could not have been leveled off in the last 12,000 years. What caused the absence of differences in populations of the Beringian region separated by 12,000 years and the preservation during this period of time of differences formed during no more than 33,000 generations? It can be suggested that populations forced southward by the Canadian ice sheet and isolates of the Alaskan refuge living in periglacial zone were in substantially different ecological circumstances and experienced the differential influence of directed selective factors.

The sharp differences in subspecies of *C. parryi* in North America, according to frequencies of transferrins, contrast with the clinal character of variability of frequencies of Tf in *C. parryi* subspecies of northeastern Asia, but are reminiscent of the pattern of distribution of Tf allelic frequencies in populations of the Siberian species *C. undulatus*, which lived there through the entire Pleistocene epoch (one million generations). We are inclined to see the evidence of a gradual pattern of evolution and of the antiquity of habitation of *C. undulatus* in Asia and of *C. parryi* in America in these facts.

Possibility of Gradual Dispersion of Some Types of Chromosomal Mutations: Cases of *Myospalax* and *Microtus arvalis*. We understand now that not all types of chromosomal mutations have reproductive consequence. Such

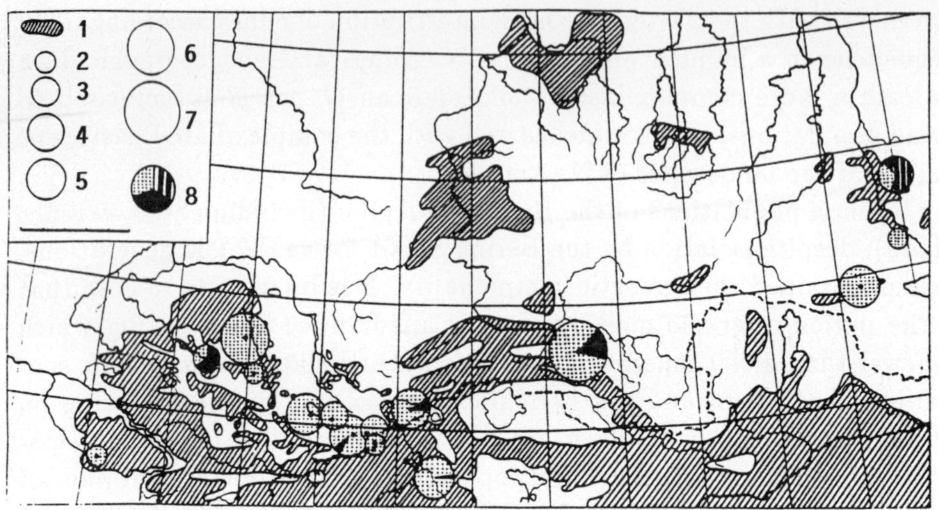

Figure 3: Complicated semiisolated pattern of distribution of *Citellus undulatus* in the central part of the area and genogeography of transferrin alleles (Vorontsov et al., 1978). (1) area, (2-7) sizes of samples: - (2) 1-5 specimens; (3) - 6-10; (4) 11-20; (5) 21-40; (6) 41-76; (7) more than 100. Black sector: Tf-Sh, dotted: Tf-10; vertical lines: Tf-11.

chromosomal mutations as duplications and deletions of heterochromatic chromosomal arms played a very small role in reproductive isolation.

Broad chromosomal variability was discovered in Asian burrowing rodents *Myospalax myospalax* (Lyapunova et al., 1974; Vorontsov and Martynova, 1976). *Myospalax* (Myospalacinae, Cricetidae), endemical burrowing animals of mountain steppes of Asia, have a similar ecology to West Palaeartic mole rats from the Spalacidae family. Intrapopulational and interpopulational chromosomal polymorphism involve 9 chromosomal pairs. Frequency of heterozygosity in most studied populations is great. Chromosomal polymorphism in *Myospalax* is connected with duplications and deletions of heterochromatic arms. Chromosomal pairs without heterochromatic arms are constant. We have no evidence that heterozygotic animals have reduced fertility. Perhaps the chromosomal polymorphism of *Myospalax myospalax* has an adaptive value, as it has the chromosomal polymorphism in populations of black rat *Rattus rattus* (2n=42), from Japan.

The common vole, (*Microtus arvalis* Pall., 2n=46), is a well studied

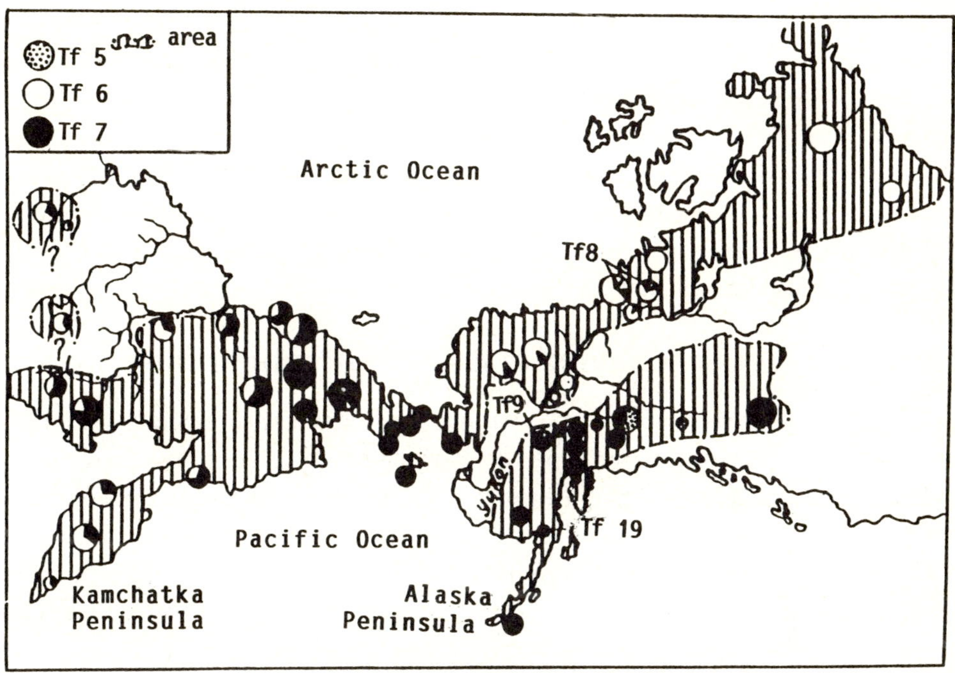

Figure 4: Map of distribution of Arctic long tailed ground squirrel *Citellus (=Spermophilus) parryi* Richardson (Sciuridae) and transferrin allelic frequencies (Vorontzov et al., 1980). This species probably originated in N. America and migrated across the Bering land bridge into Siberia at the beginning of the last (Wisconsin) Glaciation. Note the similarity of Tf-7 frequencies in both sides of Bering Strait. In the American range of distribution five alleles of Tf are found (Tf-5, Tf-6, Tf-7, Tf-9, Tf-19); in the Asian one only two (Tf-6 and Tf-7). This is the result of a bottleneck effect of the dispersion of gene flow across the Bering land bridge.

west-palaearctic species from the cytogenetical point of view. Polymorphism in the fifth chromosomal pair was described as a rare case in populations from Spain (Gamperl, 1982) and from the Volga Basin (Vorontsov et al., 1984). This pair shows a submetacentric morphology in which acrocentrics are rare. One hundred common voles from the Ukraine all have typical Sm/Sm structure of this pair (Zagorodnyuk and Teslenko, 1986). In the Armenian Transcaucasian mountains chromosomal polymorphism is common and the frequency of the acrocentric variant is greater than 0.1. The ratio of homo- and heterozygotic animals corresponds to the Hardy-Weinberger rule (Lyapunova et al., 1988a).

As noted above, we must separate two types of chromosomal polymorphisms: (1) one that leads to the origin of cytogenetic differences and reproductive isolation between populations, subspecies, sibling-species and species *in statu nascendi*; (2) balanced chromosomal polymorphism as a form of intraspecific diversity and evolution without the origin of reproductive isolation (Lyapunova, 1983). The second type of chromosomal polymorphism based on duplications/deletions of heterochromatin has the same evolutionary fate as classical gene mutations.

CHROMOSOMAL SPECIATION AS A FORM OF SUDDEN SPECIATION

Most parts of chromosomal mutations, such as translocations and inversions of euchromatic parts of the karyotype, interrupt the gene flow between individuals and populations with different chromosomal structure. As a secondary result of such reproductive separation, genetically and reproductively isolated populations with different karyotypes and with identical gene pools have diversification not only in chromosomal morphology, but also in allelic frequencies.

White (1978, p.8) noted that "structural rearrangements of chromosomes, such as inversions and translocations, have played a special and perhaps a primary role in the origin of many or even most species". Chromosomal speciation plays an exceptionally important role in the evolution of many mammalian groups, particularly the rodents (except perhaps the gerboas and rare kayologically stable genera such as *Clethrionomis*, Microtinae), the primates and the perissodactyls, which are characterized by exceptionally higher evolutionary rates. Speciation rates in mammals are directly correlated with the tempo of chromosomal rearrangements (Wilson et al., 1975). In plant speciation the most important role in punctuated ways of diversification is attributed to various forms of polyploidization (autopolyploidy, allopolyploidy, polyploid complexes, etc.) (Grant, 1971; Stebbins, 1982). The importance of other types of chromosomal rearrangements in speciation of plants is not clear, because the methods of analysis by differential staining of chromosomes are not as developed in plants as in cytogenetically well studied groups such as mammals or Diptera.

Robertsonian translocations represent a very important basis for chromosomal speciation. In the last twenty years, cases of broad Robertsonian variability - "Robertsonian fans" - of karyotypes have been described in different mammalian taxa. The first Robertsonian fan was discovered in African pigmy mice of the genus *Leggada* (Matthey, 1970). Similar Robertsonian fans were described for the domestic mouse *Mus domesticus* (Gropp et al., 1972; Capanna, 1982), for the subterranean microtine rodent *Ellobius talpinus* superspecies (Lyapunova et al., 1974, 1980, 1985; Vorontsov, 1980), for the insectivore *Sorex araneus* (Fredga, Nawrin, 1977) and for the semisubterranean microtine rodent *Pitymys daghestanicus* superspecies (Lyapunova et al., 1988b).

Chromosomal Evolution of the Genus Ellobius. Mole voles ("slyepushonka") of the genus *Ellobius* are unique material for cytogenetic studies. The broadest chromosomal intrageneric variability was described for this genus: *E. lutescens* has 2n=17, *E. fuscocapillus* 2n=36 and *E. talpinus* superspecies has a typical diploid number of 54 (Lyapunova and Vorontsov, 1978). The broadest variability of sex chromosomal mechanisms is also typical for this genus (Vorontsov et al., 1980b): *E. fuscocapillus* has a typical XX-XX system; *E. talpinus* superspecies has isomorphic sex chromosomes with identical differential staining of sex elements in males and females, *E. lutescens* has an odd chromosomal number and XO-XO system in both sexes (Vorontsov, 1973; Lyapunova and Vorontsov, 1978; Vorontsov et al., 1980b). *E. lutescens* has 50% of zygotic mortality as a result of the elimination of XX and OO sex-chromosomal constitutions (Lyapunova et al., 1975). X-chromosomes of *E. fuscocapillus* and *E. lutescens* are homologous. Differential staining demonstrates the existence of homologous regions in most parts of the chromosomes of *E. talpinus*, *E. fuscocapillus* and *E. lutescens*.

***Ellobius talpinus* Superspecies: a Karyotypic Stability in Most Parts of its Distribution Area and the Broadest Variability in the Seismic Active Region.** A very interesting example of chromosomal speciation is provided by the *E. talpinus* superspecies, which has been studied in our laboratory. Over most of its large area of distribution from the Ukraine in the west to Mongolia in the east, the karyotype of this superspecies is relatively stable: 2n=54, NF=54 in the western part of its range (*E. talpinus* s. str.), and 2n=54, NF=56 in eastern part of its range (*E. tancrei*). This small

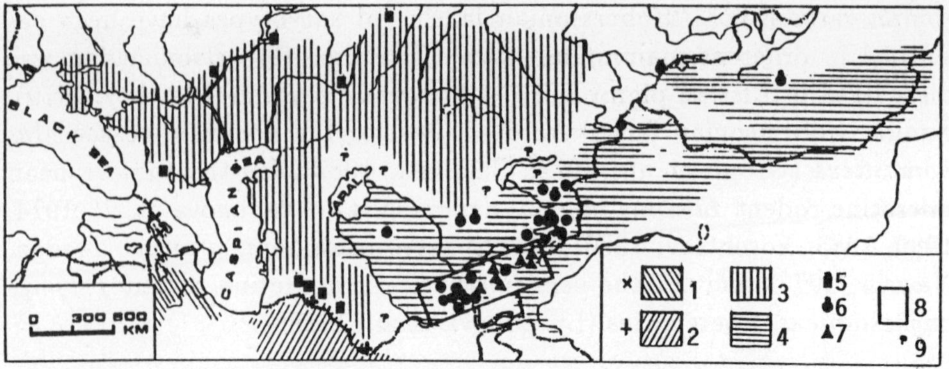

Figure 5: Geographical distribution of the genus *Ellobius* (Lyapunova et al. 1978, with additions). Symbols indicate localities from where chromosomal data were examined. (1) - *E. lutescens* (2n=17); (2) *E. fuscocapillus* (2n=36); (3-7) *E. talpinus* superspecies, (3) range of *E. talpinus* s.str.; (4) range of *E. tancrei* group: (5) *E. talpinus* s.str. (2n=NF=54); (6) *E. tancrei* s.str. (2n=54, NF=56); (7) *E. alaicus* semispecies (2n=52, NF=56); (8) Pamiro Alay zone, where a Robertsonian fan was discovered 2n=54, 53, 52, 34, 33, 32, 31 in *E. tancrei*, but NF=56 is constant.

level of chromosomal variability in the *E. talpinus* superspecies in most parts of its distribution area contrasts with a large karyotype variability in one valley of the Pamiro-Alay mountains (valley of Surkhob-Vakhsh rivers in Tadzhikistan). In the Surkhob-Vakhsh valley (Fig. 5), we discovered a full Robertsonian fan of karyomorphs: twenty four karyomorphs of the *E. tancrei* taxonomical group (all karyomorphs have the same number of chromosomal arms, NF=56) distributed in one valley with all possible chromosomal numbers between 54, 53, 52, 51... and 34, 33, 32 and 31 (Lyapunova et al., 1980, 1984). (Fig. 6).

These karyomorphs are indistinguishable in gross morphology and neither can be distinguished on the basis of electrophoretic studies (Lyapunova et al., 1980). Nevertheless, reproductive isolation, i.e., decrease in the fertility of hybrids between karyomorphs with different chromosome numbers, exists as a result of chromosomal differences (Lyapunova and Yakimenko, 1985). But, as was demonstrated in experiments with hybridization between different karyomorphs, reproductive barriers are not absolute.

We have studied the synaptonemal complex (SC) of F_1 hybrids of *E.*

tancrei karyomorphs with 2n=54 and 2n=34, hybrids are heterozygous for ten Robertsonian translocations and have 2n=44 (Kolomiets et al., 1986; Bogdanov et al., 1986). During meiosis in these hybrids the appearance in meiocytes of 7 SC-bivalents and 10 SC-trivalents should be expected. Electron microscopic studies of pachytene spermatocytes completely confirmed this expectation in relation to SC-bivalents, but the picture turned out to be more complicated in relation to SC-trivalents. In each spermatocyte at the zygotene stage several SC-trivalents (from two to five) combine into a chain by forming an SC between monosomic arms of acrocentrical autosomes belonging to adjacent trivalents (Kolomiets et al., 1986). Partial arrest of meiosis at pachytene as a result of the formation of an SC-chain is a cytogenetical basis of reproductive isolation of different karyomorphs (Fig. 7).

Why is the distribution of chromosomal mutations in the species range not homogenous? Why did we discover only two karyomorphs of the *E. talpinus* superspecies in most parts of its distribution and 24 karyomorphs in only one valley of the Pamiro-Alay mountains? Why have we not such a broad chromosomal variability in other valleys of the Pamiro-Alay and Tian-Shan mountains? We know only of one distinguishing peculiarity of the Surkhob valley: it contains the longest tectonical boundary fault of the whole region. The zone where the Robertsonian fan was found wholly overlaps with the 500-1000 year-old seismical zone of earthquakes of intensity 9 or more. The Vakhsh-Surkhob valley is where the Indostan plate of Gondwana submerges under the Paleoasian plate. Such plate mobility is connected with tectonical and seismic activity.

Problems of the correlation between the seismicity level and degree of chromosomal variation were discussed earlier (Vorontsov and Lyapunova, 1984). Seismicity correlated with concentration of various mutagenic factors (X-rays, radon water, salts of heavy metals, etc) in tectonically active regions. If our idea about the correlation between seismicity and genetical variation is correct we can predict a possibly high level of mutability and genetic variation in tectonic regions not only in *Ellobius* but in other organisms. The study of genetic variability in barley (*Hordeum bulbosum*) from the Pamiro-Alay mountains revealed the highest genetic variability in populations from the Surkhob valley, "which may be connected with the seismic activity of this region" (Fedorenko et al., 1988).

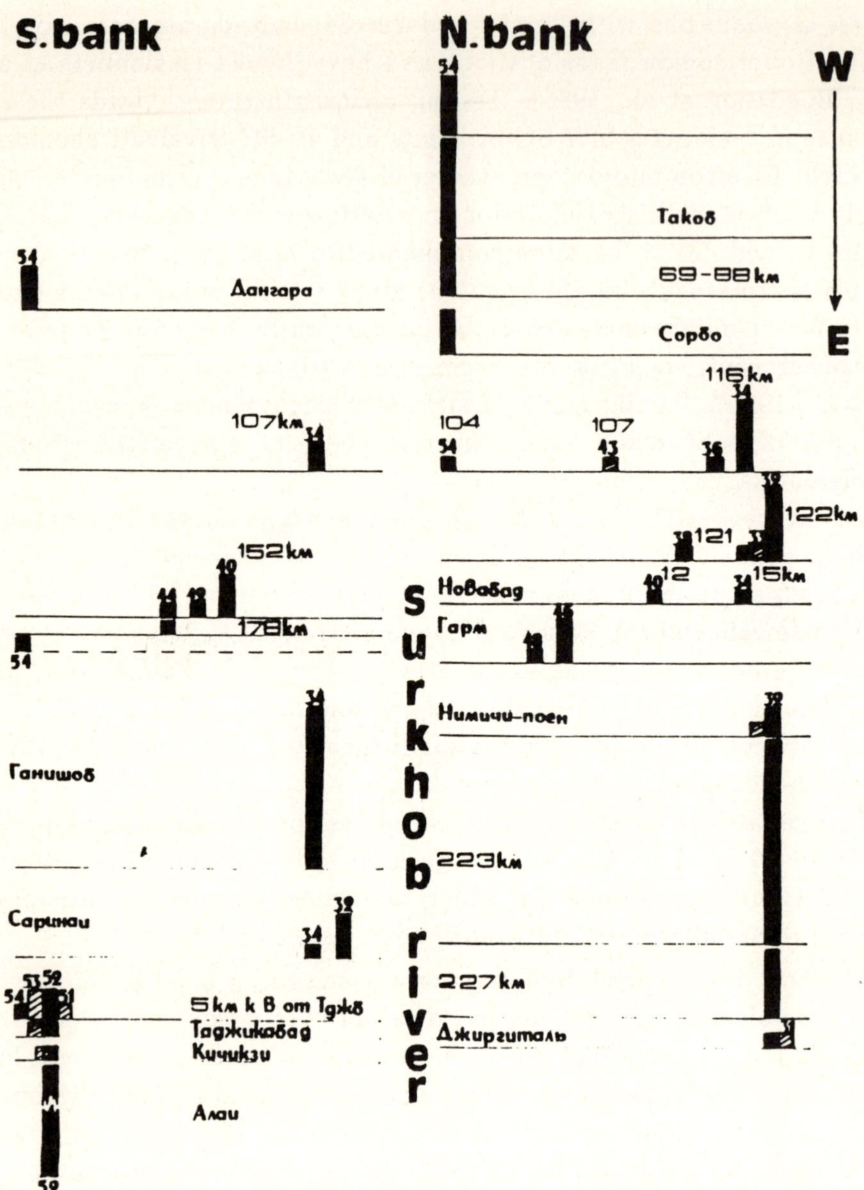

Figure 6: Distribution of different chromosomal Robertsonian karyotypes in some parts of the Surkhob valley. Some forms with intermediate 2n have a hybridogenous origin.

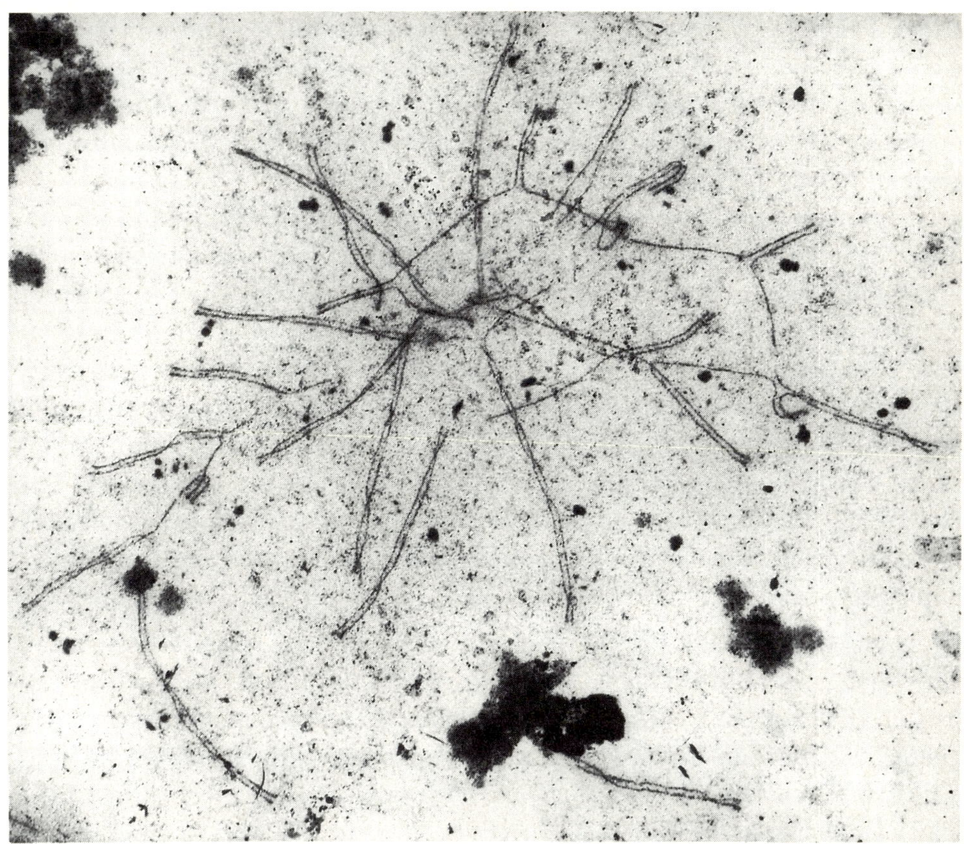

Figure 7: Disturbance of meiosis in F_1 hybrids between two Robertsonian karyomorphs of *Ellobius tancrei* group with 2n=54 and 2n=34. Electron microscope photo by O.L. Kolomiets. Fragment of pachytene plate with 7 SC-bivalents and complicated SC-chain with partial conjugation in heterochromatic parts of chromosomes, are shown.

We also studied chromosomes of the *Ellobius tancrei* group from six localities of Tian-Shan and Easren Kazakhstan. In two localities of seismically active regions, specimens with 2n=53 heterozygous for Robertsonian translocations were discovered. G-staining showed nonhomology of the Robertsonian metacentrics in animals from different populations, which indicates an independent origin of the 53-chromosome karyotype at different points of the area (Lyapunova et al., 1985) and confirms the hypothesis of a relationship between chromosomal variability and seismic activity put forth on the basis of investigating the wide Robertsonian polymorphism of mole vole populations in the high-seismic zone of Pamiro-Alay.

Other Evidence of Broad Chromosomal Variability of Mammals in Seismic Active Regions. In our special publication about explosive chromosomal speciation in seismic active regions (Vorontsov and Lyapunova, 1984) we demonstrated a correlation in many cases between broad chromosomal variability and high levels of seismicity.

It is interesting that karyotypes of closely related species have a different sensibility to mutagenic factors. In *Mus musculus* superspecies, *M. musculus* s. str., *M. hortulanus*, *M. abbotti*, *M. spretus* have a constant karyotype with 2n=40. In contrast, four semispecies of *M. domesticus* produced many Robertsonian translocations in many territories. Robertsonian translocations have been reported in the Orkney Islands, Scotland, Bavaria, Switzerland, Italy, Sicily, Spain, Yugoslavia, Greece and two localities from India (Capanna, 1980, 1982; Gropp et al., 1982; Larson et al., 1984). We have enough data to compare the karyotype stability or instability with the level of seismic activity only from the Apennines, Alps, Sicily and Dalmatia (Fig. 8).

A similar situation of differences in chromosomal variability between various closely related taxons was discovered in our laboratory with Caucasian semifossorial voles from the genus *Pitymys* (Lyapunova et al., 1988b). Earlier, we noticed an increased correlation in chromosomal variability of *P. subterraneus* populations from seismic active territories of the Carpathians and the Balkans. In the Caucasus, *P. majori* superspecies diverged into two taxonomical groups: *P. majori* s.str., 2n=54, NF=60, with a constant karyotype in all studied localities; and *P. daghestanicus*, with NF=58 and with different diploid numbers (2n=54, 53, 52, 46, 42, 40, 38). The enforcing of geographical isolation in the Great Caucasus is more intense than in the not so higher mountains of the Transcaucasus, but most parts of the karyomorphs of *P. daghestanicus* were discovered in the seismic active Transcaucasus, which contrasts with the Great Caucasus, that shows a much lower level of tectonic activity.

In mole rats (*Spalacidae*) - fossorial rodents with restricted mobility - we discovered two chromosomal variants for five species of the genus *Spalax* s.str. only in the large territories of southeastern Europe, (Lyapunova et al., 1974), but in the genus *Microspalax* the chromosomal variation is large in some parts of this area. In the nonseismic parts of the range of the *M. ehrenbergi* superspecies, large geographical barriers such as the

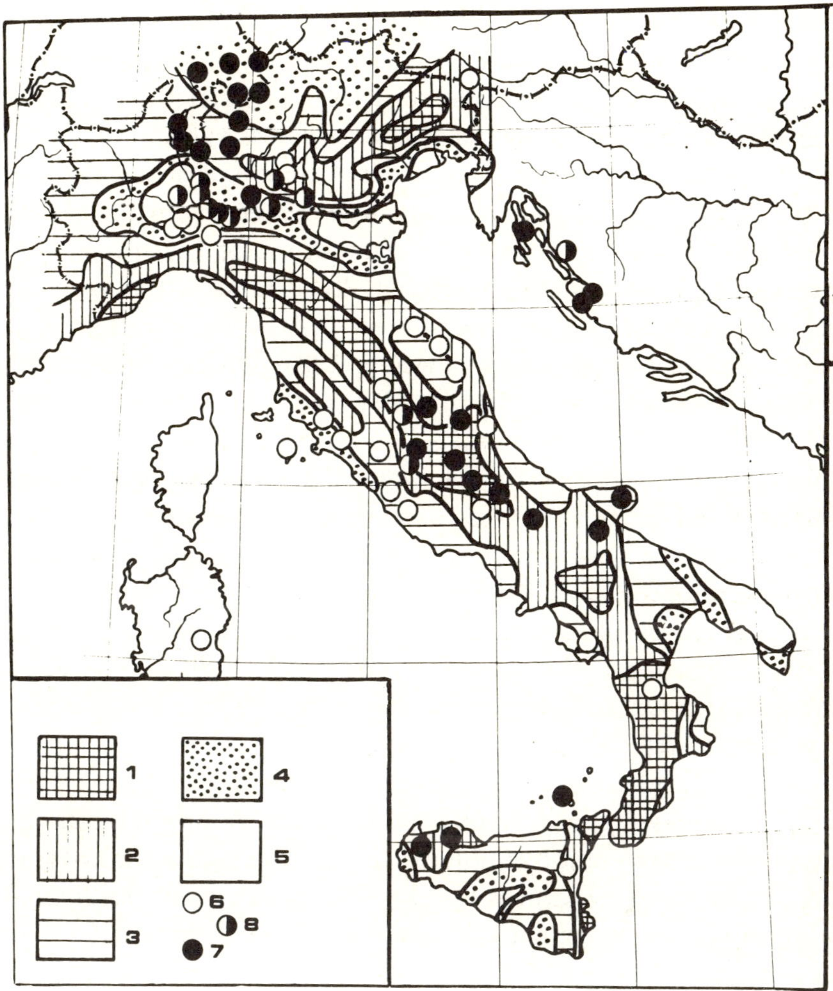

Figure 8: Seismic zones of the Apennine peninsula and the Alps and distribution of normal and Robertsonian karyomorphs of the *Mus domesticus* taxonomical group (localities according to: Capanna, 1980; Gropp et al., 1982). From Vorontsov and Lyapunova, 1984. (1) seismical intensity *9*; (2) intensity *8*; (3) intensity *7*; (4) intensity *6*; (5) intensity *5*; (6) standard karyomorph, 2n=40; (7) Robertsonian karyomorphs, 2n=22-30; (8) hybrids between standard and Robertsonian karyomorphs.

Nile River don't separate different karyomorphs. An outbreak of chromosomal speciation in mole rats of the *M. ehrenbergi* superspecies has been noted in Israel and in the adjacent territories of Palestine (Wahrman et al., 1969). Here the zone of intensive chromosomal variability is confined to the zone of rift running from the Great African Lakes and the Red Sea across the Dead Sea, Jordan River and Lake Tiberiad. In the seismically active Balkans, a wide chromosomal variation within the species complex *Microspalax leucodon* has been observed. Ten karyomorphs were discovered in Yugoslavia (Savic and Soldatovic, 1979), ten other karyomorphs were also obtained in Bulgaria (Peshev, 1981; Peshev and Vorontsov, 1982) and a large chromosomal diversity was also discovered in seismically active territories of Rumania, Greece and Turkey.

American pocket gophers (genus *Thomomys*) utilize in the Nearctic region the same ecological niches as *Ellobius* in the Palaearctic. *Thomomys* also has a broad chromosomal variability as *Ellobius*. Thaeler (1968, 1983), who studied pocket gophers of the *Thomomys talpoides* complex, found wide chromosomal variability in these fossorial rodents in two seismically active regions: Montana, Wyoming and Utah, on the one hand, and Washington, Oregon and California, on the other, while the numbers of karyomorphs are significantly lower in Idaho and Nevada, which also coincide with a region with lower levels of seismicity .

We also discussed the intriguing problem of the origin of the human karyotype (Vorontsov and Lyapunova, 1984). The high seismical region of the Great African Rift is a possible place for the origin of the *Homo* karyotype. Here, it seems proper to remember that our karyotype (2n=46) differs from those of the gorilla and two species of chimpanzee (2n=48 in all three African species), man's next of kin, by a limited number of chromosomal rearrangements (Seuañes, 1979). We have very solid grounds in believing that 2n=48 was the ancestral diploid number in this phylogenetic line. In the extremely complex and multifactorial process of antropogenesis the chromosomal mutations ensuring reproductive isolation of man's ancestors must not have played the least role whatsoever. Recent findings of early hominid fossils suggest that this chromosomal aberration from 2n=48 to 2n=46 occurred in the seismically active region of the Great African Rift (Vorontsov and Lyapunova, 1984).

CONCLUSIONS

Only 30 years ago evolutionary biologists believed that chromosomal mutations were very rare events. Now, we understand the broad distribution of the chromosomal variability not only in higher plants, but in natural populations of many groups of animals. Of course, seismicity is not the single exogenous factors of karyotype variation. Viral infections play a great role in chromosomal mutations (Vorontsov, 1973). The importance of chromosomal variability in speciation was demonstrated in a brilliant book by White (1978).

A very common objection against the chromosomal way of speciation is: How can a male find a female with the same chromosomal mutation? If the new mutation originated in a primordial germinative cell, the same chromosomal mutation is received by half of the gametes - ancestors of this cell. If the germinative cell of an animal with 2n=40 has a Robertsonial translocation (2n=39), half of the gametes from this cell will receive n=19 and the other half a normal haploid number, (n=20). The joining of a mutant gamete (n=19) with a normal gamete (n=20) would produce many heterozygous animals with a diploid karyotype of 2n=39. In the first meiotic division the level of fertility of such animals with chromosomal heterozygosity decreases as a result of partial nondisjunction and the production of gametes with an unbalanced karyotype (1M+1A; 1A+0; balanced gametes must have 1M or 2A), but they can produce offspring with the binomial proportion of karyotypes:

$$1\ (2n=40) : 2\ (2n=39) : 1(2n=38).$$

One fourth of the progeny (F_2 generation) will receive a new karyologically balanced karyotype (2n=38) with partial reproductive isolation from the ancestral 2n=40 chromosomal form. These two chromosomal species *in statu nascendi* can have the same gene pools at this moment of karyological separation.

Now we understand that the isolation mechanisms of evolution are not absolute. We have evidence that a very limited gene flow can penetrate across the barriers of isolation mechanisms of evolution. We discovered interspecific gene flow between *Mus musculus* and *M. domesticus* in the Transcaucasus. Many soviet zoologists (V. Kapitonov, A.A. Nikolsky, D.I. Bibikov and N.A. Formosov) have demonstrated a zone of natural hybridization between two species of marmots in Central Kazakhstan, *Marmota bobak* and *M. baibacina* and also a hybrid zone between *M. baibacina*, and *M. sibirica* in Altay and W. Mongolia. We have evidence of natural hybridization between two absolute real species of ground squirrels as *Citellus suslicus* (2n=34) and *C. pygmaeus* (2n=36). Nadler discovered the hybrid zone between the "muflon" sheep *Ovis orientalis* (2n=54) and the urial sheep *O. vignei* (2n=58). In a narrow hybrid zone wild sheep with 2n=54, 55, 56, 57, 58, were collected. This zone was predicted on the basis of classical taxonomic analysis of this group by Nasonov (1922). We have evidence that a very limited gene flow can penetrate across the interspecific barriers of isolation mechanisms.

The new molecular data concerning the possibility of horizontal gene transfer among taxa by viral factors switches the understanding of the speciation processes over to another level.

The most important genetic result of the different forms of isolation is disruption of panmixia between populations. Coluzzi (1982) used some illuminating terms for various forms of speciation by separation of gene flow: (1) speciation by passive isolation, and (2) speciation by active isolation. The concept of geographical speciation involves the first form of isolation. Now we have much evidence that a second way of speciation can be started from the origin of isolation by chromosomal and other "sudden" forms of speciation.

REFERENCES

Bogdanov YuF, Kolomiets OL, Lyapunova EA, Yanina IYu, Mazurova TF (1986) Synaptonemal complex and chromosome chains in the rodent *Ellobius talpinus*, heterozygous for ten Robertsonian translocations. Chromosoma 94(2): 94-102

Capanna E (1980) Chromosome rearrangement and speciation process in *Mus musculus*. Folia Zoologica (Brno) 29: 43-57

Capanna E (1982) Robertsonian numerical variation in animal speciation: *Mus musculus* an emblematic model. In: Barigozzi C (ed) Mechanisms of Speciation Alan R Liss New York, pp 155-177

Chaline J (ed) (1983) Modalités, Rythmes, Méchanismes de l'Évolution Biologique Gradualisme Phylétique ou Équilibres Ponctués? CNRS Paris, pp 3-337

Coluzzi M (1982) Spatial distribution of chromosomal inversions and speciation in Anopheline mosquitoes. In: Barigozzi C (ed) Mechanisms of Speciation Alan R Liss New York, pp 143-154

Dobzhansky Th (1924) Die geographische und individuelle Variabilitat von *Harmonia axyridis*. Biologisches Zentralblatt 44: 401-421

Dobzhansky Th (1937) Genetics and the Origin of Species. Columbia Univ Press New York

Eldredge N, Gould S (1972) Punctuated equilibria, an alternative to phyletic gradualisme. In: Schopf T (ed) Models in Paleobiology Freeman San Francisco, pp 82-115

Fedorenko OM, Schevchenko VA, Mitin AN (1988) Genetical variability of mountain populations of *Hordeum bulbosum* L on the four esterase loci. Genetika (Moscow) 24(1): 110-117

Fredga K, Nawrin J (1977) Karyotype variability in *Sorex araneus* L. Chromosomes Today v.6. George Allen & Unwin London, pp 153-161

Gamperl R (1982) Die Chromosomen von *Microtus arvalis* (Rodentia, Microtinae). Zeitschrift fur Saugetierkunde 47(6): 356-363

Grant V (1971) Plant Speciation. Columbia Univ Press New York

Gropp A, Winking H, Zech L, Muller HJ (1972) Robertsonian chromosomal variation and identification of metacentric chromosomes in feral mice. Chromosoma 39: 265-288

Gropp A, Winking H, Redi C, Capanna E, Britten-Davidian J, Noak G (1982) Robertsonian karyotype variation in wild house mice from Rhaeto-Lombardia. Cytogenetics and Cell Genetics 34: 67-77

Hoffmann RS, Nadler Ch, Lyapunova EA, Vorontsov NN (1974) Evolutionary relationships of Holarctic ground squirrels (tribe *Marmotini*). In: Kratochvil J, Obrtel R (eds) Symposium Theriologicum II Proc Intern Sympos on Species and Zoogeography of European Mammals Academia Praha, pp 11-18

Kolomiets OL, Lyapunova EA, Mazurova TF, Yanina IYu, Bogdanov YuF (1986) Participation of heterochromatin in formation of synaptonemal complex chains in animals heterozygous for multiple Robertsonian translocations. Genetika (Moscow) 22(2): 237-280 Russian with English summary English translation: Soviet Genetics Consultants Bureau NY Plenum Publ Corp (ISSN 0038-5409)

Komai T (1956) Genetics of Ladybeetles. In: Demerec M (ed) Advances in Genetics v. VIII Academic Press New York, pp 155-188

Larson A, Prager EM, Wilson AC (1984) Chromosomal evolution and morphological change in vertebrates: the role of social behavior. Chromosomes Today v.8 George Allen & Unwin London; pp 215-228

Lyapunova EA, Vorontsov NN (1970) Chromosomes and some issues of the evolution of the ground squirrel genus *Citellus* (Rodentia, Sciuridae). Experientia 26(8): 1033-1038

Lyapunova EA, Vorontsov NN, Martynova LYa (1974) Cytogenetical differentiation of burrowing mammals in the Palearctic. In: Kratochvil J, Obrtel R (eds) Symposium Theriologicum II Proc Intern Symp on Species and Zoogeography of European Mammals Academia Praha, pp 203-215

Lyapunova EA, Vorontsov NN, Zakaryan G (1975) Zygotic mortality in *Ellobius lutescens* (Rodentia: Microtinae). Experientia 31: 417

Lyapunova EA, Vorontsov NN (1978) Genetics of *Ellobius* (Rodentia) I Karyological characteristics of 4 *Ellobius* species. Genetika (Moscow) 14(1): 2012-2024 (in Russian with English summary) English translation: Soviet Journal of Genetics Plenum Publ Co (ISSN 0038-5409)

Lyapunova EA, Vorontsov NN, Korobitsina KV, Ivanitskaya EYu, Borisov YuM, Yakimenko LV, Doval VYe (1980) A Robertsonian fan in *Ellobius talpinus*. Genetica (Hague) 53/54: 239-247

Lyapunova EA (1983) Hybridization of the different chromosomal forms of mammals in nature and in experiment: evolutionary aspects. In: Atayan RR (ed) Pamyaty NV Timoffeeva-Ressovskogo (In Memoriam of NV Timoffeeff- Ressovsky) Academy of Sciences of Armenian SSR Erevan, pp 115-132 (In Russian)

Lyapunova EA, Ivnitsky SB, Korablev VP, Yanina IYu (1984) A complete Robertsonian fan of the chromosomal forms in the mole-vole superspecies *Ellobius talpinus*. Doklady Academii Nauk SSSR 274(5): 1209-1219 (In Russian) English translation: Doklady AN SSSR Biological Sciences (Proc Acad Sci of the USSR) 274(1-6): 87-90 Plenum Publ Co

Lyapunova EA, Yakimenko LV (1985) Genetics of *Ellobius* (Rodentia) IV Decrease in the fertility of hybrids between the forms of *Ellobius talpinus* superspecies with different chromosome numbers. Genetika (Moscow) 21(12): 1960-1969 (in Russian with English summary) English translation: Soviet Journal of Genetics Plenum Publ Co (ISSN 0038-5409)

Lyapunova EA, Yadav JS, Yanina IYu, Ivnitsky SB (1985) Genetics of *Ellobius* (Rodentia). III. Independent origin of the Robertsonian translocations of chromosomes in different populations of the superspecies *Ellobius talpinus*. Genetika (Moscow) 21(9): 1503-1506 (in Russian with English summary) English translation: Soviet Journal of Genetics Plenum Publ Co 21(9): 1184-1187 (ISSN 0038-5409)

Lyapunova EA, Akhverdyan MR, Teslenko SV (1988a) Chromosomal and protein polymorphism in populations of *Microtus arvalis* from Transcaucasus. In: Rodents (Grysuny) Theses of VII All-Union Conference v.1 Sverdlovsk, pp 77-78 (in Russian)

Lyapunova EA, Akhveryan MR, Vorontsov NN (1988b) Robertsonian fan of the chromosomal variability in the Caucasian subalpine voles (*Pitymys*, Microtinae, Rodentia). Doklady Academii Nauk SSSR 298(2): 480-483 (in Russian) English translation: Doklady AN SSSR Biological Sciences (Proc Acad Sci of the USSR) 298(1-6 Plenum Publ Co

Matthey R (1970) L'eventail Robertsonien chez le *Mus* (*Leggada*) africain du groupe *minutoides-musculoides*. Revue Suisse de Zoologie 77: 625-629

Mayr E (1963) Animal Species and Evolution. Cambridge Univ Press Cambridge (Massachusets)

Nadler Ch, Hoffmann RS, Sukernik RI, Vorontsov NN (1974) A comparison of a biochemical and morphological evolution in Holarctic ground squirrels (*Spermophilus*). In: I International Theriological Congress Theses 2: 52-53 Nauka Moscow (in English)

Nasonov NV (1922-1923) Geographicheskoye rasprostranenie gornych baranov Starogo Sveta (Geographical Distribution of the Wild Sheeps of the Old World). Academy of Sciences Petrograd (in Russian)

Peshev DT (1981) On the karyotypes in some populations of the mole rat (*Spalax leucodon* Nordmann) in Bulgaria. Zoologischer Anzeiger (Jena) 208(2): 129-136

Peshev DT, Vorontsov NN (1982) Chromosomal variability of the mole rat *Nannospalax leucodon* Nordmann complex in Bulgaria. In: Myllymaki A and Pulliainen P (eds) Third Intern Theriological Congress Abstracts of papers Helsinki, pp 190

Savic I, Soldatovic B (1979) Distribution range and evolution of chromosomal form in the Spalacidae of the Balkan Peninsula and bordering regions. Journal of Biogeography 6: 363-374

Seuañes HN (1979) The phylogeny of human chromosomes. Springer Berling Heidelberg New York, pp 1-189

Stebbins L (1982) Perspectives in evolutionary theory. Evolution 36(6): 1109-1118

Tan CC (1946) Mosaic dominance inheritance of colour pattern in the lady-bird beetles *Harmonia axyridis*. Genetics 31: 195-210

Thaeler CS Jr (1968) Karyotypes of sixteen populations of the *Thomomys talpoides* complex of pocket gophers (Rodentia, Geomyidae). Chromosoma (Berlin) 25(2): 172-183

Thaeler CS Jr (1983) Chromosome variation in the *Thomomys talpoides* complex. Acta Zoologica Fennica Proc III Intern Theriol Congress (Helsinki)

Timoffeeff-Ressovsky NV, Vorontsov NN, Yablokov AV (1969) Kratkyi Ocherk Teoryi Evolutsii (An outline of evolutionary concepts; in Russian). Nauka (Moscow), pp 3-407 Kurzer Grundriss der Evolutionstheorie (1975) Gustav Fisher Verlag Jena, pp 3-360 (in German)

Vorontsov NN (1960) The palaearctic species of *Cricetinae in statu nascendi.* Doklady Academii Nauk SSSR 132(6): 1448-1451 (in Russian) English translation: Doklady AN SSSR Biological Sciences (Proc Acad Sci of the USSR) 132(1-6) Plenum Press German translation: (1961). Palaarktische Zwerghamsterarten (*Cricetinae-Rodentia*) *in statu nascendi.* Sowjetwissenschaft Naturwissenschaftliche Beitrage N2: 157-162 Verlag Kultur und Fortschritt Berlin

Vorontsov NN, Lyapunova EA (1969) Structure of chromosomes of *Citellus undulatus* and history of getting of the areals of *C undulatus* and *C parryi.* Doklady Academii Nauk SSSR 187(1): 207-210 (in Russian) English translation: Doklady AN SSSR Biological Sciences (Proc Acad Sci of the USSR) 187(1-6) Plenum Press

Vorontsov NN, Lyapunova EA (1970) Chromosome numbers and speciation in the Sciurid rodents (*Sciuridae*: *Xerinae* et *Marmotinae*) of Holarctic region. Bulletin of Moscow Society of Naturalists Biological series 75(3): 112-126 (in Russian with English summary)

Vorontsov NN (1973) The evolution of the sex chromosomes. In: Chiarelli B, Capanna Ed (eds) Cytotaxonomy and vertebrate evolution Academic Press London, pp 619-657

Vorontsov NN, Martynova LYa (1976) Population cytogenetics of *Myospalax myospalax* Laxm (*Rodentia, Myospalacinae*). Doklady Academii Nauk SSSR 230(2): 447-449 (in Russian) English translation: Doklady AN SSSR Biological Sciences (Proc Acad Sci of the USSR 230(1-6) Plenum Press

Vorontsov NN, Frisman LV, Nadler ChF, Lyapunova EA, Hoffmann RS, Fomichova II (1978) Population genetics and genogeography of wild mammals I Genogeography of transferrins and variants of glucose-6-phosphate dehydrogenase in populations of Palaearctic long-tailed ground squirrel *Citellus* (=*Spermophilus*) *undulatus*. Genetika (Moscow 14(5): 805-817 (in Russian with English summary) English edition: Soviet Journal of Genetics Plenum Press

Vorontsov NN, Frisman LV, Lyapunova EA, Mezhova ON, Dersdyuk VA, Fomichova II (1980a) The effect of isolation on the morphological and genetical divergence of populations. Genetica (Hague) 52/53: 361-371

Vorontsov NN, Lyapunova EA, Borisov YuM, Dovagl VE (1980b) Variability of sex chromosomes in mammals. Genetica (Hague) 52/53: 361-372

Vorontsov, NN (1980) Synthetic theory of evolution: its sources, basic postulates and unsolved problems. Zhurnal Vsesoyuznogo Khimicheskogo obshchestva imeni D.I. Mendeleeva 25(3): 293-312 (in Russian) English translation: Soviet Mendeleev Journal of the Chemistry 25(3): 29-60 Plenum Press

Vorontsov NN (1983) Genetics and geography. In: Ataya RR (1983) Pamyati NV Timofeeva-Ressovskogo (Lectures in Memoriam of NV Timofeef- Ressovsky). Publication by Academy of Sciences of Armenian SSR Erevan, pp 200-236

Vorontsov NN, Lyapunova EA (1984) Explosive chromosomal speciation in seismic active regions. In: Bennet M, Gropp A, Wolf U (eds) Chromosome Today 8 Allen & Unwin London, pp 279-294

Vorontsov NN, Lyapunova EA, Belyanin AN, Kral B, Frisman LV, Ivnitsky SB, Yanina IYu (1984) Comparative-genetic methods of diagnostics and estimation of the degree of divergence for sibling species of common voles *Microtus arvalis* and *M epiroticus*. Zoologichesky Zhurnal 63(10): 1555-1556 (in Russian with English summary)

Vorontsov NN, Blekhman AV (1986) Phenogeography and genogeography of elytral colour in natural population of lady-bird beetles *Harmonia* (=*Lais*) *axyridis* Pall (*Coleoptera*, *Coccinellidae*). Doklady Academii Nauk SSSR 286(1): 205-208 (in Russian) English translation: Doklady AN SSSR Biological Sciences (Proc Acad Sci of the USSR) 286(1-6) Plenum Press

Vorontsov NN (1987) Adaptability and neutralism in evolution. In: Zhuchenko AA (ed) Ecological Genetics and Evolution Shtiinca Publ Kishinev, pp 74-102 (in Russian with English summary)

Vorontsov NN, Frisman LV, Nadler ChF, Hoffmann RS, Serdyuk VA (1987) Population genetics and genogeography of wild mammals. VI. Genogeography of transferrins in Amphiberingian populations of Arctic long-tailed ground squirrel *Citellus parryi* Richardson. Genetika (Moscow) 23(4): 725-737 (in Russian with English summary) English edition: Soviet Journal of Genetics 23(4): 495-504

Vorontsov NN (1988) Gradual and sudden speciation: "either-or" or "and-and". In: Kolchinsky EI, Polyansky Yul (eds) Darwinism: Istoryi i Sovremennost (Darwinism: History and Modern Trends) Nauka Leningrad, pp 87-104 (in Russian)

Wahrman J, Goitein R, Nevo E (1969) Geographic variation of chromosome forms in *Spalax*, a subterranean mammals of restricted mobility. In: Benirshke K (ed) Comparative Mammalian Cytogenetics Springer Berlin Heidelberg New York, pp 30-48

White MJD (1978) Modes of speciation. Freeman & Co San Francisco

Wilson AC, Bush GL, Case SM, King MC (1975) Social structuring of mammalian population and rate of chromosomal evolution. Proc Nat Acad Sci USA 72: 5061-5065

Zagorodnyuk I, Teslenko SV (1986) Sibling species of the *Microtus arvalis* superspecies in the Ukraine. I. Occurence of *Microtus subarvalis*. Vestnik zoologii (Kiev) 3: 34-39

B.2. Chromosomal

Karyotypic Repatterning as One Triggering Factor in Cases of Explosive Speciation

O.A. Reig

GIBE, Departamento de Ciencias Biológicas, Facultad de Ciencias Exactas y Naturales, Universidad de Buenos Aires, Pabellón 2, 4°. Piso, Ciudad Universitaria Nuñez, 1428 Buenos Aires, Argentina

INTRODUCTION

As well as in the economy of several countries the wealth is unevenly distributed among individual people, there is a notorious inequality in the distribution of richness in species among different taxa of organisms in nature. In the economy of nature there are a few privileged taxa sharing most of the richness in species, and a reduced middle class including a much lesser number, whereas the great majority of taxa is extremely poor.

Liberals would argue that this is a confirmation of the naturalness of social inequality. As most invocations to biology to explain social phenomena, this assertion can be blamed of an abusive resorting to analogy to infer general explanations. But as the inequality of wealth among people can be explained more legitimally by the laws of economics, it is likewise necessary to look at biological laws to explaning the general pattern of taxic unbalance in diversity. A recent discussion on the fact that there are more species in the order of passerine birds, than in all the remaining 32 avian orders (Kochmer and Wagner, 1988; Fitzpatrick, 1988; Raikow, 1986, 1988; Vermeij, 1988), shows that biologists do not fully agree in explaining this pattern.

This problem, which has been frequently referred to as the hollow curve problem of species distribution (Chamberlain, 1934; Anderson, 1974), was widely neglected in the Modern Synthesis, and, in general, it has been scarcely a topic of interest among modern evolutionists. However, the universality of occurrence of the hollow curve distribution of taxic diversity determines that no version of the evolutionary theory can feel confortable without trying to understand its causes.

My current involvement with the study of diversity of South American mammals faced me insistently with the perplexing fact of an extreme inequality of species richness among different genera of the same family of rodents. I studied in some detail the sigmodontine cricetids, and the caviomorphs of the families Octodontidae and Echimyidae, focusing on chromosomes, evolutionary genetics, classical morphological systematics and paleontological history. I found among them several cases in which taxa showing an explosive cladogenesis also showed a high rate of karyotypic repatterning, contrasting with the much lower rate of chromosomal change shown by their relative depauperate taxa of the same rank, which suggested to me an originally causative function of chromosomal repatterning in active speciogenesis. I compared these results with the known behavior of the same variables in the Holarctic family of voles and lemmings (Arvicolidae), which has been well studied in their systematics, fossil record and comparative cytogenetics (Anderson, 1985; Chaline, 1987; Chaline and Graf, 1988; Gromov and Polyakov, 1977). The similarity of pattern found reinforced my conclusion based on South American rodents.

In this paper I shall survey those studied cases aiming to substantiate an explanation of intrafamilial taxic inequality by appeal of different rates of chromosomal repatterning. The role of karyotypic repatterning in triggering reproductive isolation, and thus, as a potential causal factor of speciation, has been a subject of recent renewed interest (Capanna, 1982; Patton and Sherwood, 1983; Imai, 1986; King, 1987; Sites and Moritz, 1987). Some of these studies shed doubts on the feasibility of chromosomal rearrangements as one of the acting factors in promoting cladogenetic speciation. These doubts turns necessary, therefore, to briefly discuss the current status of theory on this critical subject.

THE BEARING OF CHROMOSOMAL REPATTERNING IN SPECIATION

The bearing of chromosomal structural changes in promoting speciation was a well established concept in early attempts to merge cytogenetic evidence with evolutionary theory (see, among others, Hollingshead and Babcock, 1930; Levitzky, 1931; Navashin, 1932; Darlington, 1937). However, this concept played a minor role in the forging of the Modern Synthesis, under which the concept that speciation was mainly the outcome of gradual adaptive differentiation of populations separated by an extrinsic barrier dominated. Under this paradigmatic allopatric or geographic model, adaptive divergence leading to speciation was primarily expressed by changes in structural genes, chromosome restructuring being mainly considered either as an intrapopulational devise acting in the reorganization of genetic variation in polymorphic systems (i.e. paracentric inversions in *Drosophila*), or as a post-speciational adaptive change.

However, the concept of the steady and slow outcome of speciation by adaptive divergence was complemented since the fifties by the idea that in semi-isolated peripheral population speciation may occur rapidly through radical genetic changes mostly triggered by stochastic processes (Mayr, 1954), a process which was lately baptized as peripatric speciation (Mayr, 1982). Parallely, work on cytogenetis and speciation in the genus *Clarkia* (Onagraceae) by H. Lewis and associates focused again attention on rapid rates of speciation strongly correlated with structural rearrangements (Lewis and Roberts, 1956; Lewis and Raven, 1958). This was the beginning of an increasing move to complement the dominant concept of gradual and allopatric speciation by accepting that speciation was a pluralistic process.

By the sixties, Lewis (1962, 1966) stressed the idea that "catastrophic" bursts of chromosomal structural changes can favour a highly speedy speciation process in *Clarkia*. Based mostly on this evidence, Grant (1963) introduced the concept of "quantum speciation" defined later as "the budding off of a new and very different daughter species from a semi-isolated peripheral population of the ancestral species in a cross-fertilizing organism" (Grant, 1971: 114).

The concept that structural changes can be a driving factor of speciation was independently developed by Michael White and collaborators in the study of the Australian morabine grashoppers. This evidence furthered the notion of the stasipatric model of speciation (White et al., 1967; White, 1968, 1974). With the appearence of White's book (1978a) on speciation the proposal of the stasipatric model, and, more generally, of chromosomal speciation, seemed to receive a definite accolade.

In his attempt to emphasize the genetic mechanisms of speciation, Templeton (1980a, 1980b) contributed later to sharpen the concept of chromosomal speciation. He recognized the emergence of reproductive isolation throughout the fixation of a chromosomal mutation leading to hybrid sterility or semi-sterility as one of the mechanisms of transilient speciation as opposed to divergent adaptive speciation, coining the term "chromosomal transilience" for such a mechanism. He stressed the non-adaptive nature of chromosomal transilience in its first stages.

Further developments of the theory of chromosomal speciation showed the complex nature of the processes. The recent survey by Sites and Moritz (1987) singled out the following six different modes: stasipatry (White, 1968), invasive (White, 1982), primary chromosomal allopatry (King, 1981), chain process (White, 1978b), cascade (Hall, 1983), monobrachial centric fusions (Capanna, 1982, Baker and Bickham, 1986), and recombinational breakdown (Shaw, 1981; Shaw and Coates, 1983).

Besides, various theoretical studies (Bengtsson and Bodmer, 1976; Bush et al., 1977; Lande,1979, 1985; Hedrick, 1981, Walsh, 1982; Templeton, 1980a, 1980b, and others) led to an assessment of the necessary conditions which are theoretically required to lead to the fixation of chromosomal rearrangement triggering reproductive isolation.

These studies concluded that rearrangements leading to effective reproductive isolation in heterokaryotypic hybrids through meiotic malfunction, require to be fixed high negative heterosis of the hybrids, high rates of chromosomal mutation, and a population structure of very small-sized ($NE<10$), highly inbred, socially structured demes with no gene flow. These conditions seemed to be too unlikely to occur together in natural populations. Hence, some authors (Futuyma and Mayer, 1977; Templeton, 1981) claimed that the extremely stringent restrictions turned chromosomal transilience a mere theoretical possibility without actual occurrence in nature.

These alleged theoretical unlikelihood is paradoxical as confronted with the many cases of a strong correlation of high rates of speciation with high rates of chromosomal changes (see a review in White, 1978a). In fact, several factors have been called upon to override the restrictions posed by the theoretical models (Reig, 1983; Sites and Moritz, 1987). Among them, meiotic drive, adaptive advantage of the newly created homokaryotypes, strong inbreeding, multiple succeeding mutations, accumulation of several slightly underdominant rearrangements, group selection, are all factors which may have contributed to promote the fixation of chromosomal mutations even with more permissive deme sizes and gene flow than those required by the theoretical models (Bush, 1981, Hedrick, 1981; King, 1982; Walsh, 1982; White,1978b).

Doubts on the role of structural chromosomal changes in speciation have been also addressed from a discussion of the isolating effects of hybrid sterility (John, 1981; Patton and Sherwood, 1983. Sites and Moritz, 1987). These criticisms certainly involve a healthy appeal to the need to be cautious in attributing prime isolating effects to any chromosomal differences between species in the lack of direct experimental evidence (King, 1987). In fact, recent detailed studies demonstrated that in some cases rearrangements may be nearly neutral in their meiotic effects (Searle, 1986; Sites et al, 1987). However, these cases cannot geopardize the several instances in which effective isolation in chromosomally differentiated forms has been well documented (see review in White, 1978a, and, among others, Lopez Fernández et al., 1984, Shaw, 1981; Shaw and Coates, 1983; Wahrman et al., 1985). Among them, those produced by monobrachial centric fusions have been recently singled out as a very effective isolating factor (Capanna, 1982, Barker and Bickham, 1986; Moritz, 1986).

The feasibility of chromosomal speciation has been also bolstered by the demonstration that the initial fixation and establishment of major heterotically negative chromosomal arrangements is theoretically very likely under Wright's shifting balance model, in conditions of rapid deme extinction and colonization (Lande, 1984).

Moreover, in all theoretical models, one of the limiting factors for the establishment of chromosomal mutations comes from the known regular rates of chromosomal mutation. However, there is now evidence that this rate may be highly raised by introgressive hybridization (Shaw et al., 1983; Naveira and Fontdevila, 1985), and under the effect of the dysgenesis syndrom (see review in Kidwell, 1986), which strongly raise the rate of mutations, including chromosomal ones, by the agency of transposable elements (Fontdevila, 1987). One convincing piece of evidence on the effect of transposons in producing a high rate of chromosomal rearrangements has been recently documented by Engels and Preston (1984).

Thus, there are several mechanisms which can explain the fixation and spread of underdominant chromosomal mutations overriding the constraints of some current population-genetical models, and supporting an cxplanation by the primary agency of chromosome repatterning of the many cases of strongly positive correlation between rates of speciation and rates of chromosomal evolution.

Most of the analytical studies on the relation between chromosomal rates and speciation rates have been pursued on broad comparative studies (Bengtsson,; Wilson et al, 1975, Bush et al, 1977; Aguilera, 1980; Imai, 1983, 1986). This approach may obscure the pattern of chromosomal change in relation to species diversity at lower taxonomic level. Our aim is to analize that pattern in four family-level groups of rodents, in particular connection with the hollow-curve pattern of taxic diversity found in those groups. But let us first briefly review the current problems in explaining the hollow curve distribution of species in higher taxa.

THE EXPLANATION OF THE HOLLOW-CURVE

As mentioned in the Introduction, one of the most remarkable features of the distribution of intrataxonomical diversity of organisms is its characteristic asymetry. For any taxonomic major group, there are only a few of their subordinated taxa that include most of the species (speciose taxa), whereas the remaining species are distributed into a smaller number of paucispecific (< 10 species) and a large array of monotypic taxa of the same subordinated rank. This pattern was not overlooked by Darwin, who stressed the fact that in plants and beetles there were large and small genera, and that species of the former were much more variable in morphology (Darwin, 1959: 55-59). But it was the unorthodox evolutionist J.C. Willis (1922) who first developed a deep interest in this pattern upon his studies on the numbers of species per genera in the flora of Ceylon. He found that for any family, most genera were monotypic, and the number of those which were increasingly polytypic were increasingly smaller, thus that if frequency of species is arithmetically plotted against number of genera, a characteristic hollow curve is obtained. He extended this observation to the geographical distribution of species, founding that most species were endemics showing a very restricted distribution, and the number of those showing an increasing area of distribution decreased abruptly with the increase of the area. The belief that the two phenomena were related lead him to formulate his famous theory of "age and area" (Willis, 1922), claiming that the older a taxon, the greater its area of distribution. He also found that the same distribution holds for the number of genera within a family. He claimed that this distributional pattern was so widespread in nature that it might be "the expression of some definite law which is behind evolution and distribution, and does not agree with current views about these subjects" (Willis, 1940:35). Consequently, Willis advocated an explanation of evolution alternative to the Darwinian selection theory, stressing the difficulties of selection and the adaptationist programme to explain the hollow curve distributions. To him, the widespread occurrence of the hollow curve among many groups of organisms suggested the action of an evolutionary mechanisms able to generate species at regular rates, and proposed mutation as a driving factor in evolution (Willis, 1940).

Indeed, the unequal distribution of species into higher taxa is an ubiquituous and striking fact. More than 56% of living species of birds belong to one of the 32 extant avian orders, the Passeriformes (Bock and Farrand, 1980). A similar proportion holds for bony fishes of the order Clupeiformes as compared with the remaining 33 orders of Actinopterygii (Romer, 1966). Besides, 42 % of the living species of mammals belong to one (Rodentia) out 22 living orders and within that order, muroids amount 64% of the species, the remaining being distributed in 19 different families (Anderson and Jones, 1984). The pattern is repeated when taxa of family level are focused. Chamberlain (1934), Williams (1964), and Anderson (1974) gave ample evidence of this generalized pattern.

The ubiquituous observation of this pattern deserves explanation, and this explanation may unveil, as Willis claimed, a hidden mechanism of evolution, still poorly known, and difficult to explain by the canonical factors of divergence by selection and adaptation. Why, in a given cladogenetic event, most of the species evolve to the level of a genus remaining monotypic or making depauperate genera, whereas another one or two or a few more species experience explossive branching within highly speciose genera? It seems obvious that an explanation to the effect of species diversity within genera of a family level cladogenesis should be valid to explain the unequal distribution of genera among families of a given order, of families among orders of a given class, and so forth, as these higher rank phenomena are supposed to be the result of continuing divergence from species within a genus. Thus, the explanation would extend to make understandable general features of cladogenetic evolution.

As already said, Willis claimed that the pattern cannot be explained by the action of natural selection, and endorsed a mutational explanation. Large mutations creating family and genus characters would be at the origin of a radiation, diversity of species being then the result of further mutational differentiation upon which selection acted as a secondary agent allowing it to survive. Sewall Wright (1941) was one of the few to takle the issue, maintaining that mutation alone cannot be an explanation of the hollow curve.

The intriguing character of the differential taxic species distribution led some early authors to claim that the hollow curve was an artifact of the subjective nature of systematic practice (Chamberlain, 1934). In his failure to find key adaptive characters responsible of the species richness of passerine birds, Raikow (1986) advocated recently a revival of the same view. However, the size of the genera and of higher level taxa as clusters of related species is now susceptible of objective assessing, and the application of quantitative techniques widely confirmed previous intuitive classifications, and with them the objective nature of differently sized genera. Thus, Wiles et al. (1983) showed that the size of genera as based on morphological data are statistically equivalent among different groups of terrestrial vertebrates.

Many readers would doubtless believe that the biological naturalness of taxic inequality in diversity is so obvious that it is unnecessary to belabor its confirmation. Suffice it to add, however, that refined techniques, as the assessment of genetic distances by different methods, and cytogenetic studies, generally confirmed also the closer relatedness of species belonging to large taxa, and the distinctiveness of species of monotypic or depauperate genera. Thus, biochemical data confirmed the cohesiveness of the Passeriformes (Prager and Wilson, 1980), and, as we shall see later, of the speciose genera *Akodon* (Fig. 6), *Microtus* (Fig.9) and *Proechimys*, just to cite a few examples. Indeed these biochemical cohesiveness of large taxa also suggest that the hollow curve of taxic diversity has to be thought of as connected with particular properties of the organisms involved.

However, Yule (1924) and Anderson and Anderson (1975) claimed that the hollow curve results from a stochastic process, and that the distribution of the sizes of genera in nature is what is to be expected from random rates of speciation and extinction. This conclusion resulted from statistical simulations based on random rates of species extinction and origination, which easily generated characteristic hollow curves. Thus, the distribution of species within genera would have not resulted from deterministic processes, and species of multispeciose genera are not to be supposed to differ from species of depauperate genera in evolutionary properties. This explanation was supposed to thoroughly disolve the hollow curve pattern as an evolutionary problem.

However, this indeterministic approach to evolutionary phenomena lays upon a fundamental mistake. It implies the notion that species, as units of organic diversity, could behave as identical and independent particles, as the particles in microphysics and chemistry. In the latter, electrons, atoms and molecules are identified by their spaciotemporal position and not for their distinctive characteristics. Hoffman (1981, 1983) called attention to the error of treating species as if they were particles of identical nature. Contrarily to physical and chemical particles, species must be recognized in their intrinsic individual characteristics, which are the result of deterministic processes. This makes a fundamental ontological difference. Identical particles of the microcosmos of physics depend intrinsically in their behavior of statistical laws. The uniqueness of the components of organic diversity only permits to deal with the study of their dynamics by appeal to causal laws. Thus, simulation models may result in diversity patterns which may be quite similar to real patterns found in nature, but this is a statistical artifact irrelevant to explaining the latter as a result of stochastic processes. The diversity pattern shown by clades both in their evolutionary history or in the distribution of species among higher rank taxa can only be the result of causal factors linked to the different properties or the different evolutionary strategies of species.

Consequently, several attempts have been made to look for causal factors of inequality of taxic size. The literature on this problem is rather extensive, and a detailed review is hardly needed; some of the studies on this subject often deal together with taxic, biogeographic and ecological diversity, which may not be necessarily linked. Ecological extrinsic factors, as favorableness (Terborgh, 1973) or environmental heterogeneity, have been invoked by several authors, including classical studies by Mayr (1963), Simpson (1964) and MacArthur (1972), and these factors have been more recently stressed by Cracraft (1982). However, the mere fact that in one and the same geographical locality one can find species packages of several congeneric species together with monospecific genera, undermines the general application of this factor and certainly call for intrinsic properties of the involved species.

An ample field of research is open, thus, to explore the bearing of intrinsic properties of the taxa involved in cladogenetic bursts. It has been almost canonical to advocate that adaptive breakthroughs (aromorphs, key innovations, neosemes; see Reig, 1987b) are a triggering factor in the origin of a successful cladogenesis. Thus, the distinctive gnawing masticatory apparatus of rodents (Morris, 1965), the complexity of vocal and hearing organs and related neuroanatomical features in anurans (Ryan, 1986), the complex syrinx coupled with large brain size and learning abilities in passerine birds (Wiles et al., 1983; Fitzpatric, 1988), the pharingeal jaws in cichlid bony fishes (Liem, 1973), are all anatomical, physiological or behavioral innovations invoked to explain the speciogenic success of the corresponding taxa. Some skeptical authors have claimed that the concept of key innovation is post-hoc in nature, and thus, unable to be tested as a causal factor of cladogenesis (Raikow, 1986). However, if a synapomorphy, be it physiological, morphological or behavioral, is demonstrated by a careful functional analysis to enhance diversity of adaptations, and to be correlated with speciation rates, the hypothesis that the synapomorphy is important in triggering a burst of new species cannot be sweepingly rejected just because we call the synapomorphy "key" after the clade possessing it is demonstrated to have radiated (Vermeij, 1988).

Body size has been frequently invoked as a factor of taxic differential diversity. Van Valen (1973) documented that there is a negative correlation between size of organisms and species diversity, and the issue has been recently developed by Kochmer and Wagner (1988). It seems a matter of common knowledge that the smaller the organism, the more responsive it would be to the exploitation of the grain of the resource space, and thus, that small-sized organisms have larger possibilities to adaptively diverge in a given environment that bigger ones. However, some studies documented also the fact than in several groups, the negative correlation between body size and diversity does not hold always (Van Valen, 1975; Janzen, 1977; May, 1978).

Several studies have stressed other intrinsic biological factors, as elaborate mating systems in Hawaiian drosophilids and African cichlid fishes (Dominey, 1984), life-history strategy in volutid snails (Hansen, 1980, 1983; Jablonsky, 1982) and brachiopods (Valentine and Jablonsky, 1983), eurytopy and niche breadth in terrestrial vertebrates (Miller, 1956), African Bovidae (Vrba, 1984), and in Veneridae and other families of marine pelecipods (Jackson, 1974), etc. A recent paper emphasized ecological energetics, as expressed by body size, trophic level, and metabolic rate, as important factors in determining the number of species per genus in different classes of arthropods, land vertebrates and plants (Glazier, 1987). Most of those studies afford good pieces of explanation, as they can lead to predictions able to be tested (Fitzpatrick, 1988). They assess the importance of some factors, but they do not eliminate the potential bearing of alternative plausible ones. Certainly, a cogent explanation of taxic differential diversity is likely to have to appeal at a network of related or independent causes. However, these studies are still in its infancy, and we are far from having a thorough explanation of the discussed pattern.

Most of the above intrinsic factors postulated to explain differential taxic size hinge on the concept of adaptive evolution as directed by natural selection. Non-adaptive factors of stochastic origin may have also a role to play, and this may be the case of karyotypic repatterning. Several authors (Bush et al., 1977; Bengtsson, 1980; Aguilera, 1980; Imai, 1983) stressed that karyotypic changes may be linked to differential taxonomic diversity patterns. Giving the stochastic nature of the process leading to fixation of chromosomal change, this factor qualify as a stochastic intrinsic one. I submit that the effort to study their effect in detail in closely related genera within one family may afford worthy insights.

THE CASES

Octodontid and Echimyid Caviomorphs

The Caviomorpha are an endemic taxon of infraordinal rank of Neotropical rodents comprising about 19 % of living species of South American mammals (Reig, 1981, 1986). Twelve living and two fossil families are

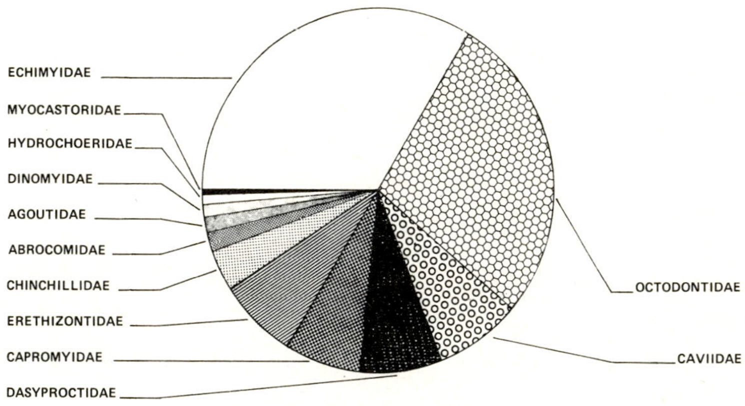

Figure 1. Pie-diagram of species diversity in the different families of living Caviomorpha. Data from Honacki et al. (1982), Woods (1982), and more recent additions.

recognized. The living families comprise 44 extant genera, encompassing a total of ca. 170 living species. But 61 % of the living species belong to any of two major families, Echimyidae (32.9 %) and Octodontidae (including Ctenomyidae) (28.2 %) (Fig. 1). Caviomorph rodents show a fairly good fossil record. The group appeared in Deseadean (early Oligocene) times, probably invading South America from Africa (Reig, 1981, 1986). They rapidly diversified in most of the known living families (Patterson and Pascual, 1972).

Karyotypic information and interpretations on chromosomal evolution have been summarized by George and Weir (1974). However, a great deal of unreviewed new information have been gathered since. The relevant new data will be summarized below.

Echimyids

Caviomorph rodents of the family Echimyidae are at present mostly forest dwellers of tropical and subtropical biomes of South and Middle America. The more primitive living forms are the Eumysopinae (formerly Heteropsomyinae, see Patton and Reig, in press). They are terrestrial rat-sized animals inhabiting the surface of forests, as the speciose and general-

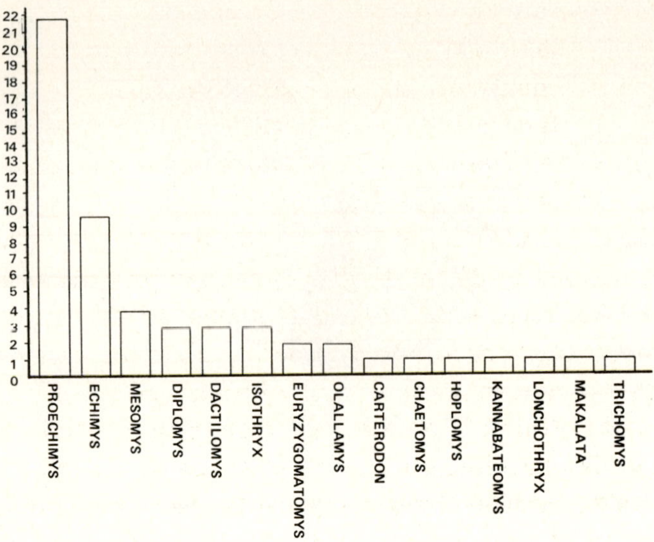

Figure 2. Hollow-curve distribution of living species diversity in extant genera of the family Echimyidae. Data from Honacki et al., (1982), Woods (1982), Gardner and Emmons (1984), and more recent additions.

ized spiny rats of the genus *Proechimys*, and the monotypic and more spiny *Hoplomys*. One monotypic genus (*Trichomys= Cercomys*) lives in drier and more open land, whereas three other, the monospecific *Euryzygomatomys*, and *Clyomys* (which may be congeneric) and the likewise monotypic *Carterodon*, are also dwellers of open land showing various degrees of specialization in adaptation to subterranean life. Forest dwelling linked to arboreal adaptations are shown by members of the Echimyinae: the paucispecific *Echimys, Mesomys, Lonchotrix, Isothrix, Diplomys*, and the single species of *Makalata*. This specialized way of life is even more advanced in the larger Dactylomyinae, *Dactylomys, Kannabateomys* and *Olallamys (=Thrinacodus*, see Emmons, 1988) and Chaetomyinae (*Chaetomys*), all of which are represented by one to three species.

As it results from the previous description, living echimyids show a typical hollow-curve pattern of distribution of species diversity, with one (*Proechimys*) out of the 15 recognized living genera amounting 42 % of the species of the family (Fig. 2). A recent allozymic study of genetic distances (Patton and Reig, in press) demonstrated the cohesiveness of *Proechimys* as a genus, and its distinction as regards other depauperate taxa as *Makalata, Isothrix*, and *Mesomys*.

Fossil echimyids are known since the Deseadean Oligocene (Wood and Patterson, 1959, 1982). The assignement of various fossil genera to living subfamilies is still doubtful, and an extinct subfamily, Adelphomyinae is recognized (Patterson and Pascual, 1968). Undoubted eumysopines are known since the Chasicoan Upper Miocene, including the fossil *Pattersomys* and *Chasichimys (= Trichomys?)* and one representative of the extant *Trichomys*. Fossil forms from the Huayquerian Late Miocene previously described by Rovereto (1914) as members of the extinct *Eumysops* have been referred to the living eumysopine genera *Proechimys, Trichomys*, and *Carterodon* (Bond, 1977). However, and at least as *Proechimys* is concerned, these assignements are not convincing, and it is more prudent to think of *Proechimys* as a genus of unknown longevity, probably rooted some time during the Pliocene to a *Trichomys*-like ancestry. The splitting of the lineage leading to the frondose diversification of spiny rats may have ocurred somewhere outside the present Pampean Region, as in that area only flourished during Pliocene times one extinct paucispecific genus of echimyids, *Eumysops* (Kraglievich, 1965).

Karyotypic information on echimyids as for chromosome number and morphology is rather good for *Proechimys*, with all of the 22 recognized species known in their beta karyotype (see resumé in Reig, 1980; Gardner and Emmons, 1984). Spiny rats are a typical example of a genus with a high rate of chromosomal heterogeneity, and chromosomal changes have been invoked to explain their species richness (Reig, 1980; Reig et al., 1980). Their chromosomal complement varies from 2n=14 to 2n=65 (Table 1), and in the case of the superspecies *P. guairae*, different karyotypes ranging in an orderly succession from 2n=42 to 2n=62, are spaced parapatrically, as to suggest that chromosomally different forms evolved from succeeding peripheral isolates (Reig, 1980), following the primary chromosomal allopatry speciation mode (King, 1981). This process took place without significant morphological changes, the chromosomally distinct forms of the Rassenkreis being practically symmorphic allospecies or semispecies. Genetic distances are also very small, suggesting a highly speedy speciation process (Benado et al., 1979). Besides, four chromosomally well differentiated species occur in sympatry in Balta, Perú, and the morphological differences among them have been hard to assess (Patton and Gardner, 1972). It is also significant that all species of *Proechimys*

Table 1. Chromosome data of *Echymidae*.

Species	2N=	FNa=	Source
Proechimys sp.	14-16	18	(1)
Proechimys steerei	24	42	(2)
Proechimys canicollis	24	44	(3)
Proechimys amphicoricus	26	44	(4)
Proechimys quadruplicatus	28	44	(2)
Proechimys cuvieri	28	50	(5)
Proechimys brevicauda	28-30	48-50	(2)
Proechimys gularis	30	48	(2)
Proechimys semispinosus	30	50-54	(2)
Proechimys oris	30	52-56	(2,6)
Proechimys decumanus	30	54	(2)
Proechimys oconnelli	32	52	(2)
Proechimys simonsi	32	58	(2)
Proechimys guyannensis	40	54	(3)
Proechimys sp (Balta)	40	56	(7)
Proechimys poliopus	42	76	(8)
Proechimys guairae	44-50	72	(8)
Proechimys mincae	48	68	(2)
Proechimys sp. (Barinas)	62	74	(8)
Proechimys trinatatis	62	80	(6)
Proechimys urichi	62	88	(6)
Proechimys iheringi	62-65	117-124	(9,10)
Clyomys laticeps	34	60	(9 ,10)
Euryzygomatomys spinosus	46	92	(9)
Trichomys apereoides	30	54	(10)
Echimysblainvilli	50	94	(10)
Echimyssp.	90	112	(9)

(1) Barros, 1978; (2) Gardener & Emmons, 1983; (3) Aguilera et al, 1979. (4) Reig et al. 1978; (5) Petter, 1978; (6) Reig et al, 1979; (7) Patton & Gardner, 1972; (8) Reig et al., 1980. (9) Yonenaga, 1975; (10) Souza, 1981. Note: look for complete references in this paper's bibliography.

are roughly similar in habitat preference, food, and limiting factors, so that ecological differentiation and adaptation does not seem to have been an important factor in their speciation. Behaviour differences among different chromosomal forms have been noticed, suggesting reinforcement of chromosomal isolation by ethological mechanisms (Reig, unpublished results).

Cytogenetic data are much scarcer for the remaining genera (Table 1). Karyotypes have been described from 2 out of the 10 usually recognized species of *Echimys*. With diploid numbers of 2n=50 and 2n=90, the two species differ greatly in their karyotypes, suggesting that these arboreal spiny rats also experienced strong chromosomal speciation. Chromosomal formulae have also been published for the closely related monotypic burrowing eumysopines *Cliomys laticeps* (2n=34) and *Euryzygomatomys spinosus (= guiara)* (2n=46) , which may belong to a single genus. The chromosomal complement and the banding pattern was also described for *Trichomys apereoides*, which shows 2n= 30. These data suggest that chromosomal rearrangements were also acting in echimyids in the differentiation of the main ecologically distinct genera.

Octodontids

Octodontid caviomorphs are typical inhabitants of the southern cone of South America. They are dwellers of mesic and arid open land biomes. The more generalized living subfamily, Octodontinae, is characterized by ever-growing eight-shaped molar teeth. This subfamily includes ecologically sharply different genera: burrowing rat-like herbivorous rodents as the three species of degus (*Octodon*); generalized surface dwellers of xeric biomes as the single species of *Octomys* and the choz-chori (*Octodontomys gliroides*); a monotypic genus with advanced adaptations to salt dry environment (*Tympanoctomys barrerae*); species with increased specialization to subterranean life as the three living ones of *Pithanotomys (= Aconaemys*, see Reig, 1986); and fully fledged fossorials as the monotypic coruro (*Spalacopus cyanus*) (see Contreras, 1987 for the ecology and biogeography of octodontines).

The highly speciose tuco-tucos are all members of the genus *Ctenomys*, which, together with an array of poorly known Upper Miocene and Pliocene-Early Pleistocene fossil genera *(Actenomys, Eucoelophorus, Megactenomys, Paleoctodon, Praectenomys, Xenodontomys*) make the subfamily Ctenomyinae (Pascual et al., 1978), which some authors prefer to treat as a full family. There are not, however, consistent reasons for this extreme separation of the two groups, as they are really closely related, which is better reflexed in joining them in one single family. The Ctenomyinae are characterized by simplified kidney-shaped ever growing cheek teeth, and by advanced adaptations to an herbivorous fossorial life. These adaptations attain a level that in octodontines is only reached by *Spalacopus* (Reig, 1970), and are surprisingly parallel to those displayed by the North American geomyid pocket gophers. Affording another sharp example of the hollow curve (Fig. 3), about fifty living and some ten

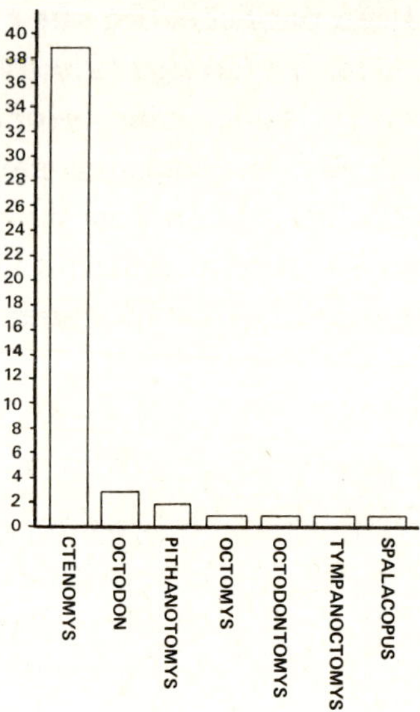

Figure 3. Hollow-curve distribution of living species of the extant genera of the family Octodontidae. Data from Honacki et al (1982), and Woods (1982). More recent additions rise to near 50 the number of species of *Ctenomys* (Reig et al., in prep.).

extinct species are recognized in *Ctenomys*, the single extant genus of ctenomyines, (Reig et al., in prep.). They are all homogeneus in their main fossorial adaptations, though they show detailed tuning to different environmental factors as soil texture and color, which is reflected in different size and morphologies distinguishing their chiefly allomorphic different species.

Fossil octodontids have been described from the Deseadan Oligocene of Argentina and Bolivia (Wood and Patterson, 1951, 1982), but those remains (*Platypittamys, Migraveramus*) are more likely to belong to the Echimyidae or elsewhere. More typical octodontids are the Acaremyinae (*Acaremys, Sciamys*) known from the Santacruzean Miocene of Patagonia. True octodontines, with eight-shaped and hypsodont molars occur later in the Upper Miocene Chasicoan beds (*Chasicomys*). During the late Miocene and the Pliocene, an extinct genus, *Pseudoplataeomys*, was rather common in deposits from Central and NW Argentina. It may be complexive, and some of their species may be ancestral to some of the living octodontine genera, the fossil history of which is poorly known. *Pitanothomys* (including *Aconaemys*) is the only living genus with a known fossil history, two or three extinct species being recorded from the Late Miocene, the Pliocene and the early Pleistocene of Argentina (Rovereto, 1914). An early Pleistocene fossil genus ancestral of the living *Octodon* is now under description by Reig and Quintana. The history of *Octomys, Tympanoctomys, Spalacopus* and *Octodontomys* is completely unknown so far.

The fossil history of Ctenomyines is interesting. They appear in the Chasicoan Miocene as represented by a still undescribed lower jaw assigned to *Paleoctodon* (including *Proctenomys*, Reig and Quintana, unpublished results). During the Late Miocene the same genus thrived in the western pre-Andean and in the Pampean region. During the Pliocene a limited radiation started, involving the extinct genera *Xenodontomys, Actenomys, Praectenomys* and *Eucoelophorus*, the latter surviving into the early Pleistocene, when the poorly known *Megactenomys* is also present. All members of this radiation are monotypic or paucispecific and each represents alternative specializations towards a fossorial life. The major ctenomyine radiation starts, however, in the early Pleistocene with *Paractenomys*, a subgenus of *Ctenomys* which is probably a derivative of the Bolivian *Praectenomys*. By the middle Pleistocene more than ten fossil species of true *Ctenomys* are known in fossil deposits of Argentina, Uruguay and Bolivia, anticipating the deployment of the present about 50 different species. Thus, *Ctenomys* represents an extensive radiation which took place within the limits of a single genus and started 1.8 MYA, and which was preceded by a previous phase of ctenomyine radiation involving three main different genera which flourished for 2 to 6 MY without experiencing much species differentiation (Fig. 4).

Cytogenetical information of octodontids is fairly good (Table 2). Of the octodontines, only *Octomys* and *Tympanoctomys* are still unknown in their karyotypes. The ecologically strikingly distinct *Octodon* and *Spalacopus* share a basically identical karyotype of 2n=58, FN=112 chromosomes (Fernández, 1968; Reig et al., 1972). A very similar karyotype of 2n=56, FN=108 is present in *Pithanotomys (=Aconaemys) fuscus* (Venegas, 1974), and in the related but family distinct *Abrocoma* (George and Weir, 1974). Thus, both by the out-group and commonality criteria,

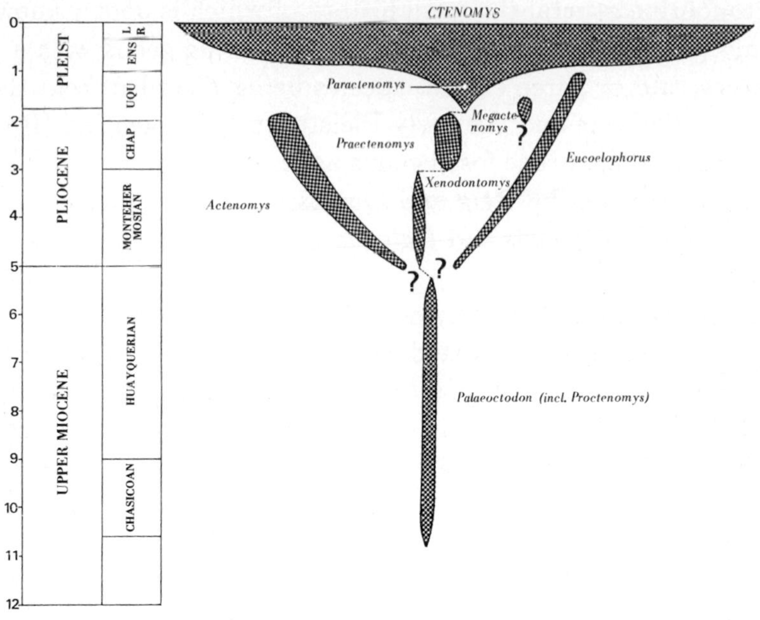

Figure 4. Tentative phylogeny of the Ctenomyinae, showing the explosive cladogenesis of *Ctenomys* since the early Pleistocene. From unpublished results by Reig and Quintana.

this may be considered as being the primitive karyotype for the family. These data also suggest that diversification of the main adaptively different lineages of octodontines took place without major chromosomal repatterning, and show that chromosomal invariance in them is related to absence or poverty in species differentiation. The lineage leading to *Octodontomys gliroides* is an exception to the first generalization, as this species shows a karyotype of 2n= 38, FN=62 chromosomes quite different from those of other octodontines (George and Weir, 1972).

But *Ctenomys* departed completely from the prevailing conservatism found in octodontine chromosomal evolution. Since the early work by Reig and Kiblisky (1969) the available information on *Ctenomys* cytogenetics advanced at a good pace lately, and now karyotypes, including many banded ones, are known for most of the recognized species (34 different karyotypes out of about 50 recognized species, Table 2). A part

Table 2. Chromosome data of the *Octodontidae*.

Species	2N	FN	Source
Ctenomys steinbachi	10	16	(1)
Ctenomys occultus	22	40	(2,4)
Ctenomys maulinus	26	50	(3)
Ctenomys opimus	26	52	(2,3)
Ctenomys fulvus	26	52	(3)
Ctenomys robustus	26	52	(3)
Ctenomys colburni (L. Blanco, Chubut Arg.)	28	46	(4)
Ctenomys sp. (Lonquimay, Chile)	28	52	(3)
Ctenomys tucumanus	28	56	(2,4)
Ctenomys boliviensis	4-46	44	(1)
Ctenomys magellanicus	4-36	68	(2,3)
Ctenomys knightiii (Cardones, Amaicha, Tucuman, Arg.)	36	68	(4)
Ctenomys fodax (Pto. Madryn, Chubut, Arg.)	38	46	(4)
Ctenomys latro	40-42	50	(2,4)
Ctenomys sp. (San Miguel, Corrientes, Arg.)	42	80	(4)
Ctenomys argentinus	44	54	(4)
Ctenomys torquatus	44-46	76	(5)
Ctenomys porteousi	47-48	78	(2,6)
Ctenomys azarae	47-48	78	(6)
Ctenomys mendocinus	47-48	80	(6)
Ctenomys australis	48	80	(2,6)
Ctenomys flamarioni	48	76	(9)
Ctenomys talarum	46-50	84	(2, 6, 10, 13)
Ctenomys minutus	46-55	74-78	(8)
Ctenomys roigii	48	80	(4)
Ctenomys conoveri	48-50	62	(1,4)
Ctenomys rionegrensis? (E. Ríos , Arg.)	48-56	72-74	(2,4)
Ctenomys rionegrensis (Uruguay)	50	72	(5)
Ctenomys yolandae	50	78	(4)
Ctenomys haigi	50	?	(9)
Ctenomys sociabilis	56	?	(9)
Ctenomys perrensis	50-56	84	(4)
Ctenomys pearsoni	56-70	84	(2,3,11,12)
Ctenomys sp. Km14 (Saladas, Corrientes, Arg.)	58	84	(4)
Ctenomys tuconax	58-61	84	(2-4)
Ctenomys sp. (San Roque, Corrientes, Arg.)	62	84	(4)
Ctenomys d'orbignyi	70	84	(4)
Octodontomys gliroides	38	62	(14)
Octodon degus	58	112	(15)
Spalacoppus cyanus	58	112	(16)
Pithanotomys (Aconaemys) fuscus	56	108	(17)

(1) Anderson et al (1987) Amer. Mus. Novitates 2891:1—20. (2) Reig and Kiblisky (1969) Chromosoma 28:211. (3) Gallardo (1979) Arc. Biol. Med. Exp. 12: 71. (4) This paper: Ortells, Barros, Reig and Contreras, unpublished data. (5) Freitas and Lessa (1984) J. Mammal. 65: 637. (6) Massarini et al., in press. (7) Freitas and Travi (1982) Resumos, Cienc. e Cult. 34: 755 (8) Freitas and Mattevi (1986) Resumos, Cienc. e Cult. 38 : 917. (9) Gallardo, 1988, pers. comm., IV Jorn. Arg. Mastozool. (10) Ortells et al. (1984) Rev. Mus. Arg. C. Nat. "B. Rivadavia" (Zool) 13 : 479. (11) Kiblisky et al. (1977) Mendeliana 2:85. (12) Novello and Lessa (1986) Z. Saugetierk. 51: 378: (13) Vidal—Rioja (1985) Caryologia 38: 169. (14) George and Weir (1972) Chromosoma 37: 53. (15) Fernandez (1968) Arch. Biol. Med. Exper. 5: 33. (16) Reig et al. (1972) Biol. J. Linn. Soc. 4: 29. (17) Venegas (1974) Bol. Soc. Biol. Concepción 47 : 207.

American and Eurasian Arvicolidae.

Sigmodontine Cricetids of the Tribe Akodontini

With about 52 genera and more than 250 recognized species living south of Panamá (Reig, 1986, 1987a), cricetids of the subfamily Sigmodontinae (Reig, 1980b) form a large array of mice representing one of the most extensively and complexly diversified group of mammals of South America. They amount by themselves 22 % of the living South American mammals. With a few exceptions, their genera are easily grouped in seven distinct tribes, namely Oryzomyini, Ichthyomyini, Akodontini, Scapteromyini, Wiedomyini, Phyllotini, and Sigmodontini. These tribes are of quite different size, as 87.1 % of the living species belong to one of three tribes: Oryzomyini (44.2%), Akodontini (24.9%) and Phyllotini (18.1%). This pattern suggests that the group as a whole differentiated in South America through three main cladogenetic events. Morphology, biogeography, and comparative cytogenetics suggest that these three main radiations took place in different areas of the Andean region as succeeding cladogenetic bursts (Reig, 1986, 1987b).

Even when we have not yet arrived to a satisfactory understanding of their diversity and relationships, the Akodontini are probably the group of sigmodontines for which more information is available. Recent surveys based on chromosomal work and classical systematics allowed to recognize no less than 62 living species distributed in 10 genera (Reig, 1986). More recent work in progress by J.L. Patton, Ph. Myers, and the author (Reig, 1987) and collaborators, increased this figure to 71 living species and 11 genera.

Predominantly central Andean in distribution, akodontine mice are, however, widely dispersed in South America. Although many of their species are versatile in ecology and habitat preferences, several other ones displayed a limited radiation. A few depauperate or monospecific genera show burrowing adaptations (*Chelemys, Notiomys* and *Geoxus*, see Pearson, 1984), and the monotypic *Blarinomys* is openly fossorial (Matson and Abrewaya, 1977). Some others specialized in divergent feeding adaptations: *Oxymycterus* and *Lenoxus* of the extant genera of the tribe Akodontini of to an animal diet, species of *Necromys (= Bolomys)* to a greater herbivory. The most speciose genus, *Akodon*, is the most gener-

of the new information is still unpublished or it has been merely advanced in annual meetings (See references in Table 2). Karyotypes vary strongly in *Ctenomys*, with diploid numbers ranging from 2n=10 to 2n=70. A few species show karyotype conservatism (Gallardo, 1979; Massarini et al, in press), but most of them are characterized by distinct chromosome number and structure. Species groups behave differently as regards the involvement of heterochromatin in karyotype repatterning. In most of them it is scarcely present in the form of centromeric heterocromatin. Those species show strong eucromatic karyotype repatterning in the form of Robertsonian rearrangements or inversions. Other group of species show whole arm heterochromatic blocks which greatly vary polymorphycally (Massarini et al., in press). Intercallary GC-rich bands also vary among species (Barros, Massarini and Reig, unpublished). Thus, the highly explossive cladogenesis of *Ctenomys* may have been triggered by chromosome repatterning. The particular population structure of tuco-tucos is likely to have favoured rapid fixation of chromosomal rearrangements. Certainly, species of *Ctenomys* are highly territorial, are mostly organized in rather small semi-isolated colonies, and show low vagility (Pearson, 1957; Busch et al., 1989). It has been noticed (Reig, 1970) that another population structure, with high-numbered and wandering colonies, is correlated in *Spalacopus* with karyotype homogeneity in a large area. Territorial populations of *Ctenomys* in a similar surface are represented by tens of chromosomally isolated species.

Cricetid and Arvicolid Muroids

Mice, rats and voles (superfamily Muroidea of the infraorder Myomorpha, sensu Simpson, 1945) are the more numerous group of mammals. In the extant fauna, muroids amount by themselves 64 % of species of rodents and 26.7 % of the species of mammals. Their 1122 living species are distributed in about 15 taxa of family rank (Anderson and Musser, 1984), but 87 % of them belong to just three highly speciose families: Muridae (41 %), Cricetidae (34.8 %), and Arvicolidae (11.2 %). Muroids mostly radiated since the Miocene, and are therefore a very suitable group to study explosive cladogenesis. We shall limit ourselves to survey here two well known groups of muroids: South American cricetids of the tribe Akodontini of the subfamily Sigmodontinae (Reig, 1980b), and North

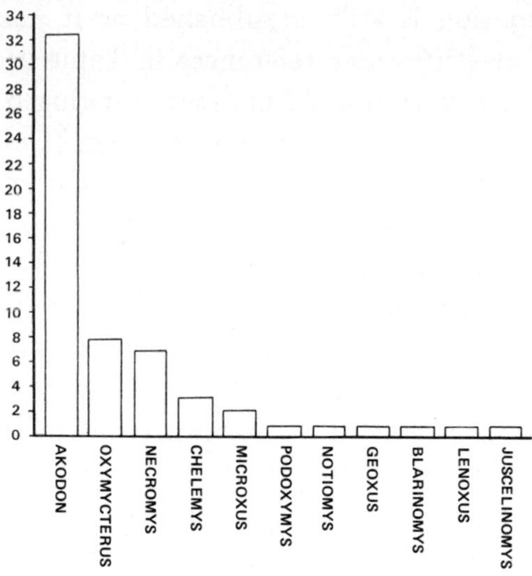

Figure 5. Hollow-curve distribution of living species of the extant genera of the tribe Akodontini of South American sigmodontine cricetids. Data from Reig (1987a), and more recent additions.

alist and the most varied in habitat distribution.

Intrageneric species distribution within the Akodontini is typically hollow-curved as more than 60.6 % of the species belong to a single genus, *Akodon.* Next in rank of species diversity come the paucispecific genera *Oxymycterus* (8 species) and *Necromys* (6 species), whereas the remaining 13 species are distributed in 7 depauperate or monotypic genera (Fig. 5). The recent allozyme genetic study by Apfelbaum and Reig (1989) confirmed the cohesiveness of Akodon as a genus as regards *Oxymycterus*, but failed to separate *Necromys* from *Akodon* (Fig. 6). A parallel study by Patton et al. (1989) also confirmed the closest biochemical relationships of species referred to *Akodon.*

The fossil history of akodontins is sketchly known from the fossil record of the Pampean region (Reig, 1978, 1987a). By the Montehermosian Pliocene, 3.8 MYA, *Necromys* was already differentiated from its relative *Akodon.* In the immediatly later Chapadmalalan beds, a representative of the subgenus *Abrothrix* of *Akodon* was present, coexisting with an ex-

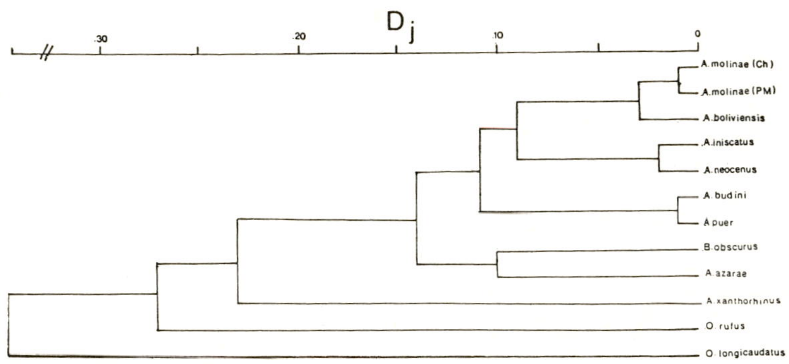

Figure 6. UPGMA phenogram generated by a matrix of Nei's genetic distances based on allozyme frequencies of *Akodon (A)*, *Necromys* (= *B. obscurus*), and *Oxymycterus rufus*, with *Oligoryzomys longicaudatus* as an out-group. From Apfelbaum and Reig (1989).

tinct specialized *Necromys*-like genus, *Dankomys*. Representatives of both *Abrothrix* and *Dankomys* have been found in the Early Pleistocene, Vorohuean beds. In the overlaying Sanandresian early Pleistocene, various extinct species of *Akodon* sensu stricto occur, and in the Middle Pleistocene Ensenadan beds, remains of the living *Akodon iniscatus* and *cursor*, an unidentified species of *Necromys*, and an extinct species of *Akodon*, have been documented (Reig, 1987a). Genetic distance studies based on allozyme differentiation proved that the absolute mean time of divergence among species of *Akodon* was of 3.04 MY (Apfelbaum and Reig, 1989). Hence, the paleontological and molecular information strongly suggest that *Akodon* is another clear case of explosive cladogenesis which led to the differentiation of about 43 species from the Upper Pliocene to the Recent.

The Akodontini have been the group of South American cricetids more intensively and extensively studied in their cytogenetics. Gross chromosomal morphology is known in 43 out of 71 recognizable species, including 29 out of 43 species of *Akodon* (see Table 3 and references therein). Moreover, 21 species have been analyzed in their banding pattern (see references in Vitullo et al., 1986).

Work in progress in my laboratory by Rosa Liascovich and María Alicia Barros afforded detailed information on phylogenetic relations of akodontines based on banding comparisons. The emerging picture indicates that the paucispecific and ecologically differentiated genera *Oxymycterus* and

Table 3. Chromosome data of *Akodontini.*

Species	2 N	FN	Source
Akodon (Akodon) sp. ("arviculoides")	14-16	20	(1,2)
Akodon (Akodon) urichi	18	32	(3,4)
Akodon (Akodon) mollis	22-23	43-44	(5)
Akodon (Akodon) cursor montensis	24	40	(6,7)
Akodon (Akodon) tolimae	24	44	(8)
Akodon (Akodon) orophilus	26	40	(9)
Akodon (Akodon) puer (Akodon (Akodon) coenosus)	34	40	(10,11)
Akodon (Akodon) iniscatus	34	42	(12)
Akodon (Akodon) dolores	34-40	44	(130
Akodon (Akodon) azarae	38	38	(3,14)
Akodon (Deltamys) kempi	37-38	38	(15)
Akodon (Hypsimys) budini	38	42	(10)
Akodon (Akodon) simulator	38-42	42	(16)
Akodon (Akodon) aerosus	40	40	(9)
Akodon (Akodon) albiventer	40	40	(3)
Akodon (Akodon) alterus	40	40	(17)
Akodon (Akodon) boliviensis tucumanensis	40	40	(3,11)
Akodon (Akodon) neocenus	40	40	(3)
Akodon (Akodon) toba	42-43	44	(18)
Akodon (Akodon) molinae	42-43	44	(13)
Akodon (Akodon) serrensis	44	44	(7, 19)
Akodon (Akodon) nigrita	52	52	(20, 21)
Akodon (Akodon) olivaceus	52	56	(22, 23)
Akodon (Akodon) andinus	52	56	(24)
Akodon (Abrothrix) hershkovitzi	52	56	(25)
Akodon (Abrothrix) illuteus	52	56	(26)
Akodon (Abrothrix) Longipilis	52	56	(3, 22, 23)
Akodon (Abrothrix) sanborni	52	56	(23)
Akodon (Abrothrix) xanthorhinus	52	56	(3)
Akodon (Chroeomys) jelskii	52	58	(9)
Necromys amoenus	34	34	(9)
Necromys lactens	36	38	(27)
Necromys lasiurus	34	34	(28)
Necromys lenguarum	34	34	(9)
Necromys obscurus benefactus	34	34	(10)
Necromys temchuki	34	34	(10)
Chelemys macronyx	52	56	(29)
Geoxys valdivianus	52	56	(29)
Microxys "bogotensis"	35-37	48	(30)
Oxymycterus angularis	54	58	(20)
Oxymycterus iheringi	52	58	(21)
Oxymycterus nasutus	54	58	(10)
Oxymycterus paramensis	54	58	(10)
Oxymycterus roberti	54	58	(20)
Oxymycterus rufus platensis	54	58	(10)

(1) Yonenaga (1972) Cytogenetics 11: 488. (2) Yonenaga et al. (1975) Cytogen. and Cell Gen. 15: 388. (3) Bianchi et al. (1971) Evolution 21: 724. (4) Reig et al. (1971) Cytogenetics 10: 99. (5) Lobato et al. (1982) Genetica 57: 199. (6) Cestari and Imada (1968) Cienc. e Cult. 20: 758. (7) Liascovich and Reig (1988) J. Mamm. (in press) (8) C. Ramirez, pers. comm. (9) Gardner and Patton (1976) Occ. Pap. Louis. St. Univ. 49: 1. (10) Vitullo et al. (1986) J. Mamm. 67 :69. (11) Barquez et ai. (1980) Ann. Carnegie Mus. 49: 379. (12) Barros et al. (1989) Z. Saugetierk. in press. (13) Bianchi et al. (1979) Genetica 50: 99. (14) Bianchi et al (1969) Can J. Genet. Cytol. 11: 233. (15) Sbalqueiro et al. (1984) Cytogenet. Cell Genet. 38: 50. (16) Liascovich pers. comm. (17) Liascovich and Reig, in prep. (18) Ph. Myers, pers. comm. (19) Sbalqueiro (1986) Cienc. e Cult. (20) Yonenaga (1975) Caryol. 28: 269. (21) Liascovich and Reig, unpublished. (22) Spotorno and Fernandez (1976) Mamm. Chrom. Newslett. 17: 13. (23) Gallardo (1982) Experientia 38: 1485. (24) Vitullo & Espinosa, pers. comm. (25) Patterson et al. (1984) Fieldiana (Zool) 23: 1. (26) Liascovich et al (1988) J. Mamm. 69, 4. (27) Ortells and Reig, in prep. (28) Maia and Langguth (1981) Z. Saugetierk. 46: 241. (29) Pearson (1984) J. Zool. 202: 225. (30) Barros & Reig (1979).

Necromys evolved without major changes in karyotype evolution, all of the species of these genera showing the same karyotype of 2n=54 or 2n=34, respectively. However, recent unpublished results by Ortells, Liaskovich and Reig demonstrated, that small departures from the uniform karyotype of their genera are found in *Necromys lactens* (2n=36, FN=38) and in *Oxymycterus iheringi* (2n=52, FN=52). Quite different is the case of species of the explosive evolving genus *Akodon*. Among them, the karyotype is highly variable, the diploid numbers ranging from 2n=14 to 2n=52 (Table 3). Robertsonian changes, tandem fussions and inversions played the central role in karyotypic repatterning within *Akodon*. Additionally, intercallary heterochromatic bands proved to be different among species with chiefly similar karyotypes (Barros et al.,1989). However, a quick glance at Table 3 shows that karyotype species-specificity is not universal among species of Akodon. Five species of the subgenus *Abrothrix* share with species of the subgenera *Akodon (A. olivaceus, A. andinus, A. nigrita)* and *Chroeomys (A. jelskii)*, a similar 2n=52, karyotype. Following the out-group criterium, this karyotype is the primitive for the genus (Reig, 1987a), as it is also present in *Chelemys* and *Geoxus*. A few species of Andean and Central Argentinian species also share karyotypes of 2n=40, FN=40 quite similar in gross morphology. The highest range of rearrangement was experienced, indeed, in the group of species which evolved in the Andean axe north of the present Puna, in north-eastern Argentina and Brasil, and in a group of species of central Argentina and Bolivia (Reig, 1987a).

Arvicolids

Voles and lemmings have been classically grouped as a subfamily Microtinae of the Cricetidae (Simpson, 1945). Most recent authors prefer to use Gray's oldest name Arvicolinae for them (Carleton and Musser, 1984), and there is an increasing tendency to rise this taxon to full family level, as Arvicolidae (Honacki et al.,1983; Chaline, 1987). Living representatives are grouped in 17 or 19 genera comprising 125-128 species (Honacki et al.,1983; Carleton and Musser, 1984). They are common and widespread components of the mammal fauna of Eurasia and North America, where they inhabit a variety of habitats within temperate, boreal, montane and arctic biomes.

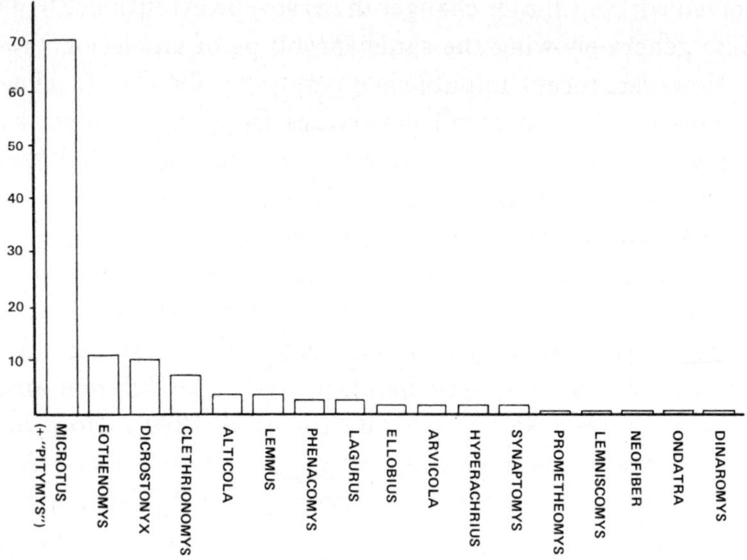

Figure 7. Hollow-curve distribution of living species of the extant genera of the family Arvicolidae. Data from Carleton and Musser (1984).

Arvicolids are mainly terrestrial, pastoral mice adapted to feeding on herbaceous vegetation. This is at a fully expressed state in the ever-growing, rootless and plicidentate molar teeth of the highly speciose genus *Microtus*. With about 70 recognized species (Carleton and Musser, 1984; Honacki et al, 1983), *Microtus* encompasses by itself ca. 56% of the species of the family (Fig 7). An allozyme study by Graf (1982) clearly demonstrated the biochemical cohesiveness of *Microtus* as a genus, ascertaining its objectiveness as a taxon (Fig. 8). Most of its species are generalists for the family standards. The related paucispecific genera *Clethrionomys* and *Eothenomys*, are also pastoral, but are more primitive, as they retain roots in their cheek teeth. The remaining genera are paucispecific or monotypic, and they show different degrees of ecological specializations. This is the case of the pastoral and terrestrial genera *Lemmus, Dicrostonyx* and *Synaptomys*, which experienced advanced adaptations to life in arctic biomes. The four living species of *Ellobius* and the single

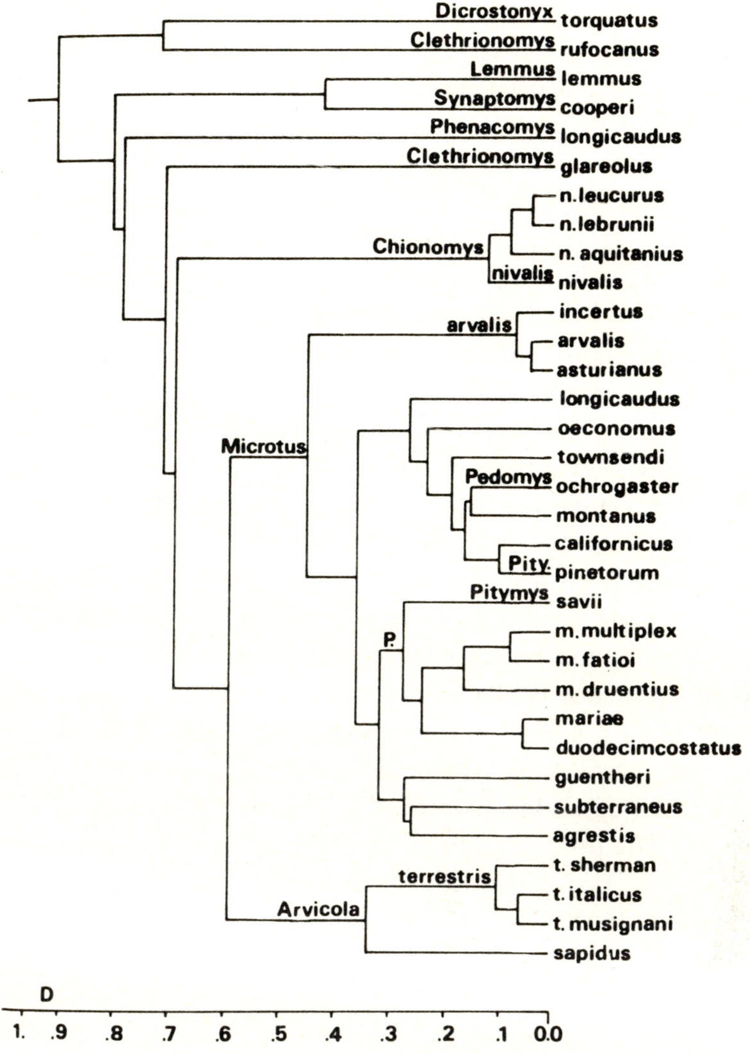

Figure 8. UPGMA Phenogram of allozyme genetic distances of arvicolid taxa. From Graf (1982).

species of *Prometheomys* show semifossorial adaptations. Less advanced modifications for underground life are expressed also in the two recognized species of Asian genus *Hyperacrius*. The monotypic genus *Neofiber* and the two species of *Arvicola* show adaptations to a semiaquatic life, which are more advanced in the single living species of muskrats (*Ondatra*). Two paucispecific genera, *Lagurus* and *Eolagurus*, show adaptations to xeric biomes, whereas the three living species of *Phenacomys* are adapted to arboreal life. Therefore, a general pattern can be recognized in arvicolids: ecological specializations are sharper in monospecific or paucispecific genera, whereas highly speciose genera are more generalists.

Arvicolids are rather well known in their paleontological history both in Eurasia and North America (see resumé in Carleton and Musser, 1984; also Chaline and Mein, 1979; Chaline, 1987; Kretzoi, 1969; Martin, 1979; Michaux, 1971; Repenning, 1968, 1983). The first undoubted arvicolids have been found in the late Miocene of Europe (Kretzoi, 1969; Mein, 1976), but it is from the earlier Pliocene that a good fossil record is abundant, starting with the genus *Promimomys* of Europe and North America. This genus is taken as the common root of a first Pliocene and early Pleistocene cladogenesis (Fig. 9), comprising several extinct genera and the origin of several ecologically specialized living ones, as the muskrats and *Neofiber* and, through the extinct *Mimomys*, of the water voles, as well as of *Clethrionomys, Eothenomys, Hyperacrius, Dinaromys* and *Lagurus*. The precise origin of the specialized cold-adapted lemmings is still unknown, but it can be traced back to the upper Pliocene. One of the lineages of the earlier cladogenetic phase related in origin also to *Mimomys*, acquired hypsodont, ever-growing molar teeth by the end of the Pliocene, ca. 1.9 MYA. This key adaptation was in the origin of the subgenus *Allophayomys* of *Microtus*, from which a new, extensive cladogenetic phase started extending throughout all Holarctica and giving rise to the voles taxic explosion.

Cytogenetics of arvicolids has been worked out in some detail, covering a great deal of the living species (Matthey, 1957, 1977). Table 4 summarizes the known information on diploid and fundamental numbers in a sample of the chromosomally known taxa. Detailed G-banding and C-banding comparisons are also available for a good share of the taxa (Mody, 1987a, 1987b; Chaline, 1987; Chaline and Mein, 1979).

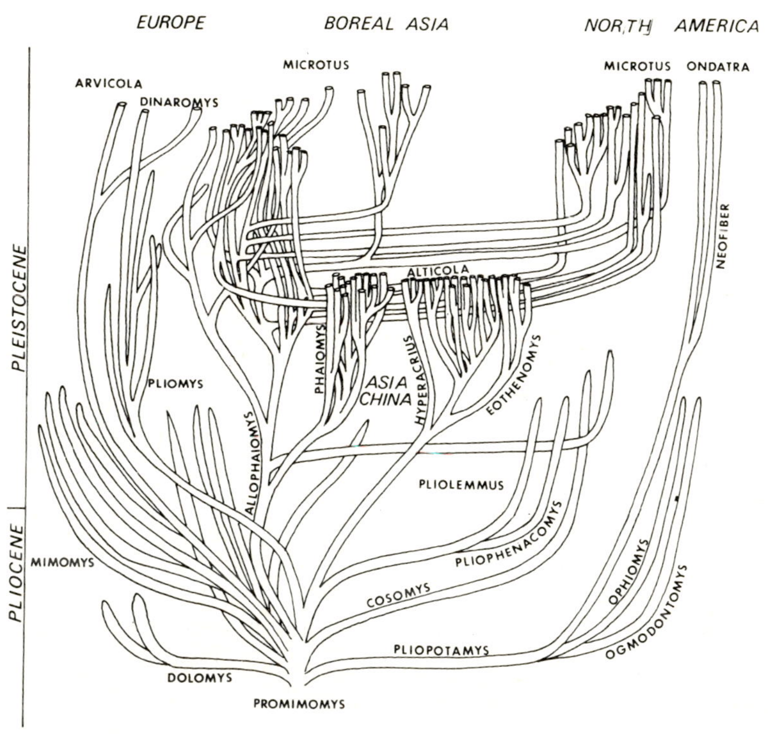

Figure 9. Phylogeny of the Arvicolidae, showing the two-phase cladogenesis of the family, and the explosive pleistocene radiation of *Microtus*. From Chaline and Mein (1979).

In karyotypic differentiation, arvicolids represent a highly multiform group or rodents, diploid numbers ranging in them from 17 to 64 (Table 4). Banding comparisons showed that chromosome repatterning within the family included centric fussions, tandem fusions, and pericentric inversions as the most abundant rearrangements (Mody, 1987b). Matthey (1957, 1973) maintained that a karyotype of 2n=56, made of 27 acrocentric autosomal pairs, was the primitive karyotype for arvicolids. This karyotype is present without major variations in species of *Clethrionomys, Eothenomys, Alticola, Dinaromys, Phenacomys*, and *Eolagurus*. *Ondatra* and *Lagurus*, with 2n= 54, and *Neofiber alleni*, with 2n=52, represent minor departures from the same karyotype. This fact suggests that most of the genera rooted with the earlier arvicolid cladogenetic phase evolved

Table 4. Chromosome data of *Arvicolidae*.

Species	2N	FN	Source
Clethrionomys californicus	56	56	(1)
Clethrionomys gapperi	56	56	(1)
Clethrionomys glareolus	56	56	(1)
Clethrionomys rufocanus	56	56	(1)
Clethrionomys rutilus	56	56	(2)
Eothenomys melanogaster	56	56	(3)
Dinaromys bogdanovi	56	56	(3)
Phenacomys intermedius	56	56	(1)
Dicrostonyx torquatus	56	56	(1)
Dicrostonyx kilangmiutak	47	56	(4)
Dicrostonyx groenlandicus	46	52	(4)
Dicrostonyx richadsoni	42-44	50	(4)
Dicrostonyx rubricatus	32-42	55	(4)
Dicrostonyx exsul	34	54	(4)
Dicrostonyx unalascensis	34	54	(4)
Dicrostonyx nelsoni	30	54	(4)
Dicrostonyx vinogradovi	28	28	(4)
Eolagurus luteus	56	56	(2)
Lagurus lagurus	54	54	(5)
Lagurus curtatus	54	54	(1)
Alticola argentatus	56	56	(2)
Ondatra zybethica	54	56	(2)
Neofiber alleni	52	54	(1)
Synaptomys borealis	54	56	(5)
Synaptomys cooperi	50	48	(1)
Lemmus amurensis	50	48	(1)
Lemmus lemmus	50	48	(1)
Lemmus sybiricus	50	48-52	(1)
Arvicola sapidus	40		(2)
Arvicola terrestris	36	64-72	(1)
Myopus schisticolor	32	60	(1)
Ellobius alaicus	52	56	(6)
Ellobius talpinus	32-54	56	(6)
Ellobius fuscopsilatus	36	56	(6)
Ellobius lutescens	17	32	(1)
Microtus oregoni	17-18	32	(1)
Microtus montanus	22-24	42-44	(1)
Microtus canicaudus	24	44	(1)
Microtus kikuchi	30		(6)
Microtus montbelli	30	54	(1)
Microtus oeconomus	30-31	54	(1)
Microtus gregalis	36	54	(7)
Microtus arvalis	46	64-82	(1)
Microtus mexicanus	44-48	54-58	(1)
Microtus pennsylvanicus	46	50	(1)
Microtus middenforti	50	54	(7)
Microtus agrestis	50	50	(7)
Microtus subterraneus	52	56	(1)
Microtus ochrogaster	52	64	(1)
Microtus guentheri	54	52	(1)
Microtus juldashi	54	52	(1)
Microtus epiroticus	54	54	(1)
Microtus californicus	54	62	(1)
Microtus nivalis	54	62	(1)
Microtus abbreviatus	54	72	(7)
Microtus miurus	54	72	(7)
Microtus carruthersi	54-58	60	(1)
Microtus richardsoni	56	58	(1)
Microtus pinetorum	62	62	(1)
Microtus duodecimocostatus	62	72	(1)
Microtus longicaudus	64	108	(1)

(1) Moodi (1987; (2) Matthey (1983); (3) Chaline (1987); (4) Rausch and Rausch (1975); (5) Matthey (1976); (6) Lyapunova et al. (1980); (7) Fedyk (1979). Note: look for complete references in this paper's bibliography.

without any major involvement of chromosomal repatternings. In most of these cases, the genera are paucispecific or monotypic.

Quite other is the case of the species involved in the second phase radiation, comprising the many species grouped in *Microtus*. After analyzing the karyotypic rate of change in 18 genera of rodents, Maruyama and Imai (1981) found that *Microtus* showed the highest rate. The primitive karyotype of 2n=56 is only found in *Microtus richardsoni*, and several species show a very similar 2n=54 karyotype. However, most other species show highly derived karyotypes with either lower or higher numbers, leading to the condensed karyotype of 2n=17-18 of *Microtus oregoni*, and to the 2n=64 karyotype of *M. longicaudatus*, respectively. Hence, there are strong reasons to suppose that chromosome repatterning played an important role in shaping the perplexing taxic diversity of the rapidly differentiating genus *Microtus*.

Karyotypic repatternings played their speciational role in two other paucispecific to scarcely polytypic genera, the phylogenetic rooting of which is still dubious: the mole-voles, *Ellobius*, and the arctic lemmings, *Dicrostonyx*. The former exemplifies a clear trend towards condensation of the genome and a high heterogeneity in karyotypes, which varies from 2n=17 to 2n=54 (Matthey, 1958; Lyapunova et al., 1981). The arctic lemmings show the primitive karyotype of 2n=56 in one species, but seven other different karyotypes are possessed by the other species, varying from 2n=47 to 2n=28 (Rausch and Rausch, 1972, 1975).

DISCUSSION

In this paper I have attempted to show by way of example that in the context of family and subfamily level cladogeneses, the genera showing an explosive pattern of species diversification are also endowed of high rates of chromosomal repatternings. In my discomfort with the reliability of the available measures of chromosomal evolution (see Wilson et al., 1975; Bush et al. 1977; Imai et al., 1983; Bianchi and Merani, 1984), I have not attempted to investigate quantitatively rates of chromosomal repatterning. However, in all four examined cases, the intuitive grasp of a high correlation between genera showing a high rate of speciation and a high rate of karyotypic change is striking.

By contrast, in those cladogeneses the general pattern in most of their remaining paucispecific or among the monotypic genera, is one of chromosomal stability. With the exceptiom of the two chromosomally known species of *Echimys*, of the three monospecific genera of echimyids among them, and of *Dicrostonyx* and *Ellobius* among arvicolids, all other cases show that depauperate genera are chromosomally homogeneous or show little intrageneric or intergeneric variation in their karyotype. Particullarly significant are the cases of the low rate of chromosomal diversification in most of arvicolids derived from the first phase of the cladogenesis of the family (*Clethrionomys, Eothenomys, Dinaromys, Phenacomys, Alticola, Ondatra, Synaptomys, Lemmus, Eolagurus, Lagurus*) as compared with *Microtus*; of the genera *Necromys* and *Oxymycterus, Chelemys* and *Geoxus*, as compared with *Akodon* among akodontines, and of the depauperate octodontid genera *Octodon, Spalacopus* and *Pithanotomys*, as compared with *Ctenomys*. The lack of an adequate fossil record precludes attempting to interpret within an evolutionary framework the contradictory cases of *Dicrostonyx, Ellobius* and echimyids other than *Proechimys*, but they may be considered as exceptions of the more general rule illustrated by the other cases.

The underlying theory of the bearing of chromosomal repatterning as a triggering factor of reproductive isolation, and the founding of high rates of that repatterning in rapidly and broadly diversifying genera among family-level cladogeneses within which there are typical hollow-curve patterns of taxic diversity, with the depauperate taxa showing in them low rates of chromosomal evolution, clearly suggest a causal connection. Species rich genera may have evolved their differential higher diversity because they are unstable in their karyotypic genomic structure. In other words, chromosomal instability may be taken as an intrinsic biological property of speciose genera versus the chromosomal conservatism of the depauperate genera of the same clade. Needless to say, this conclusion only holds for taxa like the studied ones, in which the chromosome complement behaves as an evolutionary variable. There is ample evidence of other taxa, as marsupials, whales, pinnipeds, frogs, turtles, and many others, in which the intrataxonomical karyotype stability precludes any explanation of their evolution in terms of chromosomal changes.

As for the cases like the studied ones in which chromosomal diversification is an important evolutionary variable, the important question is whether an autonomous dynamics of chromosome repatterning may elicit by itself species diversification. In might be argued that the triggering factors of the latter are quite different, and that karyotypic variability is a secondary consequence of species divergence. In other words, that speciose taxa show karyotypic heterogeneity because they are diverse, and that depauperate taxa do not show that heterogeneity because they did not diverge. One frequent assumption of this argument is that chromosome rearrangements are a secondary consequence of adaptive divergence.

The hypothesis of the secondary, adaptive outcome of karyotype evolution would predict that species showing different karyotypes may show contrasting adaptations. However, contrasting adaptations in the analized cases are not shown within the speciose and karyotypically heterogeneus genera, but among the remaining depauperate or monotypic genera endowed of chromosomal stability. All species of *Akodon, Proechimys, Microtus* and *Ctenomys* are homogeneus in ecology and main adaptations. Taking apart *Ctenomys*, which shares with its relative *Spalacopus* an advanced adaptation to fossorial life, the other qualify, also, among the more generalist members of their own clade. By contrast, the depauperate, geologically more long lived, and less variable in chromosomes, arvicolid genera *Lagurus, Eolagurus, Phenacomys, Ondatra* etc, are ecological specialists showing contrasting adaptations. Likewise, the older chromosomally homogeneus depauperate genera *Necromys*, and *Oxymycterus*, and the burrowing *Chelemys* and *Geoxus* among akodontines, are much more specialized in diet or habitat than the speciose *Akodon*. Thus, in those cases, there are sound reasons to conclude that their adaptive divergense took place in the absence of any major chromosomal repatterning. The contrast with the lack of significant ecological divergence in the speciose and chromosomically heterogeneus genera *Microtus, Proechimys, Akodon* and *Ctenomys*, is remarkable. It strongly suggests that speciogenic karyotypic rearrangements do not necessarily obey to the action of adaptive evolution. In fact, most models of fixation of chromosomal repatternings show that stochastic processes as inbreeding and drift should be unavoidable acting forces in speciation by chromosomal transilience. Selection may operate after the fixation of a chromosome rearrangement, but not necessarily at its inception.

In any case, we are faced to the phenomenon that within family level cladogeneses, one or a few genera are more susceptible of chromosomal alterations than others. Particular features of the population structure of species of those genera may explain in some cases the fixation and spread of those alterations. But these explanations are not relevant to explain the origin of differential higher rates of structural mutations. Attempts have been made to explain the enhancement of chromosomal mutations by extrinsic factors, as seismic activity (Vorontzov and Lyapunova, 1984) or exposure to ultraviolet light and/or cosmic rays induced by geomagnetic reversals (Tsakas and David, 1986). However, these studies are at their infancy, and we must admit our ignorance of the factors which may govern changes in the speed of chromosomal and, in general, genomic alterations. Certainly, it is to be expected that advance of research in this area should be most rewarding in the near future, and, as Willis hoped, may unveil a hidden factor of evolution.

ACKNOWLEDGEMENTS

This paper was presented as a lecture at the Symposion on Evolutionary Biology of Transient Unstable Populations, held at Bagnoles, Spain (25-28 July, 1988). I greatly acknowledge the organizer, Dr. Antonio Fontdevila, for giving me the opportunity to develop these ideas to a distinguished audience, and to the Comissió Interdepartamental de Recerca i Innovació Tecnològica (CIRIT) of the Generalitat de Catalunya, for providing funds to cover my travel expenses. Esteban Hasson, James L. Patton, and John A. W. Kirsch contributed with comments or with data. My associates Marisol Aguilera, Liliana Apfelbaum, María Alicia Barros, Julio R. Contreras, Silvia Dahinten, Lidia Daleffe, Orlando A. Scaglia, and my mammal-team students Marcelo Ortells, Alicia Massarini, Rosa Liascovich, Silvia Blaustein and Carlos Quintana are greatly acknowledged by their collaboration in obtaining a great deal of the original information. This information resulted from research financed by CONICIT grant PID 30853000/85, given to the author.

REFERENCES

Aguilera M (1980) Análisis intragenérico de la evolución cromosómica en algunos grupos de mamíferos. In OA Reig (ed) Ecology and Genetics of Animal Speciation: 191-209. Equinoccio Edit Univ Simón Bolivar Caracas

Anderson S (1974) Patterns of faunal evolution. Q Rev Biol 49:311-332

Anderson S (1985) Taxonomy and systematics. In RH Tamarin (ed) Biology of New World Microtus. Spec Publi 8 Amer Soc Mamm Austin Texas

Anderson S, Anderson CS (1975) Three Monte Carlo models of faunal evolution. Am Mus Nov 2563: 1-6

Anderson S, Jones JK (1984) Orders and families of recent mammals of the World. John Wiley & Sons New York pp 1-689

Apfelbaum LI, Reig OA (1989) Allozyme genetic distances and evolutionary relationships in species of akodontine rodents (Cricetidae Sigmodontinae) Biol J Linnean Soc (in press)

Avise JC (1978) Variances and frequency distributions of genetic distance in evolutionary phylads Heredity 40:225-237

Avise JC, Ayala FJ (1975) Genetic change and rates of cladogenesis. Genetics 81 :757-773

Barros MA, Liascovich RC, Gonzalez L, Lizarralde MS, Reig OA (1989) Banding pattern comparisons between *Akodon iniscatus* and *Akodon puer* (Rodentia, Cricetidae) Z Saugetierk (in press)

Barros RS (1978) Variabilidade cromosómica en Proechimys e Oryzomys. Doctorate Thesis, University of Sao Paulo, Sao Paulo

Baker RJ, Bickham (1986) Speciation by monobrachial centric fusions. Proc Natl Acad Sci USA 83: 8245-8248

Benado M, Aguilera M, Reig OA, Ayala FJ (1979) Biochemical genetics of chromosomal forms of Venezuelan spiny rats of the *Proechimys guairae* and *Proechimys trinitatis* superspecies. Genetica 50:89-97

Bengtsson BO (1980) Rates of karyotype evolution in placental mammals. Hereditas 92:37-47

Bengtsson BO, Bodmer WF (1976) On the increase of chromosomal mutation under random mating. Theor Popul Biol 9:260-281

Bianchi NO, Merani MS (1984) Cytogenetics of South American akodon rodents (Cricetidae) X. Karyotypic distances at generic and intrageneric levels. J Mammal 65:206-219

Bond M (1977) Revisión de los Echimyidae (Rodentia, Caviomorpha) de Edad Huayqueriense (Plioceno medio) de las Provincias de Catamarca y Mendoza. Ameghiniana 14:312

Busch C, Malizzia AI, Scaglia OA, Reig, OA (1989) Spatial distribution and attributes of a population of Ctenomys talarum Rodentia: Octodontidae) J Mammal 70:204-208

Bush GL (1975) Modes of animal speciation. Ann Rev Ecol Syst 6: 339-364

Bush GL, Case SM, Wilson AC, Patton JL (1977) Rapid speciation and chromosomal evolution in mammals. Proc Natl Acaad Sci USA 74:3942-3946

Capanna E (1982) Robertsonian numerical variation in animal speciation: *Mus musculus* as an emblematic model. In C Barigozzi (ed) Mechanisms of speciation:155-177 AR Liss Inc New York

Carleton MD, Musser GG (1984) Muroid Rodents. In S Anderson and JK Jones (eds) Orders and families of recent mammals of the World:289-379. Wiley & Sons New York

Chaline J (1987) Arvicolid data (Arvicolidae, Rodentia) and evolutionary concepts. In MK Hecht B Wallace and GT Prance (eds) Evolutionary Biology 21:237-310

Chaline J, Mein P (1979) Les rongeurs et l'évolution. Doin Editeurs Paris

Chaline J, Graf JD (1988) Phylogeny of the Arvicolidae (Rodentia): biochemical and paleontological evidence. J Mammal 69:22-33

Chamberlain JC (1924) The hollow curve of distribution. Amer Nat 58:350-374

Chambers SM (1987) Rates of evolutionary change in chromosome numbers in snails and vertebrates. Evolution 41:166-175

Coen E, Strachan T, Brown S, Dover G (1983) On the limited independence of chromosome evolution. In PE Branham and MD Bennett (eds): Kew Chromosome Conference 2:295-303. Allen and Unwin, London

Contreras, LC, Torres-Murua JC, Yaez JL (1987) Biogeograpy of octodontid rodents: an ecoevolutionary hypothesis. In B D Patterson and R M Timm (Eds.): Studies in Neotropical mammalogy. Essays in honor of Philip Hershkovitz. Fieldiana Zoology (NS) 39:401-411

Cracraft J (1982) A non-equilibrium theory for the rate-control of speciation and extinction and the origin of macroevolutionary patterns. Syst Zool 31:348-365

Darlington CD (1937) Recent advances in cytology. 2nd Edit Blakinston: Philadelphia

Darwin C (1859) On the origin of species by means of natural selection, or the preservation of favoured races in the struggle for life. John Murray London

Dominey WJ (1984) Effects of sexual selection and life history in speciation: species flocks in African cichlids and Hawaiian Drosophila. In AA Echelle and I Kornfield (eds): Evolution of species flocks, 231-249. University of Maine, Orono, Maine

Emmons LH (1988) Replacement name for a genus of South American rodent (Echimyidae) J Mammal 69:421

Engels WR, Preston CR (1984) Formation of chromosome rearrangements by P factors in Drosophila. Genetics 107: 657-678

Fedyk S (1979) Chromosome of *Microtus* (Stenocranius) *gregalis majora* and phylogenetic connections between subarctic representatives of the genus Microtus Schrank, 1898. Acta Theriol 15:143-152

Fernandez R (1968) El cariotipo de *Octodon degus* (Rodentia, Octodontidae) Arch Biol Med Exper 5:33-37

Fitzpatrick JW (1988) Why so many passerine birds? A response to Raikow. Syst Zool 37:71-76

Fontdevila A (1987) The unstable genome: An evolutionary approach. Genét Ibér: 315-349

Futuyma DJ, Mayer GC (1980) Non-allopatric speciation in animals. Syst Zool 29: 254-271

Gardner AL, Emmons LH (1984) Species groups in Proechimys (Rodentia, Echimyidae) as indicated by karyological and bullar morphology. J Mammal 65:10-25

George W, Weir BJ (1972) The chromosomes of some octodontids with special reference to Octodontomys (Rodentia, Hystricomorpha) Chromosoma Berlin 37: 53-62

George W, Weir BJ (1974) Hystricomorph chromosomes. In IW Rowlands and BJ Weir (eds) The biology of hystricomorph rodents pp 79-108. Academic Press New York

Glazier DS (1987) Energetics and taxonomic patterns of species diversity. Syst Zool 36:62-71

Graf JD (1982) Génetique biochimique, zoogéographie et taxonomie des Arvicolidae (Mammalia, Rodentia) Rev Suisse Zool 89:749-787

Grant V (1963) The origin of adaptations. Columbia University Press New York

Grant V (1971) Plant speciation. Columbia University Press New York and Oxford

Gromov IM, Poliakov IA (1977) Polievki (Microtinae) Fauna SSSR, Mliekopitauyushchie 3 (8):1-504. Nauka Moskvá-Leningrad

Hall WP (1983) Modes of speciation and evolution. In AGJ Rhodin and K Mitaya (eds) Advances in herpetology and evolutionary biology:643-679 Mus Comp Zool, Harvard Univ Cambridge Massachusetts

Hansen TA (1980) Influence of larval dispersal and geographic distribution on species longevity in neogastropods. Paleobiology 6:193-208

Hansen TA (1983) Modes of larval development and rates of speciation in Early Tertiary neogastropods. Science 220: 501-502

Hedrick PW (1981) The establishment of chromosomal variants. Evolution 35:322-332

Hoffman A (1981) Stochastic versus deterministic approaches to paleontology: the question of scaling or metaphysics? N J Geol Paleont Abh 162: 80-96

Hoffman A (1983) Paleobiology in the crossroad: a critique of some paleobiological research programs. In M Grene (ed) Dimensions of Darwinism:241-271 Cambridge University Press London

Hollingshead L, Babcock EB (1930) Chromosomes and phylogeny in Crepis. Univ Calif Publi Agr Sci 6:1-53

Imai HT (1983) Quantitative analysis of karyotype alteration and species differentiation in mammals. Evolution 37:1154-1161

Imai HT (1985) Modes of species differentiation and karyotype alteration in ants and mammals. In K Iwatsuki, P H Raven, and W J Bolk (eds.): Modern aspects of species:87-105 University of Tokyo Press Tokyo

Jackson JBC (1974) Biogeographic consequences of eurytopy and stenotopy among marine bivalves and their evolutionary significance. Amer Nat 108:541-560

Jablonsky D (1983) Evolutionary rates and modes in Late Cretaceous gastropods: Role of larval ecology. Proc III North American Paleontol Convention 1:257-262

Janzen DH (1977) Why are there so many species of insects? Proc XV Internat Congr Entomol Washington DC:84-94

John B (1981) Chromosome change and evolutionary change: a critique. In WR Atchley and D Woodruff (eds) Evolution and speciation:23-51 Cambridge Univ Press London

Kidwell MG (1986) Molecular and phenotypic aspects of the evolution of hybrid dysgenesis systems. In S Karlin and E Nevo (eds) Evolutionary processes and theory:169-198 Academic Press Inc Harcourt Brace Jovanovich Publ Orlando etc

King M (1981) Chromosome change and speciation in lizards. In WR Atchley and D Woodruff (eds) Evolution and speciation: 262-285. Cambridge Univ Press London

King M (1982) A case for simultaneous multiple chromosome rearrangements. Genetica 59:53-60

King M 1987 Chromosomal rearrangements, speciation and the theoretical approach. Heredity 59:1-6

Kochmer JP, Wagner RH (1988) Why are there so many kinds of passerine birds? Because they are small. A reply to Raikow. Syst Zool 37:68-69

Kraglievich JL (1965) Spéciation phylétique dans les ronguers fossiles du genre Eumysops Amegh. (Echimyidae, Heteropsomyinae) Mammalia (Paris) 29:258-267

Kretzoi M (1969) Skizze einer Arvicoliden-Phylogenie-Stand 1969. Vert Hung 11:155-193

Lande R (1979) Effective deme sizes during long-term evolution estimated from rates of chromosomal rearrangement. Evolution 33:234-251

Lande R (1985) The fixation of chromosomal rearrangements in a subdivided population with local extinction and colonization. Heredity 54:323-332

Levitzky GA (1931) The karyotype in systematics. Bull Appl Bot Genet Plant Breed 27: 220-240

Lewis H (1962) Catastrophic selection as a factor in evolution. Evolution 16: 257-271

Lewis H (1966) Speciation in flowering plants. Science 152: 167-172

Lewis H, Raven PH (1958) Rapid evolution in Clarkia. Evolution 12: 319-336

Lewis H, Roberts MR (1956) The origin of *Clarkia lingulata*. Evolution 10: 126-138

Liem KF (1973) Evolutionary strategies and morphological innovations: Cichlid phariyngeal jaws. Syst Zool 22:425-441

Lopez-Fernandez C, Rufas JS, Garcia de la Vega C, Gonzalvez J (1984) Cytogenetic studies on *Corthippus jucundus*, Fisch (Orthoptera) III The meiotic consequences of a spontaneous centric fusion. Genetica 63:3-7

Lyapunova EA, Vorontzov NN, Korobitsyna KV, Ivanitskaya EY, Borisov, YM, Yakimenko LV, Dovgal VY (1981) A Robertsonian fan in Ellobius talpinus. Genetica 52/53:239-247

MacArthur RH (1972) Geographical ecology. Harper and Row, New York

Martin LM (1979) The biostratigraphy of arvicoline rodents in North America. Trans Nebraskan Acad Sci 7:91-100

Maruyama T, Imai HT (1981) Evolutionary rate of the mammalian karyotype. J Theoret Biol 90:111-121.

Massarini AI, Barros MA, Ortells MO, Reig OA (In press) Evolutionary biology of cenomyine rodents (Octodontidae) III Chromosomal polymorphism and small karyotypic differentiation

Massoia E (1985), El estado sistemático de algunos muroideos estudiados por Ameghino en 1889 con la revalidación del género Necromys (Mammalia, Rodentia, Myomorpha) Circular Informativa Asociación Paleontológica Argentina Nr 14:4

Matson JO, Abrawaya JP (1977) Blarinomys breviceps. Mammalian Species 74:1-3

Matthey R (1957) Cytologie comparée, systematique et phylogenie des Microtinae (Rodentia, Muridae) Rev Suisse Zool 64:39-71

Matthey R (1958) Un nouveau type de determination chromosomique du sexe chez les mammifères *Ellobius lutescens* Th. et *Microtus (Chilotus) oregoni* Bachm. (Muridés-Microtinés) Experientia 14:240-241

Matthey R (1973) The chromosome formulae of Eutherian Mammals. In AB Chiarelli and E Capanna (eds.), Cytotaxonomy and Vertebrate Evolution:531-616 Academic Press: London and New York

Matthey R (1977) Cytogénétique des Microtinae. La position du genre Synaptomys Mammalia 37:325-329

May RM (1978) The dynamics and diversity of insect faunas. In LA Mound and N Waloff (eds) Diversity of insect faunas: 188-204. Blackwell Oxford

Mayr E (1954) Change of genetic environment and evolution. In J Huxley, AC Hardy, and EB Ford (eds) Evolution as a process: 157-180. G Allen and Unwin London

Mayr E (1963) Animal species and evolution. Belknap Press Cambridge Massachusetts

Mayr E (1982) Speciation and macroevolution. Evolution 36: 1119-1122

Michaux, J (1971) Arvicolinae (Rodentia) du Pliocène terminal et du quaternaire ancien de France et d'Espagne. Paleovertebrata 4:138-214

Miller AH (1956) Ecological factors that accelerate formation of races and species of terrestrial vertebrates. Evolution 10: 262-277

Modi WS (1987a) C-banding analysis and the evolution of heterochromatin among arvicolid rodents. J Mammal 68:704-714

Modi WS (1987b) Phylogenetic analysis of chromosomal banding patterns among the Nearctic Arvicolidae (Mammalia: Rodentia). Syst Zool 36:109-136

Moritz C (1986) The population biology of Gehyra (Gekkonidae): chromosomal change and speciation. Systematic Zool 35:46-67

Morris D (1965) The mammals. A guide to the living species. Harper and Row New York and Evanston

Navashin MS (1932) The dislocation hypothesis of evolution of chromosome numbers. Zeitschr Abst Verergungsl 63:224-231

Naveira, H, Fontdevila A (1985) The evolutionary history of Drosophila buzzatii. IX High frequency of new chromosome rearrangements induced by introgressive hybridization. Chromosoma 91:87-94

Patterson B, Pascual R (1968) New echimyid rodents from the Oligocene of Patagonia, and synopsis of the family. Brevioraa Mus Comp Zool Harvard Univ Nr 301:1-14

Patterson B, Pascual R (1972) The fossil mammal fauna of South America. In A Keast, F Erk, and B Gall (eds) Evolution, mammals and southern continents :247-310. State Univ of New York Press Albany

Patton JL (1987) Species groups of spiny rats, genus Proechimys (Rodentia, Echimyidae). In BD Patterson and R M Timm (eds) Studies in Neotropical mammalogy. Essays in honor of Philip Hershkovitz. Fieldiana Zoology (NS) 39:305-345

Patton JL, Gardner AL (1972) Notes on the systematic of Proechimys (Rodentia, Echimyidae), with emphasis on Peruvian forms. Occas Pap Mus Zool Louisiana State Univ 44:1-30

Patton JL, Myers P, Smith MF (1989) Electromorphic variation in selected South America. akodontine rodents (Muridae: Sigmodontinae), with comments on systematic implications. Z Saugetierk 54 (in press)

Patton JL, Reig OA (In press) Genetic differentiation among echimyd rodents, with emphasis on spiny rats, genus Proechimys

Patton JL, Sherwood SW (1983) Chromosome evolution and speciation in rodents. Ann Rev Ecol Syst 14: 139-158

Pearson OP (1959) Biology of the subterranean rodents, Ctenomys, in Perú. Mem Mus Hist Nat "Javier Prado" 9:1-56

Pearson OP (1984) Taxonomy and natural history of some fossorial rodents of Patagonia, southern Argentina. J Zool London 202: 225-237

Prager EM, Wilson AC (1980) Phylogenetic relationships and rates of evolution in birds. Proc XVIII Ornithol Congr West Berlin pp 1209-1214

Raikov RJ (1986) Why are there so many kinds of passerine birds. Syst Zool 35: 255-259

Raikow RJ (1988) The analysis of evolutionary success. Syst Zool 37:76-79

Rausch RL, Rausch VR (1972) Observations on chromosomes of *Dicrostonyx torquatus stevensoni* Nelson and chromosomal diversity in arctic lemmings. Zeit Saugetierk 73:372-384

Rausch RL, Rausch VR (1975) Taxonomy and zoogeography of *Lemmus spp* (Rodentia, Arvicolinae), with notes on laboratory-reared lemmings. Zeit Saugetierk 40:8-34

Reig OA (1970) Ecological notes on the fossorial octodontid rodent *Spalacopus cyanus* (Molina) J Mammal 51:592-601

Reig OA (1980a)A new fossil genus of South American cricetid rodents allied to Wiedomys, with an assessment of the Sigmodontinae. J Zool London 192:257-281

Reig OA (1980b) Modelos de especiación cromosómica en las casiraguas (género Proechimys) de Venezuela. In OA Reig (ed) Ecology and Genetics of Animal Speciation:149-190 Equinoccio, Edit Univ Simón Bolivar Caracas

Reig OA (1981) Teoría del origen y del desarrollo de la fauna de mamíferos de América del Sur. Monogr Naturae Mus Munic C Nat Mar del Plata 1:1-116

Reig OA (1983) Estado actual de la teoria de la formación de las especies animales. Inf Final IX Congr Latinoamer Zool Perú Octubre 1983:37-57

Reig OA (1986) Diversity pattern and differentiation of High Andean rodents. In F Vuilleumier and M Monasterio (eds) High Altitude Tropical Biogeography:404-439. Oxford University Press New York Oxford

Reig OA (1987a) An assessment of the systematic and evolution of the Akodontini, with the description of new fossil species of Akodon. In BD Patterson and R M Timm (eds) Studies in Neotropical mammalogy. Essays in honor of Philip Hershkovitz. Fieldiana Zoology (NS) 39:347-399

Reig OA (1987b) Notes on biological progress, the changing concepts of anagenesis, and macroevolution. Genét Ibér 39: 473-520

Reig OA, Aguilera M, Barros MA, Useche M (1980) Chromosomal speciation in a Rassenkreis of Venezuelan spiny rats (genus Proechimys, Rodentia, Echimyidae)

Genetica 52/53:291-312

Reig OA, Contreras JR, Ortells M (In prep.). A progress report on evolutionary history, systematics and speciation of Ctenomys (Rodentia, Octodontidae)

Reig OA, Kiblisky P (1969) Chromosome multiformity in the genus Ctenomys (Rodentia, Octodontidae). Chromosoma Berlin 28: 201-244

Reig OA, Spotorno A, Fernandez R (1972) A preliminary survey of chromosomes in populations of the Chilean burrowing rodent *Spalacopus cyanus* Molina (Caviomorpha, Octodontidae) Biol J Linnean Soc London 4:29-38

Repenning CA (1968) Mandibular musculature and the origin of the subfamili Arvicolinae (Rodentia) Acta Zool Kracov 13: 29-72

Repening CA (1983) Quaternary rodent biochronology and its correlation with climatic and magnetic stratigraphy. In WC Mahaney (ed) Correlations of Quaternary Chronologies:105-118. York University Toronto

Romer AS (1966) Vertebrate Paleontology (Third Edit). Univ Chicago Press Chicago & London

Ryan MJ (1986) Neuroanatomy influences speciation rates among anurans. Proc Natl Acad Sci USA 83:1379-1382

Searle JB (1986) Factors responsible for a karyotypic polymorphism in the common shrew, *Sorex araneus*. Proc R Soc Lond B 229:277-298

Shaw DD (1981) Chromosomal hybrid zones in orthopteroid insects. In WR tchley and D Woodruff (eds) Evolution and speciation:146-170. Cambridge Univ Press London

Shaw DD, Coates DJ (1983) Chromosomal variation and the concept of the coadapted genome -a direct cytological assessment. In PE Brandhan, MD Bennett (eds) Kew Chromosome Conference 2:207-216. G Allen and Unwin London

Shaw DD, Wilckinson P, Coates DJ (1983) Increased chromosomal mutation rate after hybridization between two subspecies of grashoppers. Science 220: 1165-1167

Simpson GG (1945) The principles of classification and a new classification of mammals. Bull Ame Mus Nat Hist 85: 1-350

Simpson GG (1964) Species densities of North American recent mammals. Syst Zool 13: 57-73

Sites JW, Moritz B (1987) Chromosomal evolution and speciation revisited. Syst Zool 36: 153-174

Sites JW, Porter CA, Thompson P (1987) Genetic structure and chromosomal evolution in the *Sceloporus grammicus* complex. Natl Geogr Res 3:343-362

Souza MJ de (1981) Caracterizacâo cromossômica en otto espécies de roedores brasileiros das famílias Cricetidae e Echimyidae. Doctorate thesis, Universidade de Sao Paulo, Sao Paulo

Souza MJ, Yonenaga Y (1983) Chromosomal variability of sex chromosomes and NORs in *Trichomys apereoides* (Rodentia, Echimyidae) Cytogenetics and Cell Genetics 23:1-5

Templeton AR (1980a) The theory of speciation via the founder principle. Genetics 94:1011-1038

Templeton AR (1980b) Modes of speciation and inferences based on genetic distances. Evolution 34:719-739

Templeton AR (1981) Mechanisms of speciation. A population-genetic approach. Ann Rev Ecol Syst 12:23-48

Terborgh J (1973) On the notion of favorableness in plant ecology. Amer Nat 107: 481-501

Tsakas SC, David JR (1986) Speciation burst hypothesis: an explanation for the variation in rates of phenotypic evolution. Génét Sél Evol 18:351-358

Valentine JW, Jablonsky D (1983) Larval adaptations and patterns of brachiopod diversity in space and time. Evolution 37:1052-1061

Van Valen L (1973) Body size and numbers of plants and animals. Evolution 27:87-94

Van Valen L (1975) Group selection, sex, and fossils. Evolution 29:87-94

Venegas W (1974) Estudio citogenético en *Aconaemys fuscus fuscus* Waterhouse (Rodentia: Octodontidae) Bol Soc Biol Concepción 47 207-214

Vermeij GJ (1988) The evolutionary success of passerines: A question of semantics? Syst Zool 37:70-72

Vitullo AD, Merani MS, Reig OA, Kajon AE, Scaglia OA, Espinosa MB, Perez-Zapata A (1986) Cytogenetics of South American akodon rodents (Cricetidae) New karyotypes and chromosomal banding patterns of Argentinian and Uruguayan forms. J Mammal 67:69-80

Vorontzov NN, Lyapunova EA (1984) Explosive chromosomal speciation in seismic active regions. Chromosomes Today 8: 279-294

Vrba ES (1984) Evolutionary pattern and process in the sister-groups Alcelaphinae-Epicerotinae (Mammalia: Bovidae). In N Eldredge and S Stanley (eds) Living fossils:62-79. Springer Verlag New York

Wahrman J, Richler C, Gamperl R, Nevo E (1985) Revisiting Spalax: Mitotic and meiotic chromosome variability. Israel J Zool 33:15-38

White MJD (1968) Models of speciation. Science 159:1065-1070

White MJD (1974) Speciation in the Australian morabine grashoppers. The cytogenetic evidence. In MJD White (ed): Genetic mechanisms of speciation in insects:57-68. Australian and New Zealand Book Co Sydney

White MJD (1978a) Modes of speciation. W H Freeman and Co San Francisco

White MJD (1978b) Chain processes in chromosomal speciation. Syst Zool 27:285-298

White MJD (1982) Rectangularity, speciation, and chromosome architecture. In C Barigozzi (ed) Mechanisms of speciation: 75-103. AR Liss Inc New York

White MJD, Blackwith E, Blackwith M, Cheney J (1967) Cytogenetics of the viatica group of morabine grashoppers. I: The "coastal" species. Australian J Zool 15:263-302

Williams CB (1964) Patterns in the balance of nature and related problems in quantitative ecology. Academic Press London New York

Willis JC (1922) Age and area. Cambridge Univ. Press London

Willis JC (1940) The course of evolution by differentiation or divergent mutation rather than by selection. Cambridge Univ. Press, London

Wilson AC, Bush GL, Case SM, King MC (1975) Social structuring of mammalian populations and rate of chromosomal evolution. Proc Nat Acad Sci USA 72:5061-5065

Wood AE, Patterson B (1959) The rodents of the Deseadan Oligocene of Patagonia and the beginning of South American rodent evolution. Bull Mus Comp Zool 120:279-248

Wood AE, Patterson B (1982) Rodents from the Deseadan Oligocene of Bolivia and the relationships of the Caviomorpha. Bull Mus Comp Zool 149:371-543

Woods C A (1982) The history and classification of South American hystricognath rodents: reflections on the far away and long ago. In M Mares and H H Genoways (eds) Mammalian biology of South America:377-392. Spec Publ Pymatuning Laboratory of Ecology No 6. Univ Pittsburg

Wright S (1941) The "age and area" concept extended. Ecology 22:345-347

Wiles JS, Kunkel JG, Wilson AC (1983) Birds, behavior and anatomical evolution. Proc Natl Acad Sci USA 80: 4394-4397

Yorenaga Y (1975) Karyotypes and chromosome polymorphism in Brazilian rodents. Caryologia 28: 269-286

Yule GU (1924) A mathematical theory of evolution, based on the conclusions of Dr JC Willis. Phil Trans Roz Soc London (B) 213: 21-87

Subject Index